Decision Making in Service Industries

A Practical Approach

T0225571

Decision Making in Service Industries

A Practical Approach

Edited by

Javier Faulin • Angel A. Juan
Scott E. Grasman • Michael J. Fry

CRC Press
Taylor & Francis Group
Boca Raton London New York

CRC Press is an imprint of the
Taylor & Francis Group, an **informa** business

CRC Press
Taylor & Francis Group
6000 Broken Sound Parkway NW, Suite 300
Boca Raton, FL 33487-2742

First issued in paperback 2017

© 2013 by Taylor & Francis Group, LLC
CRC Press is an imprint of Taylor & Francis Group, an Informa business

No claim to original U.S. Government works

Version Date: 20120703

ISBN 13: 978-1-4398-6734-1 (hbk)
ISBN 13: 978-1-138-07368-5 (pbk)

Library of Congress Cataloging-in-Publication Data

Decision making in service industries : a practical approach / editors, Javier Faulin ... [et al.].
    p. cm.
Includes bibliographical references and index.
ISBN 978-1-4398-6734-1 (hardcover : alk. paper)
1. Service industries--Management. 2. Decision making. I. Faulin, Javier.

HD9980.5.D418 2013
658.4'03--dc23   2012017890

Visit the Taylor & Francis Web site at
http://www.taylorandfrancis.com

and the CRC Press Web site at
http://www.crcpress.com

Contents

Section 3   Decision Making in Logistic Services

Section 4   Decision Making in Other Service Areas

Foreword

Today, most of the industries in developed countries are service industries. This may seem a strong affirmation, but, by looking around, one can realize that we are surrounded by service industries or goods that come with services, like a car with its insurance or a tablet with its applications.

Most of our economic activities can be classified under either of the two main industries—manufacturing or services. The manufacturing industry produces tangible products (i.e., goods), whereas the service industry produces intangible products (i.e., services). The service industry includes a myriad of activities: banking and financial services, health care services, retailing, logistics, insurance services, travel services, tourism, leisure, recreation and entertainment activities, and professional and business services (e.g., accountancy, marketing, and law).

In developed countries, the service industry generates over two-thirds of the gross national product (GNP) and their relevance is also growing in developing countries. Moreover, the larger part of employment in developed countries belongs to the service industry. Usually, the service industry requires a considerable amount of qualified labor, and even workers from manufacturing industries perform service-type jobs related to marketing, logistics, management, and so forth. It is commonly accepted that productivity in services is lower than that in manufacturing. Among the main reasons for this lower productivity is that service organizations often require a high degree of customer contact with employees or even required customer participation in the service activity. This high degree of personalization and the inherent variability of the service encounter can lead to lower productivity. These interactions make stochasticity especially relevant in service industries compared to that of manufacturing industries. Therefore, any process of decision making and efficient management in the service industry must consider the randomness implicit in the system.

The positive effect of information technology (IT) on business is well documented in business literature. The use of IT in services can also change the way services are managed, created, and offered, especially as we are becoming a knowledge-based society. In particular, to respond to the challenge of increasing productivity and improving management efficiency, modern businesses need IT and sophisticated decision support systems (DSSs) based on powerful mathematical models and solution techniques. There is no doubt that quantitative models and computer-based tools for decision making, like simulation and probabilistic methods, play a major role in today's business environment. This is especially true in the rapidly growing area of service industries. These computer-based systems can make a significant impact on decision processes in organizations by providing insights to the problems and robust methodologies for decision making under complex and uncertain environments. This will lead to improvements in firms' and organizations' productivity and competitiveness, since better or more robust decisions will be made. This is the reason why both industry and academia have become increasingly interested in using ITs and DSSs as a means of responding to the challenging problems and issues posed by the service area.

This book is an important contribution and a major milestone for the service industry. The material covered reflects the increasing role of decision-making processes under uncertainty in service industries. The editors of this volume bring together researchers from the decision-making field whose contributions present new research findings, developments, and future directions for decision making in service industries. The book covers applications of simulation, probabilistic algorithms, and hybrid approaches to several important areas of service industries from health care to logistics services. For example, the reader can find, in Section 1, a general description of the methodological issues related to decision-making processes under uncertainty in service industries; followed by several chapters describing real applications in areas such as health care, logistics, banking, and IT services. Another interesting aspect of this volume is that the methodologies presented here can be extended and generalized to many other service industries.

I want to compliment the editors and the authors of the chapters for having written such a relevant book about decision making in service industries.

Helena R. Lourenço
Department of Economics and Business
Universitat Pompeu Fabra
Barcelona, Spain

Preface

Purpose of the Book

Efficient service management plays a fundamental role in today's global world. Service industries are becoming more important in knowledge-based societies, where information technology (IT) is changing the way these services are offered and managed. At the same time, new challenges must be faced in order to provide intelligent strategies for efficient management and decision making. In real-life scenarios, service management involves complex decision-making processes that are usually affected by stochastic variables. Under such uncertain conditions, it becomes necessary to develop and use robust and flexible strategies, algorithms, and methods that contribute to increasing organizational competitiveness. The main purpose of this text is twofold. First, to provide insight and understanding into practical and methodological issues related to decision-making processes under uncertainty in service industries. Second, to provide insight and understanding into current and future trends regarding how these decision-making processes can be efficiently performed for a better design of service systems by using probabilistic algorithms and simulation-based approaches.

Generally speaking, decision making is a difficult task in business fields. Traditionally, many quantitative tools have been developed to improve decision making in production companies. This book tries to give support to better decision making in service industries. Thus, strategic, tactical, and operational problems in service companies are tackled with the help of suitable quantitative models such as optimization, heuristic algorithms, simulation, and queuing theory. Most of the common traits of the service companies' problems are related to the uncertainty of service demand. Consequently, this book presents the business analytics needed to improve decision making in service industries.

The main objectives and mission of this book are described by the following points:

- To identify and publish worldwide best practices regarding decision-making processes in service industries facing uncertainty.
- To share theoretical and applied methodologies and strategies used in service management to deal with uncertainty, including the use of stochastic algorithms.

- To forecast emerging tendencies regarding decision support systems and information systems to support decision making in service industries.

- To provide the academic community with a text that can serve as a reference for research on practical problems related to service management under uncertainty.

- To present up-to-date research on how probabilistic algorithms and simulation-based approaches are changing experiences and practices in service management.

It is important to highlight that this book presents research that provides support for decision making in a variety of service industries. We do not constrain the scope of the book to any specific type of service. Nevertheless, knowing that there are services that are more popular than others in the managerial arena, it is more probable to describe and study decision making in those services. Thus, you will find several applications in this book on health care, logistics, and financial applications.

Target Audience

The target audience of this book is wide because it can be a valuable tool for researchers, practitioners and managers, master's and doctoral students, consultants, and government officials and policy makers.

Researchers—Knowing that this book is focused on different methodological approaches to make decisions in service settings, it provides innovative ways to focus research problems in the following areas related to service operations: management science, business, finance, marketing, supply chain management and design, production and operations management, and computer science.

Practitioners and managers—Logistics and operations managers, supply chain managers, production managers and control engineers are dealing with real-life problems on a daily basis. The decision-making techniques applied to service industries presented in this book provide a valuable resource to be implemented in practical settings.

Master's and doctoral students—The book can be of interest for lecturers and students in business, finance, production, marketing, operations research, computer science, logistics, operations management, management information systems, industrial engineering and systems engineering. It should be especially useful for the teaching and learning of the following topics: information technology and service, simulation in service industries, decision making in health care, and decision making in logistics.

Consultants—This book provides new methodological approaches to solving problems in services industries that can be used by consultants and further applied to their clients.

Government officials and policy makers—In general, decision makers in both private and public sectors can benefit from this book by tailoring and assessing the methodology explained here according to their decision-making criteria.

Overview of the Book

The chapters published in this book are authored by a total of 42 researchers affiliated with higher-education institutions or research centers located in 10 countries, as follows (in alphabetical order): Australia, Brazil, China, Korea, Portugal, Romania, Spain, Turkey, United Kingdom, and United States of America. The chapters cover a range of decision-making theory in the services field in three main areas: health care, logistics, and other service areas. These areas correspond to the final three sections of the book, following an introductory section, as explained in detail next.

Section 1 contains five chapters and it is devoted to the applications of different decision-making techniques in service industries and serves as an introductory section to the interested reader.

Chapter 1 is written by Aliza R. Heching and Mark S. Squillante. This chapter describes the global IT services delivery providers who handle requests that vary with respect to required skills and performance targets, while satisfying contractual service level agreements that include monetary penalties for violations. Motivated by the authors' experience working with a global IT services delivery provider, this chapter considers these challenges representing a fundamental stochastic decision-making problem.

Chapter 2 is authored by Sanad H. Liaquat, Pragalbh Srivastava, and Lian Qi. It provides simulation-based studies for supply chains with random disruptions at the supplier and retailers. The authors evaluate the advantages of using a dynamic sourcing strategy to allocate sources of supply to customers based on supply availabilities. Similarly, they identify the conditions under which such advantages are significant, and they demonstrate using simulations that it may not be cost effective to keep perfectly reliable retailers in a supply chain when dynamic sourcing and transshipment are allowed.

Chapter 3 is written by George N. Kenyon and Kabir C. Sen. This chapter builds a model for understanding how the consumer builds a perception of service quality. Later, a framework for service design is described. The design encompasses the route from the consumer's first awareness of the service offering to possible future encounters. The key dimensions of service design in the current "information age" are identified. The chapter suggests that the provider should work at the interface of these dimensions to reach the design's sweet spot, where the effective creation of value for all parties is achieved.

Chapter 4 is written by Teresa Oliveira, Amílcar Oliveira, and Alejandra Pérez-Bonilla. This chapter reviews an application of data mining in the service industry and explores some quality-related topics. In the service industry, the main goal of data mining is to improve both the quality of the interface and the interaction between the organizational system and the respective customers and their needs. Today, due to recent technological advances, a new challenge of exploring data and extrapolating patterns has emerged. Data mining, as an exploratory data analysis tool used in large and complex data sets, allows for the discovery of new snippets of knowledge and patterns in these vast amounts of data.

Chapter 5 is authored by Elena Pérez-Bernabeu, Horia Demian, Miguel Angel Sellés, and María Madela Abrudan. It describes how decision-making tools (DMTs) and the related software (DMTS) have made changes in the service sector, focusing their analysis on banks, hospitals, and hotel applications. It should be noted that the evaluation results obtained from multiple attribute decision-making problems of diverse intensity, such as service quality evaluation, may be misleading if the fuzziness of subjective human judgment is not taken into account.

Section 2 of the book is composed of four chapters and it is devoted to the presentation of decision-making techniques in health care services.

Chapter 6 is authored by Yi-Chin Huang, Michael J. Fry, and Alex C. Lin, and examines an instance of a vehicle-routing-type problem encountered by a small chain of drugstores, Clark's Pharmacy, which provides home deliveries of medical prescriptions. The authors use several versions of the traveling salesperson problem and vehicle routing problem to analyze different scenarios for providing cost-effective home delivery of prescriptions. In most cases the authors are able to solve the presented models to optimality using max-flow formulations to identify subtours in the traveling salesperson problem.

Chapter 7 is written by Alberto Isla, Alex C. Lin, and W. David Kelton. It discusses how computer simulation can be used in community pharmacy management to improve various operational aspects while keeping costs under control. Moreover, some of the most useful performance and quality-of-service statistics that pharmacy managers can obtain from a simulation model are introduced. Finally, the chapter discusses some simulation-related projects that are currently being implemented in community pharmacies.

Chapter 8 is written by Omar M. Ashour and Gül E. Okudan Kremer. It presents the use of multiattribute utility theory (MAUT) and fuzzy analytic hierarchy process (FAHP) to improve the problem of patient prioritization in emergency department (ED) settings. Patient prioritization, which is called triage, is a classification system that is used to sort patients based on the acuity of their conditions. Inherent uncertainty in the triage process requires the use of MAUT and FAHP in this context. The uncertainty can be due to incomplete information (e.g., X-ray, blood tests), subjectivity of certain health indicators (e.g., pain level) and individual differences, inadequate understanding, and undifferentiated alternatives (e.g., many patients have the same chief complaint or the same acuity level when they arrive at the ED).

Chapter 9 is authored by John F. Kros and Evelyn C. Brown. It explores how hospitals can reduce room turnaround time. Several approaches are combined, including process-mapping techniques, capability analysis, and heuristics. The chapter also explores the practical steps the analyzed hospital took to improve room turnaround time.

Section 3 is composed of four chapters, and it is devoted to the presentation of decision-making techniques in logistic services.

Chapter 10 is authored by Renata Albergaria de Mello Bandeira, Luiz Carlos Brasil de Brito Mello, and Antonio Carlos Gastaud Maçada. This chapter seeks to answer the question of how organizations that are considering hiring logistics services providers analyze the decision to outsource. This work is an empirical study that uses qualitative and quantitative techniques to propose a set of key decision factors for logistics outsourcing. Logistics managers and decision makers can refer to this set of factors when structuring the decision-making process in their organizations and, therefore, reduce the risk of making incorrect decisions.

Chapter 11 is written by Hongyan Chen and Ruwen Qin. It analyzes issues related to initial capital investments and operating/maintenance costs necessary to build, operate, and maintain transportation infrastructures. Similarly, it discusses how these two types of costs are related and also how travel demands can vary during the long service life of the infrastructure, thus affecting service revenue.

Chapter 12 is authored by Galina Sherman, Peer-Olaf Siebers, David Menachof, and Uwe Aickelin. This chapter presents cost-benefit analysis as a tool for decision making related to port operations. It demonstrates different modeling methods, including scenario analysis, decision trees, and simulation that are used for estimating input factors required for conducting cost-benefit analysis based on a case study of the cargo screening facilities at the ferry port of Calais, France.

Chapter 13 is written by Buyang Cao and Burcin Bozkaya. It reviews the vehicle routing problem (VRP) as an optimization problem often encountered for routing and distribution in the service industry. Some examples include product delivery to a customer's address, school bus routing and scheduling, or service technician routing. In order to solve problems from the real world, practitioners have been utilizing contemporary IT techniques including GIS, GPS, and RFID in addition to traditional optimization technologies.

Section 4 is composed of three chapters and is devoted to the presentation of decision-making techniques in other service areas.

Chapter 14 is written by Magno Jefferson de Souza Queiroz and Jacques Philippe Sauvé. It addresses the challenge of prioritizing IT service investments under uncertainty. The authors identify and discuss solution requirements, they detail methods and procedures to address these requirements, and they develop a solution model. A case study and a sensitivity analysis illustrate the robustness and usefulness of the presented model.

Chapter 15 is authored by Natalie M. Scala, Jayant Rajgopal, and Kim LaScola Needy. It develops a methodology for determining inventory criticality when demands are extremely intermittent and predictive data are unavailable. The methodology is described in the context of managing spare parts inventory in the nuclear power generation industry. The authors present a scoring methodology using influence diagrams and related models that place the various spare parts into groups according to the criticality of storing each part in inventory.

Chapter 16 is authored by David Y. Choi, Uday S. Karmarkar, and Hosun Rhim. It presents a framework that models service technology, design, pricing, and competition in the specific context of retail banking. Two alternative service delivery processes, conventional and electronic, are examined and their effects on customer costs, process economics, market segmentation, and competition are analyzed. Both customer and provider processes are analyzed to derive results about consumer preferences for alternative service delivery mechanisms.

Thus, this book provides the reader with a general overview of the current decision-making processes across a full range of practical cases in service industries. In addition, the literature is enriched by the discovery of current and future trends due to the evolution of the research in this field.

MATLAB® is a registered trademark of The MathWorks, Inc. For product information, please contact:

The MathWorks, Inc.
3 Apple Hill Drive
Natick, MA 01760-2098 USA
Tel: 508 647 7000
Fax: 508-647-7001
E-mail: info@mathworks.com
Web: www.mathworks.com

Acknowledgments

The completion of this book has involved the diligent efforts of several people in addition to the authors of the chapters. The editors are grateful to the members of the Editorial Advisory Board, who are well-recognized academics in their fields, for supporting and overseeing the development of this editorial project. We also express our gratitude to Professor Helena R. Lourenço for writing the foreword of this book.

The chapter submissions for this book were subjected to a double-blind refereeing process that engaged up to three reviewers per chapter. The editors thank the reviewers for reviewing, proposing improvements, and advising the status of the chapters considered for publication in this book; their names are provided in a list of reviewers elsewhere in this book. Without the efforts of the reviewers, this book could not have been completed as scheduled. We also extend our appreciation to those authors who considered publishing in this book, but whose chapters could not be included for a variety of reasons. We trust that their work will eventually appear elsewhere in the literature.

The editors express their recognition to their respective organizations and colleagues for the moral support and encouragement that have proved to be indispensable during the preparation and execution of this editorial project. This book is part of the CYTED IN3-HAROSA Network. Javier Faulin and Angel Juan also thank the financial aid of the VERTEVALLÉE research network (funded by the Working Community of the Pyrenees by means of the grant IIQ13172.RI1-CTP09-R2), the Sustainable TransMET research network (funded by the program Jerónimo de Ayanz from the government of Navarre, Spain), and the project TRA2010-21644-C03 (funded by the Spanish Ministry of Science and Technology). Scott E. Grasman acknowledges the Missouri University of Science and Technology and Rochester Institute of Technology for accommodating his visiting positions through the CYTED IN3-HAROSA Network, which allowed for work on this book. Similarly, Michael J. Fry acknowledges the help and support of his colleagues both at his home institution of the University of Cincinnati and at the University of British Columbia, where much of the work for this book was performed during his sabbatical.

Last, but not least, the editors acknowledge the help of the Taylor & Francis team, and in particular to Cindy Carelli and Amber Donley, as their assistance and patience have been significant and heartening. Finally, the editors extend their sincere appreciation to their respective families for their support and understanding in the development of the edition of this book.

Editors

Javier Faulin, Ph.D., is a professor of operations research and statistics at the Public University of Navarre (Pamplona, Spain). He holds a Ph.D. in economics; an M.S. in operations management, logistics, and transportation; and an M.S. in applied mathematics. His research interests include logistics, vehicle routing problems, and decision making in service companies. Dr. Faulin is a member of the Institute for Operations Research and Management Sciences (INFORMS) and EURO, the Association of European Operational Research Societies; and an editorial board member of the *International Journal of Applied Management Science* and the *International Journal of Operational Research and Information Systems*. His e-mail address is: Javier.Faulin@unavarra.es.

Angel A. Juan, Ph.D., is an associate professor of simulation and data analysis in the Computer Science Department at the Open University of Catalonia (UOC). He is also a researcher at the Internet Interdisciplinary Institute (IN3-UOC). He holds a Ph.D. in applied computational mathematics, an M.S. in information technologies, and an M.S. in applied mathematics. Dr. Juan's research interests include computer simulation, applied optimization, and mathematical e-learning. He is the coordinator of the CYTED IN3-HAROSA Network, as well as editorial board member of both the *International Journal of Data Analysis Techniques and Strategies* and the *International Journal of Information Systems and Social Change*. His Web address is: http://ajuanp. wordpress.com.

Scott E. Grasman, Ph.D., is professor and department head of industrial and systems engineering at Rochester Institute of Technology. He has held previous permanent, adjunct, or visiting positions at Missouri University of Science and Technology, Olin Business School at Washington University in St. Louis, Public University of Navarre, and IN3–Universitat Oberta de Catalunya. He received his B.S.E., M.S.E., and Ph.D. in industrial and operations engineering from the University of Michigan. Dr. Grasman's primary research interests relate to the application of quantitative models, focusing on the design and development of supply chain and logistics networks. He is a member of the American Society for Engineering Education (ASEE), the Institute of Industrial Engineers (IIE), and the Institute for Operations Research and Management Sciences (INFORMS). His e-mail is: Scott. Grasman@rit.edu.

Michael J. Fry, Ph.D., is an associate professor of operations and business analytics in the Carl H. Lindner College of Business at the University of Cincinnati. He has also been a visiting professor at the Johnson School at

Cornell University and in the Sauder School of Business, University of British Columbia. He holds a Ph.D. in industrial and operations engineering from the University of Michigan. Dr. Fry's research interests include applications of operations research methods in supply chain management, logistics, and public-policy operations. He is a member of the Institute for Operations Research and Management Sciences (INFORMS) and the Production and Operations Management Society (POMS). His e-mail is: Mike.Fry@uc.edu.

Contributors

María Madela Abrudan
University of Oradea
Oradea, Romania

María Mandela Abrudan is a senior lecturer at the University of Oradea–Romania, Faculty of Economics and Chief of Management–Marketing Department. She holds a Ph.D. in management with a thesis in human resource management. Moreover, she teaches courses in human resource management, comparative management, international management, and project management. As a coordinator, author, or coauthor she has published 15 books and more than 70 scientific papers in journal and conference publications in different countries, including Poland, Slovenia, Croatia, Bosnia-Herzegovina, and Hong Kong. Since 2006, Dr. Abrudan has annually participated in the Erasmus teaching staff mobility program at different universities in Italy, Spain, Slovakia, Hungary, and Turkey. Similarly, she coordinates and participates in research projects in the field of development of human resource management, entrepreneurship, and management of local government.

Uwe Aickelin
University of Nottingham
Nottingham, United Kingdom

Uwe Aickelin is professor of computer science and an Advanced EPSRC Research Fellow at the University of Nottingham, School of Computer Science. His long-term research vision is to create an integrated framework of problem understanding, modeling, and analysis techniques, based on an interdisciplinary perspective of their closely coupled nature. His main research interests are mathematical modeling, agent-based simulation, heuristic optimization, and artificial immune systems. For more information, see: http://ima.ac.uk.

Omar M. Ashour
Pennsylvania State University
University Park, Pennsylvania

Omar Ashour is a Ph.D. candidate in the Department of Industrial and Manufacturing Engineering at Pennsylvania State University. He received his M.S. and B.S. in industrial engineering from Jordan University of Science and Technology in 2007 and 2005, respectively. He also received his M.Eng.

in industrial engineering with an option in human factors/ergonomics in 2010. His research interests include decision making, applications of engineering methods and tools in health care, and human factors.

Renata Albergaria de Mello Bandeira
Instituto Militar de Engenharia
Rio de Janeiro, Brazil

Renata Albergaria de Mello Bandeira is a professor of transportation engineering at Instituto Militar de Engenharia, Rio de Janeiro, Brazil. She earned her Ph.D. in business administration in 2009 and her masters in industrial engineering in 2006 at the Universidade Federal do Rio Grande do Sul. Dr. Bandeira is currently teaching and conducting research on supply chain management, logistics outsourcing, and operations management. She has had 7 articles published in journals and almost 30 papers in conferences. In addition, she has worked for almost 10 years as a civil engineer and project manager for the Brazilian government.

Burcin Bozkaya
Sabanci University
Istanbul, Turkey

Burcin Bozkaya earned his B.S. and M.S. degrees in industrial engineering at Bilkent University, Turkey, and his Ph.D. in management science at the University of Alberta in Canada. He then joined Environmental Systems Research Institute Inc. of California and worked as a senior operations research analyst. During this period, Dr. Bozkaya has participated in many public and private sector projects and acted as an architect, developer, and analyst in applications of GIS to routing and delivery optimization. One such project completed for Schindler Elevator Corporation was awarded with the 2002 INFORMS Franz Edelman Best Management Science Finalist Award. Currently, he is a professor at Sabanci University, Turkey. His current research interests include routing and location analysis, heuristic optimization, and application of GIS as decision support systems for solving various transportation problems. Dr. Bozkaya is a winner of the 2010 Practice Prize awarded by the Canadian Operational Research Society.

Evelyn C. Brown
East Carolina University
Greenville, North Carolina

Evelyn Brown earned a B.S. in mathematics from Furman University, an M.S. in operations research from North Carolina State University, and a Ph.D. in systems engineering from the University of Virginia. Before becoming a full-time academician, Dr. Brown worked for Nortel Networks in Research Triangle Park, North Carolina, as an operations researcher and project manager. She currently works as a professor in the Department of Engineering at East Carolina University. Prior to her employment at East Carolina, Dr. Brown taught for seven years at Virginia Tech. While at Virginia Tech, her research focused on applications of genetic algorithms, particularly to grouping problems. Dr. Brown has published her work in journals such as the *International Journal of Production Research, Computers and Industrial Engineering, OMEGA: The International Journal of Management Science*, and *Engineering Applications of Artificial Intelligence.*

Buyang Cao
Tongji University
Shanghai, China

Buyang Cao earned his B.S. and M.S. degrees in operations research at the University of Shanghai for Science and Technology, and his Ph.D. in operations research at the University of Federal Armed Forces, Hamburg, Germany. He is working as an operations research team leader at ESRI Inc. in California. Currently, he is also a guest professor at the School of Software Engineering at Tongji University, Shanghai, China. Dr. Cao has been involved and led various projects related to solving logistics problems, including Sears vehicle routing problems, Schindler Elevator periodic routing problems, a major Southern California Sempra Energy technician routing and scheduling problems, and taxi dispatching problems for a luxury taxi company in Boston. He has published papers in various international journals on logistics solutions, and he has also reviewed scholar papers for several international journals. Dr. Cao's main interest is to apply GIS and optimization technologies to solve complicated decision problems from the real world.

Hongyan Chen
Missouri University of Science and Technology
Rolla, Missouri

Hongyan Chen is a Ph.D. student of engineering management and systems engineering at Missouri University of Science and Technology (formerly University of Missouri–Rolla). Her current research focuses on the application of real options, game theory, and optimization to the design of incentive mechanisms.

David Y. Choi
Loyola Marymount University
Los Angeles, California

David Choi is an associate professor of management and entrepreneurship at Loyola Marymount University in Los Angeles. Dr. Choi has written and published articles on technology management, strategy innovation, and entrepreneurship. He received a Ph.D. in management from the Anderson School at the University of California–Los Angeles (UCLA), and an M.S. and B.S. in industrial engineering and operations research from the University of California–Berkeley.

Horia Demian
University of Oradea
Oradea, Romania

Horia Demian is a Ph.D. lecturer at the University of Oradea–Romania, Faculty of Economics, Management–Marketing Department. He holds a Ph.D. in cybernetics and economic statistics with a thesis in the field of integrated information systems. Similarly, Dr. Demian has published books as a coordinator, as well as over 30 scientific papers in journals and conference publications. Moreover, he has annually participated in the Erasmus teaching staff mobility program since 2007 at Escuela Politécnica Superior de Alcoy, Universidad Politécnica de Valencia, Campus of Alcoy in Spain. As a result of his research, Dr. Demian has developed integrated computer applications for business and has more than 15 years of experience in computer programming.

Aliza R. Heching
IBM Thomas J. Watson Research Center
Yorktown Heights, New York

Aliza Heching is a research staff member in the Business Analytics and Mathematical Sciences Department at the IBM Thomas J. Watson Research Center, Yorktown Heights, New York. She received her Ph.D. in operations from Columbia University Graduate School of Business, New York. The focus of Dr. Heching's work at IBM is on using statistical methods and mathematical modeling and optimization to solve complex problems for IBM and its clients. She has spent a number of years working with IBM's global delivery teams developing analytical tools and mathematical models that are being used to improve the performance of IBM's service delivery. Dr. Heching is a member of the editorial boards of *Manufacturing and Service Operations Management*, *Production and Operations Management*, and the *International Journal of Operations Research*.

Yi-Chin Huang
University of Cincinnati
Cincinnati, Ohio

Yi-Chin Huang graduated from Ming Lung Senior High School in 1997 and attended Soochow University, Taipei, to study economics. After graduation, she worked for two years as a financial auditor and three years as an operations manager. She earned an M.B.A. and M.S. in quantitative analysis from the University of Cincinnati. Working with Dr. Michael Fry and Dr. Alex Lin as advisors, her master's thesis focused on vehicle routing optimization for a pharmacy prescription-delivery service. She currently works for Medpace Inc., a global clinical research organization based in Cincinnati.

Alberto Isla
University of Cincinnati
Cincinnati, Ohio

Alberto Isla is a doctoral student in the pharmaceutical sciences program of the James L. Winkle College of Pharmacy, University of Cincinnati, Ohio. He received his M.Eng. from the Universidad Politécnica de Madrid (Spain), M.B.A. in finance, and M.S. in quantitative analysis from the University of Cincinnati College of Business. He has a 15-year career as an IT consultant and financial risk analyst in Europe and in the United States. Isla's research involves the application of innovative operations-management techniques such as stochastic discrete-event computer simulation and lean processing to improve standard operating procedures and quality of service in health care settings, such as community pharmacies, hospitals, and emergency rooms. His dissertation analyzes performance and error rate measures at the pharmacy of a large children's hospital after the implementation of lean six sigma methodology.

Uday S. Karmarkar
UCLA Anderson School of Management
Los Angeles, California

Uday Karmarkar is the LA Times Professor of Technology and Strategy at the University of California–Los Angeles (UCLA) Anderson School of Management. His research interests are in information intensive sectors, operations and technology management, and strategy for manufacturing and service firms. Karmarkar coordinates the business and information technology global research network centered at UCLA, with research partners in 16 countries.

W. David Kelton
University of Cincinnati   ·
Cincinnati, Ohio

David Kelton is a professor in the Department of Operations, Business Analytics, and Information Systems at the University of Cincinnati, and director of the MS–Business Analytics Program; he is also a visiting professor of operations research at the Naval Postgraduate School. His research interests and publications are in the probabilistic and statistical aspects of simulation, applications of simulation, and stochastic models. He is coauthor of *Simio and Simulation: Modeling, Analysis, Applications,* as well as *Simulation with Arena;* he was also coauthor of the first three editions of *Simulation Modeling and Analysis,* all for McGraw-Hill. He was editor-in-chief of the *INFORMS Journal on Computing* from 2000 to 2007. He served as Winter Simulation Conference (WSC) Program chair in 1987, general chair in 1991, was on the WSC Board of Directors from 1991 to 1999, and is a founding trustee of the WSC Foundation. Kelton is a fellow of both the Institute for Operations Research and Management Sciences (INFORMS) and Institute of Industrial Engineers (IIE).

George N. Kenyon
Lamar University
Beaumont, Texas

George Kenyon is an associate professor of operations and supply chain management at Lamar University. He was recently named as the William and Katherine Fouts Faculty Scholar in Business. He received his B.S. in technology from the University of Houston, an M.S. in management science from Florida Institute of Technology, and a Ph.D. in business administration from Texas Tech University. Dr. Kenyon's industry experience includes positions in engineering, manufacturing, business planning, and supply chain management. His research has been published in several noted journals, including *Quality Management Journal, Journal of Marketing Channels,* and the *International Journal of Production Economics.*

Gül E. Okudan Kremer
Pennsylvania State University
University Park, Pennsylvania

Gül Okudan Kremer is an associate professor of engineering design and industrial engineering at the Pennsylvania State University. She received her Ph.D. from University of Missouri–Rolla in engineering management and systems engineering. Dr. Kremer's research interests include multicriteria decision analysis methods applied to improvement of products and systems and enhancing creativity in engineering design settings. Her published work appears in journals such as the *Journal of Mechanical Design*, the *Journal of Engineering Design*, the *Journal of Intelligent Manufacturing*, and the *Journal of Engineering Education and Technovation*. She is a member of the Institute of Industrial Engineers (IIE), the American Society of Mechanical Engineers (ASME), and the American Society for Engineering Education (ASEE). She is also a National Research Council–US AFRL Summer Faculty Fellow for the Human Effectiveness Directorate (2002–2004), an invited participant of the

National Academy of Engineering (NAE) Frontiers in Engineering Education Symposium (2009), and a Fulbright Scholar to Ireland (2010).

John F. Kros
East Carolina University
Greenville, North Carolina

John Kros is a professor of marketing and supply chain management in the College of Business at East Carolina University, Greenville, North Carolina. Before joining academia, Dr. Kros was employed in the electronic manufacturing industry by Hughes Network Systems (HNS), Germantown, Maryland. His research areas include simulation/process analysis, quality control, and applied statistical analysis. He has published in numerous journals, including *Interfaces*, the *Journal of Business Logistics*, *Quality Engineering*, *Quality Reliability Engineering International*, *Computers and Operations Research*, and the *Journal of the Operational Research Society*. Dr. Kros has a bachelor of business administration from the University of Texas at Austin, a master of business administration from Santa Clara University, and a Ph.D. in systems engineering from the University of Virginia.

Sanad H. Liaquat
Missouri University of Science and Technology
Rolla, Missouri

Sanad Liaquat graduated from Missouri University of Science & Technology (Missouri S&T, formerly University of Missouri–Rolla) with a master's degree in information science and technology. At Missouri S&T, Liaquat also earned a graduate certificate in enterprise resource planning (ERP) systems as well as the SAP-UMR University Alliance Program certificate. Liaquat's experience includes working as a software consultant at Thought Works Inc. and as a research assistant at Missouri S&T.

Alex C. Lin
University of Cincinnati
Cincinnati, Ohio

Alex Lin is an assistant professor of pharmacy systems and administration in the Division of Pharmacy Practice and Administrative Sciences, James L. Winkle College of Pharmacy, University of Cincinnati, Ohio. Dr. Lin's research interests and publications are in pharmacy automation technologies and medication use system design using computer simulation. He has carried out research projects for many major U.S. and international chain drugstores and health care enterprises, including Fortune 500 companies McKesson, Rite Aid Pharmacy, Procter & Gamble Healthcare, and Kroger Co.

Antonio Carlos Gastaud Maçada
Military Institute of Engineering
Rio de Janeiro, Brazil

Antonio Carlos Gastaud Maçada is a professor of information technology management and operations management at Universidade Federal do Rio Grande do Sul in Porto Alegre, Brazil. He earned his Ph.D. in business administration in 2001. Dr. Maçada is currently teaching and conducting research on supply chain governance, supply chain security and risk, and information technology management. He has had over 30 articles published in journals and 130 papers in conferences. He is a visiting professor at Texas A&M, Mays Business School.

Luiz Carlos Brasil de Brito Mello
Military Institute of Engineering
Rio de Janeiro, Brazil

Luiz Carlos Brasil de Brito Mello is a professor of industrial engineering at Universidade Federal Fluminense, Niterói, Brazil. He earned his Ph.D. in civil engineering in 2007 and his master's in industrial engineering in 2003 at Universidade Federal Fluminense. Dr. Mello is currently teaching and conducting research on supply chain management, logistics, project management, and quality management. He has had over 30 articles and papers published in journals and conferences. He has also worked for more than 34 years as a senior executive in multinational companies both in Brazil and abroad in engineering, procurement, logistics, and project management.

David Menachof
Hull University
Hull, United Kingdom

David Menachof is professor of Port Logistics at the Business School, University of Hull. He received his doctorate from the University of Tennessee and was the recipient of the Council of Logistics Management's Doctoral Dissertation Award in 1993. He is a Fulbright Scholar, having spent a year in Odessa, Ukraine, as an expert in logistics and distribution. Dr. Menachof's work has been published and presented in journals and conferences around the world. His research interests include supply chain security and risk, global supply chain issues, liner shipping and containerization, and financial techniques applicable to logistics.

Kim LaScola Needy
University of Arkansas
Fayetteville, Arkansas

Kim LaScola Needy is department chair and 21st Century Professor of Industrial Engineering at the University of Arkansas. She received her B.S.

and M.S. degrees in industrial engineering from the University of Pittsburgh, and her Ph.D. in industrial engineering from Wichita State University. Prior to her academic appointment, Dr. Needy gained significant industrial experience while working at PPG Industries and the Boeing Company. Her first faculty appointment was at the University of Pittsburgh. Her research interests include engineering management, engineering economic analysis, integrated resource management, and sustainable engineering. She is a member of the American Society for Engineering Education (ASEE), ASEM, the Association for Operations Management (APICS), the Institute of Industrial Engineers (IIE), and the Society for Women Engineers (SWE), and is a licensed professional engineer in Kansas.

Amílcar Oliveira
Universidade Aberta
Lisbon, Portugal

Amílcar Oliveira is an assistant professor of statistics in the Department of Sciences and Technology at the Universidade Aberta (UAb), Portugal, and he is co-coordinator of the master on statistics, mathematics, and computation program. He holds a Ph.D. in mathematics–statistical modeling, Universidade Aberta, and an M.S. in statistics and optimization, Faculty of Sciences and Technology, New University of Lisbon. Dr. Oliveira's research interests include statistical modeling as well as mathematical e-learning. He has published several papers in international journals, books, and proceedings regarding these fields. He is an integrated member of the Center of Statistics and Applications, University of Lisbon, and a collaborated member of Laboratory of Distance Learning, Universidade Aberta.

Teresa Oliveira
Universidade Aberta
Lisbon, Portugal

Teresa Oliveira is an assistant professor in the Department of Sciences and Technology at the Universidade Aberta (UAb), Portugal, and she is the coordinator of the master on biostatistics and biometry. She holds a Ph.D. in statistics and operations research–experimental statistics and data analysis, University of Lisbon; and an M.S. in statistics and operations research, Faculty of Sciences, University of Lisbon. Dr. Oliveira's research interests include experimental design and statistical modeling as well as mathematical e-learning. She has published several papers in international journals, books, and proceedings, and she has supervised several Ph.D. and master's theses. She is an integrated member of CEAUL–Center of Statistics and Applications, University of Lisbon, and is a collaborator in UIED–Research Unit on Education and Development, FCT–New University of Lisbon; in IPM–Preventive Medicine Institute, Faculty of Medicine, University of Lisbon; and in LEaD–Laboratory of Distance Learning, Universidade Aberta.

Elena Pérez-Bernabeu
Technical University of Valencia
Alcoy, Spain

Elena Pérez-Bernabeu is a lecturer with the Department of Applied Statistics, Operational Research and Quality at the Technical University of Valencia, Campus of Alcoy, Spain. She joined the university in October 2001. Pérez-Bernabeu has a bachelor's degree in telecommunications engineering and a master's in industrial organization. She got her Ph.D. in 2008 at the Technical University of Valencia. She teaches undergraduate and graduate courses in the Department of Applied Statistics, Operational Research and Quality. She has written a book on applied statistics and edited a book on new technologies, and she has participated in many conferences. Her research lines are control charts and decision making. She is a member of the Spanish Society of Statistics and Operational Research. She is also involved in the management of her home university, and she is the vice dean for International Relations at the Alcoy campus since 2008.

Alejandra Pérez-Bonilla
Open University of Catalonia
Barcelona, Spain

Alejandra Pérez-Bonilla was born in Chile. She was a postdoctoral researcher at the Internet Interdisciplinary Institute (IN3-UOC). She also collaborated as an assistant professor with the Department of Applied Mathematics I at the Technical University of Catalonia (UPC). She is a full-time professor in EUSS–Engineering at the Business Organization Department, teaching six subjects. She holds a Ph.D. in statistics and operational research from the Technical University of Catalonia (UPC). She holds an M.S. in industrial engineering and a civil industrial engineering degree, both from University of Santiago of Chile, and an engineer in industrial organization from Ministry of Education and Science of Spain. Dr. Pérez-Bonilla's research interests include both service and industrial applications of computer simulation, probabilistic algorithms, educational data analysis, multivariate analysis and data mining, statistics, and artificial intelligence such as clustering knowledge acquisition for concept formation in knowledge-based systems. She has also been involved in some international research projects.

Lian Qi
Rutgers Business School
Rutgers University
Newark, New Jersey

Lian Qi joined Rutgers Business School in September 2008. Prior to that, he was an assistant professor and the director of the Center for ERP at Missouri University of Science and Technology (formerly University of

Missouri–Rolla). His other work experience includes application consultant at SAP. He received a Ph.D. in industrial and systems engineering from the University of Florida (Gainesville) in 2006. His main research field is supply chain design. Dr. Qi's papers have been published in journals, such as *Transportation Science, Production and Operations Management,* the *European Journal of Operational Research,* and *Naval Research Logistics.* He won the Institute of Industrial Engineers Annual Research Conference best paper award in Operations Research in 2006. Dr. Qi is also the recipient of the Junior Faculty Teaching Award at Rutgers Business School in 2011, and Outstanding Faculty Award at the School of Management & Information System, University of Missouri–Rolla, in 2007.

Ruwen Qin
Missouri University of Science and Technology
Rolla, Missouri

Ruwen Qin is an assistant professor of engineering management and systems engineering at Missouri University of Science and Technology (formerly University of Missouri–Rolla). She holds a Ph.D. in industrial engineering and operations research. Dr. Qin's current research focuses on the application of optimization, real options, and optimal control to the manufacturing and service operations management, workforce engineering, transportation, and energy strategies.

Magno Jefferson de Souza Queiroz
University of Wollongong
Wollongong, Australia

Magno Jefferson de Souza Queiroz is a Ph.D. candidate in information systems and technology at the University of Wollongong, Australia. He received a bachelor's and a master's of computer science from the Federal University of Campina Grande, Brazil. Before commencing his Ph.D. research, Queiroz worked on the development of tools applied to IT governance practices at CHESF (a major Brazilian state electric utility) and collaborated with Hewlett-Packard (HP) researchers on the development of linkage models for business-driven IT management (BDIM). His research interests include the alignment of IT and business strategy, IT investment decision making, and the economic and organizational impacts of IT.

Jayant Rajgopal
University of Pittsburgh
Pittsburgh, Pennsylvania

Jayant Rajgopal has been on the faculty of the Department of Industrial Engineering at the University of Pittsburgh since January 1986. His current research interests are in mathematical modeling and optimization, global supply chains, operations analysis, and health care applications of operations research. He has taught, conducted sponsored research, published, or consulted in all of these areas. Dr. Rajgopal has over 60 refereed publications in books, conference proceedings, and scholarly journals, including *IIE Transactions, Operations Research, Mathematical Programming, Naval Research Logistics, Technometrics*, the *European Journal of Operational Research, Operations Research Letters*, and *Vaccine*. He holds a Ph.D. in industrial and management engineering from the University of Iowa, and is a senior member of the Institute of Industrial Engineers and the Institute for Operations Research and the Management Sciences. He is also a licensed professional engineer in the state of Pennsylvania.

Hosun Rhim
Korea University Business School
Seoul, Korea

Hosun Rhim is a professor of logistics, service, and operations management at the Korea University Business School in Seoul, Korea. Dr. Rhim has published articles on service operations and supply chain management. He received a Ph.D. in management from the Anderson School at University of California–Los Angeles, and a M.S. and B.S. in business administration from Seoul National University.

Jacques Philippe Sauvé
Federal University of Campina Grande
Campina Grande, Brazil

Jacques Philippe Sauvé received a Ph.D. in electrical engineering at the University of Waterloo, Canada. He is currently a professor in the Computer Science Department at the Federal University of Campina Grande where his efforts are concentrated in the areas of advanced architectures for software systems and IT management. Dr. Sauvé has published 10 books and over 70 papers in international journals and conferences. He has been and is on the technical program committees (TPCs) of many conferences and is on the editorial board of several international journals. He is a member of the IEEE and ACM.

Natalie M. Scala
University of Pittsburgh
Pittsburgh, Pennsylvania

Natalie Scala holds a Ph.D. in industrial engineering from the University of Pittsburgh, an M.S. degree in industrial engineering from the University of Pittsburgh, and a B.S. degree magna cum laude in mathematics from John Carroll University. Prior to the doctoral program, she interned as a technical agent at the Sherwin Williams Company and worked as an analyst at First Energy Corporation. She has also interned as a summer associate at RAND Corporation. She is currently employed by the U.S. Department of Defense. Her research interests include applications of operations research and engineering management in the utility industry and sports. She is a member of the American Society for Engineering Management (ASEM), the Institute of Industrial Engineers (IIE), and the Institute for Operations Research and Management Sciences (INFORMS).

Miguel Angel Sellés
Technical University of Valencia
Alcoy, Spain

Miguel Angel Sellés is a lecturer in the Department of Mechanical and Materials Engineering at the Technical University of Valencia, at its Campus of Alcoy, Spain. He joined the university in October of 2000. Dr. Sellés has a bachelor's degree in electronics engineering, and a master's in industrial organization. He received his Ph.D. in 2009 from the Technical University of Valencia. He teaches undergraduate and graduate courses in the Department of Materials and Mechanical Engineering, and supervises and advises students. He is participating now as technical chair of the International Research Group on Tribology in Manufacturing, and is also a member of the Society of Spanish Manufacturing Engineers since its foundation in 2004. Dr. Sellés is the author of books for teaching, several book chapters, and many conference papers. His research is focused on the field of polymer science, tribology, and sheet metal forming.

Kabir C. Sen
Lamar University
Beaumont, Texas

Kabir Sen is a professor of marketing, and chair of the Management and Marketing Department at Lamar University. He received his Ph.D. in marketing from Washington University in St. Louis. Dr. Sen's research interests are in franchising, service design, health care marketing, and sports economics. He has published in academic journals such as the *Journal of Advertising Research*, *Journal of Consumer Marketing*, and *Managerial and Decision Economics*.

Galina Sherman
Hull University
Hull, United Kingdom

Galina Sherman is a Ph.D. student at Hull University, Business School. Her current research is related to supply chain management, risk analysis, and rare event modeling.

Peer-Olaf Siebers
University of Nottingham
Nottingham, United Kingdom

Peer-Olaf Siebers is a research fellow at the University of Nottingham, School of Computer Science. His main research interest is the application of computer simulation to study human-centric complex adaptive systems. This is a highly interdisciplinary research field, involving disciplines such as social science, psychology, management science, operations research, economics, and engineering.

Mark S. Squillante
IBM Thomas J. Watson Research Center
Yorktown Heights, New York

Mark Squillante is a research staff member in the Business Analytics and Mathematical Sciences Department at the IBM Thomas J. Watson Research Center, Yorktown Heights, New York, where he leads the applied probability and stochastic optimization team. He received his Ph.D. from the University of Washington, Seattle. He has been an adjunct faculty member at Columbia University, New York, and a member of the technical staff at Bell Telephone Laboratories, Murray Hill, New Jersey. Dr. Squillante's research interests concern mathematical foundations of the analysis, modeling, and optimization of complex stochastic systems, including stochastic processes, applied probability, stochastic optimization and control, and their applications. He is a fellow of the Association for Computing Machinery (ACM) and the Institute of Electrical and Electronics Engineers (IEEE), and currently serves on the editorial boards of *Operations Research*, *Performance Evaluation*, and *Stochastic Models*.

Pragalbh Srivastava
Missouri University of Science and Technology
Rolla, Missouri

Pragalbh Srivastava received his master's degree from Missouri University of Science and Technology (formerly University of Missouri–Rolla) in information science and technology. His coursework was focused on enterprise applications and data warehousing. His work experience includes working in the health care IT industry with Cerner Corporation. His area of work includes database management, data integration, data warehousing, and providing reporting solutions using various business intelligence applications.

Reviewers

Matthew Bailey
Bucknell University
Lewisburg, Pennsylvania

Barry Barrios
Northwestern University
Evanston, Illinois

Bogdan Bichescu
University of Tennessee
Knoxville, Tennessee

Didac Busquets
University of Girona
Girona, Spain

Jose Caceres
Internet Interdisciplinary Institute
Barcelona, Spain

Steven M. Corns
Missouri University of Science
    and Technology
Rolla, Missouri

Yann Ferrand
Clemson University
Clemson, South Carolina

Albert Ferrer
Technical University of Catalonia
Barcelona, Spain

Miguel A. Figliozzi
Portland State University
Portland, Oregon

Pau Fonseca
Technical University of Catalonia
Barcelona, Spain

Alvaro Garcia
Technical University of Madrid
Madrid, Spain

Esteban Garcia-Canal
University of Oviedo
Oviedo, Spain

Sergio Gonzalez
Internet Interdisciplinary Institute
Barcelona, Spain

Sabrina Hammiche
University of Rennes
Rennes, France

Michael Hewitt
Rochester Institute of Technology
Rochester, New York

Patrick Hirsch
University of Natural Resources and
    Life Sciences
Vienna, Austria

Adamantious Koumpis
Research Programmes Division
ALTEC Software
Thessaloniki, Greece

Saravanan Kuppusamy
University of Cincinnati
Cincinnati, Ohio

Fermin Mallor
Public University of Navarre
Pamplona, Spain

David Masip
Open University of Catalonia
Barcelona, Spain

Section 1

Services and Decision Making

1

Stochastic Decision Making in Information Technology Services Delivery

Aliza R. Heching and Mark S. Squillante

CONTENTS

1.1 Introduction

The services industry has come to dominate activity in most advanced industrialized economies. In the United States alone, more than 75% of the labor force is employed in the services industry with an overall output representing around 70% of total industry output (Woods 2009). The effective and efficient management of delivery decision-making processes within the services industry is, on the one hand, critically important to any services business in practice, whereas on the other hand, very complex and difficult to achieve because of various practical problem characteristics. This is especially the case for the complex decision-making processes involved in the delivery of information technology (IT) services, whose many challenges include new revenue-cost delivery models, highly uncertain and potentially volatile client demands, complex service level agreement (SLA) models, and the need to capture various practical aspects of IT services delivery (e.g., customer-specific demands customer-dedicated IT environments, and the limited ability to pool resources and knowledge).

In this chapter, we consider such a set of problem IT services delivery management (SDM) and decision making. Specifically, we investigate a real-world IT services delivery environment (SDE) in which a services provider with global delivery locations responds to requests for service from a collection of customers around the world. Service requests arriving to the system vary with respect to required skills and associated target performance metrics, resulting in multiple service classes. The request arrival rate is highly nonstationary and variable over relatively coarse time scales (e.g., hourly, daily), driven in part by customer business practices and the global customer base as well as inherent uncertainty in the request arrival process. The agents handling these requests have different scope and levels of training that render differences in both breadth and depth of agent skills. Agents are organized in virtual or colocated teams, where all agents in a team have common training. Contractual SLAs, which dictate the target performance metrics, also specify the monetary penalties that are applied in the case that SLAs are breached. One challenge to the service provider is to optimize the staffing decisions related to the different agent teams so as to optimize a financial function of managing the services delivery system.

Despite their importance in practice, the complex management and decision-making problems of IT services delivery have received very limited attention. We address this class of decision-making problems based in part on our experiences with a global IT services delivery provider (SDP) and the challenges observed while working in that SDE. Toward this end, we propose and investigate a decision-making framework consisting of a practical two-stage approach that combines methods from applied probability, queueing theory, combinatorial optimization, and simulation-based optimization. For the first stage of our approach, we develop general stochastic models of the IT services delivery decision-making processes that represent the inherent uncertainties, potential volatilities, and risks associated with real-world IT SDEs. We derive approximate analytical solutions to these stochastic models within the context of a profit-based objective and formulate the corresponding optimization as a mathematical programming problem. The resulting approximate solutions consist of queueing-theoretic formulas that capture the stochastic system dynamics and provide an efficient, nearly optimal solution of the original decision management problem by exploiting advanced algorithms for various classes of such mathematical programming problems. For the second stage of our approach, we take this approximate first-stage solution as a starting point and utilize a simulation-based optimization approach to move from the nearly optimal local solution of our analytical methodology to the globally optimal solution of the original deci-sion management optimization problem. Our analytical methodology sup-ports general optimization under a wide range of assumptions, scenarios, and workloads, thus enabling us to efficiently explore the entire design space and then leverage our simulation-based optimization in a more surgi-cal manner to obtain globally optimal solutions for portions of the perfor-mance–profit space with the greatest importance or sensitivity.

There are several important reasons for adopting such an approach to IT SDM, which develops and integrates a stochastic analytical methodology together with a simulation-based optimization methodology into a unified solution framework. Most important, from the perspective of the actual busi-ness, it is critical to obtain both an accurate and efficient solution to the prob-lems within a complex SDE. This necessitates leveraging the often opposing (accuracy–efficiency) benefits of analytical and simulation methods while mitigating their disadvantages. Analytical methods typically resort to simpli-fied versions of complex practical systems because they are usually unable to address all real-world complexities, often resulting in highly efficient, high-quality solutions to simplified problems. On the other hand, while simula-tion allows one to consider all of the complexities of real-world application environments, the black-box optimization of such simulation-based systems typically requires a significant amount of time and computational resources in order to iterate among possible solutions identified by the optimization model and their evaluation via simulation. By developing our stochastic analytical methodology (to very efficiently obtain a nearly optimal solution) and our simulation-based optimization methodology (to fine-tune the initial solution in a small number of simulation runs) and integrating the comple-mentary nature of these methodologies to obtain an optimal solution, we are realizing the goals of an accurate and efficient practical approach to IT services delivery decision management. Our framework is similar in spirit to the recent work of Dieker et al. (in preparation), although the underlying mathematical methods and application area are quite different.

Although very little attention has been paid to the complex manage-ment and decision-making problems of IT services delivery, there are a few research areas that are somewhat related and worth mentioning. The most closely related area, which has received a great deal of attention, concerns the operational management of call center environments (CCEs). Several such studies consider various mathematical models of the staffing problem in call centers with skills-based routing. Gurvich and Whitt (2007a) describe a policy for assigning service requests to agent pools depending upon state-dependent thresholds of agent pool idleness and service class queue length. Gurvich and Whitt (2007b) propose another solution by collapsing the multi-class problem into a single class problem and determining the staffing levels under the assumption that all agents can handle all service requests. Gurvich et al. (2010) propose a formulation of the skills-based routing problem by converting mean performance SLA constraints to chance-based constraints such that a service delivery manager can select the risk of failing to meet the former constraints. We refer the reader to Gans et al. (2003), Aksin et al. (2007), Balseiro et al. (in preparation), and the references therein for further details. A few recent studies have attempted to more closely incorporate the complexities of various operational decisions in CCEs using a wide range of approaches that combine simulation and optimization techniques. Atlason et al. (2008) solve a sample-mean approximation of the problem using a simulation-based analytic center cutting-plane method. Cezik and L'Ecuyer (2008) apply this approach to large problem instances and develop heuristic methods to address numerical challenges. Feldman and Mandelbaum (2010) employ a stochastic approximation approach to determine optimal staffing, where service levels (SLs) are modeled as strict constraints or costs in the objective function, and simulation is used to evaluate SL attainment. It is important to note, however, that there are many fundamental differences between CCEs and the real-world IT SDEs of interest in this chapter, as explained throughout the chapter. These fundamental differences and the actual business motivating our study led us to develop a novel decision-making solution framework that addresses the needs of real-world IT services delivery processes in practice.

The remainder of this chapter is organized as follows. We first describe the motivating IT SDE and related key business issues. We then present our formulation of the SDM problems, followed by a description of our two-stage solution framework. A representative sample of results from a vast collection of numerical experiments is presented next. We then conclude with a discussion of our experience with the motivating real-world IT SDE and the benefits of our decision-making solution framework to the business in practice.

1.2 Business Environment and Issues

1.2.1 Information Technology Services Delivery

Outsourcing of IT services is on the rise,[*] involving important shifts from the traditional motivation and operation of IT services outsourcing. Although the traditional usage was viewed as a cost savings measure, outsourcing providers are increasingly valued as strategic partners to enable standardization, process innovation, and operational efficiency. IT services delivery has moved from the traditional mode of dedicated on-site customer support to *services delivery locations* (SDLs) where agents are colocated at centralized locations from which support is provided to multiple customers. SDPs have also moved to a so-called *global delivery model* (GDM), in which SDPs respond to customer IT service requests from on-site and geographically distributed

---
[*] http://www.gartner.com/it/page.jsp?id=1322415

off-site locations. The geographical dispersion of SDLs allows SDPs to take advantage of available expertise in local markets as well as local labor rates. Additional GDM benefits include round-the-clock service due to time zone differences in the locations from which services are provided, ensured business continuity and resiliency in case of disaster, and easy scaling of customer core business components while providing required increased IT support.

In this new GDM, there are two predominant models for organizing skills across the network of SDLs: *center of competency* (CoC) and *cross competency* (xC). The CoC model organizes SDLs as centers of competency in a single IT area. All customer requests associated with the specific IT area are routed to the SDL where agents trained at varying levels of skill in this targeted area are colocated. Thus, requests arriving for a customer whose contract contains a menu of IT services may be routed to multiple SDLs depending upon the IT area required for each request. This model benefits from enabling greater economies of scale at each SDL and allowing increased technical knowledge sharing across the colocated agents within an SDL. The challenge lies in ensuring that customers experience seamless and consistent quality of care across all SDLs, as this delivery model requires the sharing of customer knowledge across SDLs. The xC model populates each SDL with a wide array of skills such that all customer requests arriving to the SDP can be routed to a single SDL. This model benefits from allowing a single SDL to gain an end-to-end view of a customer's supported systems as well as all outstanding customer requests at any point in time, enabling more holistic customer support and improved customer satisfaction. However, the xC model leaves resources skilled in each IT area more distributed, thus potentially requiring an increase in the overall number of agents for the SDP to achieve economies of scale. An SDP network of SDLs may also contain a mixture of both types of models where some SDLs serve as CoC and others as xC locations. Availability of resources and skills in the local market often is a significant factor in determining whether an SDL should follow the xC or CoC model of skills organization. These decisions may be revisited by the SDP as new SDLs are opened or as a significant number of new agents are hired and added to the delivery network.

We next define the different types of requests handled by an IT SDE. An *incident* is an event that is not part of standard operations and results in service interruption or reduced quality of service (QoS) to the customer, where a *major incident* is one with a (potentially) high impact. SDPs typically maintain separate management systems and processes dedicated to handling major incidents, often requiring cross-SDL coordination, escalation, additional resources, and increased communications. A *request for change* involves proposing a change to any component of an IT infrastructure or any aspect of an IT service. More basic user requests include risk-free services such as standard changes, new user access to an IT service, or password reset. Finally, *project* requests are typically highly complex, multistage customer requests that involve multiple agent teams to ensure successful execution and closure, requiring a longer relative duration than other requests handled by the SDP. We shall focus on the process for managing and staffing the agent teams that respond to requests other than project workload.

Requests to handle incidents or major incidents arrive to the SDE from various sources such as telephone, Web, e-mail, and monitoring systems. These requests are documented as *incident tickets* in *ticketing systems,* where the latter may be customer specific or shared by multiple customers. The ticket includes information on customer name, date and time when the request was opened, urgency of the request, target resolution time, brief problem description, and classification of the problem within a standard taxonomy of problem classifications that are used to route tickets to agent teams based upon the skills of the agent teams. Ticket target resolution time is calculated based on the customer, problem classification, and urgency. The volume of incident requests typically varies over time with predictable variability, such as seasonal patterns, observed over longer periods of time. Typically, an increased demand on the systems results in higher volume of incidents and reported problems.

Requests for change are managed through a customer change coordinator who reviews the request details and forwards the request to a designated SDP change analyst. A change request requires detailed documentation regarding the different components of the change. The change analyst reviews the request for completeness and opens a ticket in the change ticketing system. A change impacts the customer production systems and therefore, unless an emergency (addressing major system failure), is scheduled to be implemented over the weekends or during late night or at other times when affected systems usage is low.

We now discuss the decisions faced by the SDP when a potential new customer arrives to request service. The customer requests a selection of services from the menu of IT services offered by the SDP. The SDP reviews the customer requests and estimates the demand that will be driven by these requests and the skills required to support the customer. The SDP then decides whether to accept the customer into its portfolio of existing customers based upon this expected demand, forecasted skill requirements relative to available skills of the SDP, the workload arrival pattern as compared with availability of agents, and other factors. Pricing and marketing considerations are additional factors that are considered as part of this decision. The SDP will also review any special requests placed by the customer. For example, the customer may insist that nonstandard equipment be used for service provisioning or that support be provided on-site at the customer location. (These are typically more costly for the SDP, often requiring dedicated resources and customer specific technology.) Other examples of special requests include requirements for special certification or restrictions on access to the customer account.

Customer contracts also contain QoS constraints in the form of SLAs. These contractual agreements between a customer and SDP involve certain main features:

1. *Service level target* (SLT) indicating the target QoS that must be attained by the SDP, where examples include target time within which an SDP must react to customer requests (maximum waiting time in queue, stated in business hours or calendar hours), target time within which an SDP must resolve a customer request (maximum total time in system including time in service), maximum number of minutes allowed for network downtime, and so on.

2. *Time frame* over which SDP performance is measured for QoS, rather than against each individual request, such that the time frame specified in the SLA indicates this rolling measurement time horizon.

3. *Scope of coverage* in which there are often different SLAs in place for the different types of services included in the contract between the customer and the SDP.

4. *Percentage attainment* (PA) specifying the percentage of customer requests within the SLA scope whose measurement time frame must achieve the SLT.

Examples of SLAs include 90% (PA) of all severity 1 incident tickets (scope of coverage) opened each month (time frame) must be resolved within 4 hours (SLT), or 99% (PA) of all console alerts (scope of coverage) each month (time frame) must be reacted to within 15 minutes (SLT). We note that a single customer may, and typically does, have multiple SLAs in place for the different services contracted with the SDP. We also note that SLAs are contracted on a customer-by-customer basis. As such, there is significant variance between the contract terms (in particular, the SLT and PA) across the SLAs for the different customers supported by the SDP. As an example, for our motivating SDP, we often find approximately 30 SLAs for a given customer and over 500 distinct SLAs across the customer base.

Once it accepts the customer into its portfolio of existing customers, the SDP must determine which SDL will support each of the different types of requests selected by the customer from the menu of technology offerings supported by the SDP. This decision can be made dynamically, that is, each time a request arrives, the SDP can assess the number of requests outstanding at each SDL as well as the number of available agents. In practice, however, these decisions are made on a more static basis, where the SDP routes all customer requests for each of a given technology type to a selected SDL(s). In some cases, the routing decisions depend upon the complexity of the request in addition to the technology area. (The SDP may decide to staff some SDLs with less skilled agents and other SDLs with agents with different skill levels; these decisions may also be driven by limited availability of more highly skilled agents in some geographic areas.) Alternatively, routing decisions may depend upon time of day in the case that a customer contracts for 24-hour support. Support is sometimes provided by multiple SDLs where each provides service during its local regular business hours. Another consideration is legal requirements on classes of requests that are restricted from cross-ing regional borders, which vary by geography and by request class. Other explanations for more static assignment of customers to agent teams include geography *ownership* of customers from a profit and loss perspective.

This describes the organization of the network of SDLs and how workload is assigned to SDLs within this network. We next consider the local organi-zation within each SDL, in which agents are grouped into teams. Each team consists of agents with identical skills. Skills may include depth or breadth of knowledge in a technology area, language skill, and required certifications. A team of agents will support one or more customers. The SDP faces various challenging problems to address the optimal clustering of customers. This includes determining the optimal composition of the teams of agents and the optimal assignment of customers to these teams of agents. The former is not a primary subject of this chapter, whereas the latter is directly included as part of our decision-making solution framework.

When customer requests arrive to the SDP, they are routed to supporting agent teams. In cases where multiple agent teams have matching skills, there are policies for assigning requests to teams. For example, the agent teams may be located in opposite time zones, in which case the SDP will route customer requests so as to minimize the need for *off-shift* (overtime) hours in each of the agent teams. In this respect, the SDP is leveraging the GDM and considers the time at which requests arrive when determining how to route requests. The SDP may adopt a *follow the sun* approach wherein customer workload is routed to the SDL based upon the supporting technology area and working hours.

A dispatcher is assigned to each team of agents, possibly even shared across a few agent teams. The role of the dispatcher is to review requests as they arrive to the agent team and evaluate the complexity of each request. Complexity is most often a function of the number of interactions required to complete the task (e.g., people, technologies, systems). The dispatcher also assesses the urgency of the customer request and prioritizes the request in the queue of cus-tomer requests waiting to be handled. Requests are assigned to agents based upon agent availability. Due to the priority of some customer requests as well as the differences in handling times observed among different types of customer requests, the environment operates under a preemptive-resume regime; for example, one observes 3- to 4-minute mean handling times for some requests while other requests have a mean handling time of 40 minutes. These differ-ences in mean handling times are also taken into consideration (in addition to request priority) upon assigning requests to agents. The dispatcher continually monitors agent availability and whether each of the requests will meet its SLT.

1.2.2 Key Issues

The SDP faces a number of challenging decisions within the SDE. First, the number of agents with each of the different sets of skills must be determined.

This decision is usually made prior to having knowledge of the uncertain arrival rate of the workload for the different classes of requests. There is typically little ability to modify the overall staffing levels on a frequent basis, which would be equivalent to hiring and terminating agents, bringing in temporary workers, or using significant overtime. Moreover, the work performed by these agents often requires highly specialized technical knowledge, and use of temporary workers is an impractical approach. With respect to overtime, shifts in this environment are often 12 hours long and human resources regulations limit overtime usage. In other cases, there are legal restrictions on the number of hours of overtime (per month as well as a total amount of overtime per year) that agents may work. Thus, reliance on overtime is not a viable approach to manage the uncertainties in an SDE. The second challenge faced by the SDP is determining a policy for prioritizing requests in the system. This policy can be dynamic and may depend upon factors such as the volume of requests currently assigned to the agent team (both in queue and being handled) and PA to date for the current attainment review period. The agent team may be organized such that there is a single queue in which all workload is prioritized or multiple queues that are used to prioritize subsets of the workload with policies governing how customer requests are assigned to the different queues. The third challenge is determining a policy for assigning arriving customer requests from the various classes of work to the different agents. The policy may be dynamic and may depend upon factors including, for example, required skills, workload in the system for each class of work, and agent utilization. The objective is to determine minimal cost staffing levels in this SDE where staffing costs and the penalties for any SL violations are the main cost contributors.

The challenges faced by the SDP are further compounded by the nature of SLAs, which often have short targets while PA is measured against aggregate performance over a longer time horizon. Although there are many types of SLAs, we focus on the most common class found in practice and those most common in our motivating real-world SDE. Specifically, the SLA specifies a minimum level of service over a given period of time, which may specify the proportion of time that a customer network is available over a specified period or the proportion of console alerts that are responded to within 15 minutes over a specified period. Again, the SLA applies to an extended period of time rather than on a request-by-request basis. Thus, over the SLA measurement time window, the system may sometimes overachieve and sometimes underachieve on SLTs. However, the objective is to at least meet targets over the extended time horizon. Monetary penalties for missing targets can be significant and are typically dependent upon the criticality of the request class.

We note that SL attainment is measured on a customer request class basis and is measured based on performance over each review period. Thus, in some cases it may be preferable to prioritize lower priority requests if the weighted cost of missing SLAs for higher priority class requests is lower than that of the lower priority requests (weighted by the likelihood of missing SLAs for the prevailing review period). In reality, the SDP considers not only the contractual SLA but customer satisfaction as well. The customers are often dissatisfied when targets are missed or if there are high priority infrastructure problems, even if the SLAs for the prevailing review period are met. In addition, one often finds that customers expect QoS that exceeds the contractually agreed upon level of service, for example, faster response times or services beyond the contract. Such situations increase the challenges of an SDP in using SLAs as a constraint for determining required staffing levels.

Another practical consideration is that there may be contractual constraints limiting the minimum number of agents available during specified hours of the day (irrespective of workload). In other cases, there are physical constraints that may again place a lower bound on the number of agents that must be available in a team; for example, the agent team monitoring events that appear on consoles may be constrained by the physical placement of the consoles. The distance between consoles combined with a tight reaction time SLA on console alerts could require a higher number of agents than the workload appears to require.

We also observe that the workload arrival rates are often nonstationary over the entire time horizon. Some of the variability in the arrival rate of requests is explained by the business environment; for example, nights and weekends may be assigned higher volume of scheduled maintenance workload for systems where the primary customer activity occurs during business hours. CCEs typically respond to this scenario by designing highly flexible shift schedules that match the fluctuations in arrival of workload. These flexible shift schedules can be supported due to the high volume of workload and the relatively lower fragmentation in skills required to respond to the different requests that arrive to a call center. Such highly flexible shift schedules are not typically observed in IT SDEs, which can be explained by the more sparse skills matrix that is observed in an SDE. IT services delivery requires greater expertise and deeper skills relative to the volume of requests arriving for each of the different skills. Thus, the request volume is (relatively) low and does not support the associated numerous shifts that would be required to meet the highly nonstationary arrival patterns for each of the different classes of workload.

The decision-making requirements for staffing in an SDE are sensitive to fluctuations in the arrival process, especially in cases where the target times are short. These staffing levels depend upon assumptions regarding the shape and parameters of the handling time distributions and differences in these values for a given request class across agent teams. Target times are not necessarily monotonic in urgency across the request classes, yet customers anticipate that critical requests will receive the highest priority irrespective of service period-to-date attainment. There are multiple request classes and multiple agent teams, and SL attainment is measured with respect to performance across multiple time periods, creating a link across the time periods. Determining optimal staffing levels per period may result in overstaffing as it will ensure that SLTs are met in each period rather than across the attainment measurement period. This is why, in practice, our decision-making solution framework presented in Sections 1.4.1 and 1.4.2 determines the optimal staffing levels and shift schedules over the entire time horizon.

Another challenge faced in practice is the dispatching of workload to agents in a specific agent team. To make optimal assignment decisions and determine the likelihood of meeting SLTs, the dispatcher must have knowledge of the *state of the system* at all times. This includes volume of workload in queue and being handled (by request class), time remaining until target for each request in the system, set of skills associated with each agent, percentage completion for each request currently being handled, measurement period-to-date, PA for each request class, and current availability for each agent. However, in practice, the dispatcher rarely has visibility into the majority of this information. Unlike CCEs, there typically are not automated systems to capture the state of an agent. These notifications also are not always reflected (in real time) in any systems. Moreover, agents rarely update the systems while handling customer requests to reflect the current status of the request and percentage completion. Finally, in some cases, system integration issues prevent the dispatcher from receiving an automated view of the volume of requests in the system. This hinders the ability to install automated support to the dispatching role as the required data feeds are often unavailable.

1.3 Problem Formulation

Motivated by this business context, we consider an IT SDE supporting multiple types of requests arriving from multiple customers. The different types of requests arrive from independent stochastic processes to capture uncertainties in actual request patterns. These different types of requests may be correlated, but for the most part the relationship between arrivals processes of requests is not well understood or is anecdotal. For example, there is anecdotal evidence that failed requests for change that are executed over a weekend (when fewer users are using the systems) render an increase in reported incidents in the early days of the following workweek. However, due to limited data availability as well as limited understanding of the nature of the relationship between these two types of requests (weekend and start of workweek), it can be difficult to directly model the nature of the correlation between the arrival processes of these two types of requests. We generally model these effects within our stochastic analytical methodology as a sequence of stationary intervals each consisting of independent stochastic arrival processes with any correlations captured through the connections among the stochastic processes of different intervals.

The arriving requests are handled by agents who are grouped into teams where agents in the same agent team have identical (breadth and depth of) skills. For each request type and each team handling such requests, the handling time follows a general probability distribution that captures uncertainties in request handling times fitted against data. As a specific example, two agent teams may have identical breadth of skills where the first team is comprised of less experienced agents (e.g., recent hires) and the second team is comprised of more experienced agents. Each of these teams could respond to requests for which they have the skills, but the handling times for the less experienced agents may be significantly longer (depending upon the complexity of the request and the expected resolution). Requests arriving to the system can be described as belonging to one of a number of different classes of service, where service classes are differentiated by the skills required to complete the request. As such, service classes can be handled by subsets of the agent teams.

The scope of services (the types of requests) supported by the SDP varies on a customer-by-customer basis depending upon the scope of the customer contract. The delivery of all requests must adhere to some form of SLA, further refining all classes of service to include both required skills and request targets. Although there are many possible types of SLAs, we consider the most common class of SLAs found in practice as described in the previous section.

The following decisions must be made in this SDE: (1) Required number of agents with each of the different sets of skills; (2) policy for prioritizing work in the system; and (3) policy for the assignment of arriving workload from various classes to the different teams of agents. We consider a profit-based objective where revenues are gained for requests, penalties are incurred for SLA violations, and costs are incurred for the staffing levels of agent teams. More specifically, the SDP receives revenues as a function of $\mathcal{R}_{j,k}$ for class k requests that are handled by agent team j, incurs penalties as a function of $\mathcal{P}_{j,k}$ for class k requests whose waiting or response times from team j exceed a function of their SLTs $\mathcal{Z}_{j,k}$ with respect to their PA $\mathcal{A}_{j,k}$, and incurs costs as a function of $\mathcal{C}_{j,k}$ for the staffing capacity levels C_j of each agent team j. The aforementioned decisions are made in order to maximize the profit received by the IT SDP under the revenue, penalty, and cost functions, subject to business constraints.

1.3.1 Stochastic Model

The scope of services supported by the IT SDE varies on a customer-by-customer basis, but all customers are supported over a long-term time horizon comprised of different work shifts and time-varying request arrival and handling time patterns. Our system model of the IT SDE consists of partitioning this long-term time horizon into stationary intervals within which the underlying stochastic behaviors of the system are invariant in distribution to shifts in time across a sufficiently long period. We then model each stationary interval of the system and combine these stationary stochastic models to represent the entire long-term time horizon together with the overlap of different work shifts. In what follows, to elucidate the exposition, our focus will be on any given stationary interval of system behavior that is of interest.

We generally model the IT SDE as a collection of J priority queueing systems, one for each team of agents indexed by $j = 1, \ldots, J$. Requests from each of K classes, indexed by $k = 1, \ldots, K$, arrive to the system according to an independent stochastic process $A_k(t)$ with finite rate $\lambda_k = 1 / \mathbb{E}[A_k]$. Upon arrival, class k requests are assigned (routed) to the queueing system of agent team j with a mean rate denoted by $\lambda_{j,k}$ such that $\lambda_k = \sum_{j=1}^{J} \lambda_{j,k}$ and where $\lambda_{j,k}$ is fixed to be 0 whenever team j does not have the appropriate skills to handle class k requests. This results in an arrival process of class k requests to the team j queueing system that follows a stochastic process $A_{j,k}(t)$ with finite rate $\lambda_{j,k} = 1 / \mathbb{E}[A_{j,k}]$. The times required for agent team j to handle class k requests are assumed to be independent and identically distributed according to a common general distribution function $S_{j,k}(\cdot)$ with finite first three moments $\mathbb{E}[S_{j,k}^i] = \int_0^\infty t^i dS_{j,k}(t)$, $i = 1, 2, 3$, and mean service rate denoted by $\mu_{j,k} = 1 / \mathbb{E}[S_{j,k}]$. Similarly, the interarrival times for class k requests to the agent team j queueing system are assumed to have finite first three moments $\mathbb{E}[S_{j,k}^i] = \int_0^\infty t^i dS_{j,k}(t)$, $i = 1, 2, 3$. Let C_j denote the number of agents that comprise team j, and define $A_{j,k}^{(1)}, A_{j,k}^{(2)}, \ldots$ and $S_{j,k}^{(1)}, S_{j,k}^{(2)}, \ldots$ to be the sequences of interarrival time and handling time random variables (r.v.s), respectively, for class k requests at the team j queueing system where $A_{j,k}^{(n)} \overset{d}{=} A_{j,k}$ and $S_{j,k}^{(n)} \overset{d}{=} S_{j,k}$.

The requests arriving to the agent teams may vary significantly with respect to their urgency and handling times. For example, one class of requests may require agent teams to respond to alerts indicating failures on critical customer systems. Other requests may require the agents to perform checks to verify that no failures are observed on any systems. Finally, another class of requests may require agents to perform periodic maintenance or system health checks. These three examples of requests have significantly different customer impact if delayed; this dictates their priority ordering in the system. Correspondingly, our model assumes that each team employs a group-based fixed priority scheduling policy in which the k request classes are mapped into L priority groups such that requests having group index ℓ are given priority over requests having group index ℓ' for all $1 \leq \ell < \ell' \leq L$. Thus, requests in priority group 1 have absolute priority over those in group 2, which in turn have absolute priority over those in group 3, and so on. Let G_ℓ be the set of request class indices comprising priority group ℓ, and let σ be the mapping from the set of request class indices $\{1, \ldots, K\}$ to the set of priority group indices $\{1, \ldots, L\}$, where $\sigma(k) = \ell$ if and only if $k \in G_\ell$, or in other words class k is in group $G_{\sigma(k)}$.

We consider a preemptive-resume scheduling discipline, in which the handling of preempted requests is resumed from where they left off without any overhead. This assumption models the actual practice of the real-world SDE, as previously noted, where responses to requests often follow a fixed set of procedures and a preempted request is resumed at the procedure point where it was paused. Requests within each priority group are resolved in a first-come, first-served (FCFS) manner, consistent with the goal of achieving SLTs with relatively low variance (Kingman 1962). For further consistency with the real-world SDE, every agent team follows the same group-based fixed priority scheduling policy. The system dynamics are such that class k requests arrive to the team j queueing system according to the assignment policy rendering $A_{j,k}(\cdot)$, wait in the corresponding class k queue to be resolved (including possible preemptions), and leave the system upon completion according to $S_{j,k}(\cdot)$.

We consider the aforementioned profit-performance model based on the notion that revenues may be gained for handling requests and costs may be incurred for both SLA violations and agent staffing levels. Let $T_{j,k}$ denote the r.v. for the stationary response time of class k requests handled by team j, $W_{j,k}$ the r.v. for the stationary waiting time of class k requests handled by team j, and $R_{j,k}$ the r.v. for the stationary residence time of class k requests handled by team j, where $T_{j,k} = W_{j,k} + R_{j,k}$. Define $T_{j,k}^{(1)}, T_{j,k}^{(2)}, \ldots$ and $W_{j,k}^{(1)}, W_{j,k}^{(2)}, \ldots$ to be the sequences of response time and waiting time r.v.s for class k requests handled by team j, respectively, such that $T_{j,k}^{(n)} \stackrel{d}{=} T_{j,k}$ and $W_{j,k}^{(n)} \stackrel{d}{=} W_{j,k}$. SLAs can be defined for each customer class k as a function $\mathcal{S}_k(\cdot)$ of $T_{j,k}^{(n)}$ or $W_{j,k}^{(n)}$ or both. As previously noted, the SDP receives revenue as a function of $\mathcal{R}_{j,k}$ for each class k request that is handled by agent team j, incurs a penalty as a function of $\mathcal{P}_{j,k}$ for the nth class k request handled by agent team j whenever the SLA function $\mathcal{S}_k(T_{j,k}^{(n)}, W_{j,k}^{(n)})$ exceeds the corresponding SLT $\mathcal{Z}_{j,k}$ relative to the corresponding PA $A_{j,k}$, and incurs a cost as a function of $\mathcal{C}_{j,k}$ for the capacity C_j of team j. To be consistent with the real-world SDE of interest, we have the same per-class SLT and PA for every team, that is, $\mathcal{Z}_{j,k} = \mathcal{Z}_k$ and $A_{j,k} = A_k$.

1.3.2 Optimization Model

Our objective is to determine the optimal staffing capacity levels C_j^* for each agent team j and the optimal assignments (routing) of class k requests to agent team j that maximize expected profit under the described profit-performance model with respect to all customer class SLAs, given class–team assignment constraints and exogenous arrival and handling time processes $A_k(\cdot)$ and $S_{j,k}(\cdot)$. The class–team assignment decision variables are based on a simplifying modeling assumption in which assignment decisions are probabilistic such that a request of class k is assigned to agent team j with probability $P_{j,k}$, independent of all else. This approach still allows us to address important assignment features observed in practice, such as balancing agent utilization or reserving agents with unique skills. However, when other nonprobabilistic assignment mechanisms are used in practice (e.g., weighted round robin), our optimal solutions within this probabilistic modeling framework can be applied to set the parameters of these mechanisms (e.g., the per-class weights). The overall decisions additionally include the priorities attributed to each collection of requests assigned to each agent team. For now, we will assume that the priority structure has already been determined from the perspective of the business, that is, the priority of requests is determined when the customer and the SDP enter into a contractual agreement. We shall revisit the removal of this assumption in Section 1.6.

To simplify the formulation, suppose the optimal assignments of class k requests satisfy $\lambda_{j,k}/(C_j\mu_{j,k}) < 1$. Recall that $\lambda_{j,k}$ and $\mu_{j,k}$ are the mean arrival and service rates for class k requests handled by team j, respectively. For expository convenience, consider the SLAs of class k to be such that penalties are applied according to the SLT indicator function $I\{S_k(T_{j,k}^{(n)}, W_{j,k}^{(n)}) > Z_k, A_k\}$, which takes on the value 1 (0) when there is an SLA violation (otherwise). Furthermore, we initially assume that the revenue, penalty, and cost functions are linear in the total number of requests, number of SLA violations, and number of agents, revisiting these assumptions in Section 1.4.1.2. We then have the following formulation of the IT services delivery optimization problem (OPT-SDP):

$$\max_{C,P} \sum_{j=1}^{J} \sum_{k=1}^{K} \mathbb{E}\left[\sum_{n=1}^{N_{j,k}(T)} R_{j,k} - P_{j,k}I\{S_k(T_{j,k}^{(n)}, W_{j,k}^{(n)}) > Z_k, A_k\} \right] - C_{j,k}C_j \qquad (1.1)$$

$$\text{s.t.} \quad \sum_{j=1}^{J} P_{j,k}\lambda_k = \sum_{j=1}^{J} \lambda_{j,k} = \lambda_k, \quad k = 1,\ldots,K \qquad (1.2)$$

$$P_{j,k} = \lambda_{j,k} = 0, \quad \text{if } \mathcal{I}\{j,k\} = 0, \quad k = 1,\ldots,K, j = 1,\ldots,J \qquad (1.3)$$

$$P_{j,k}\lambda_k = \lambda_{j,k} \geq 0, \quad \text{if } \mathcal{I}\{j,k\} = 1, \quad k = 1,\ldots,K, j = 1,\ldots,J \qquad (1.4)$$

$$\sum_{j=1}^{J} P_{j,k} = 1, \quad k = 1,\ldots,K \qquad (1.5)$$

where $\mathbf{C} \equiv (C_j)$, $\mathbf{P} \equiv (P_{j,k})$, $N_{j,k}(t)$ is the cumulative number of class k requests assigned to team j through time t, T is the time horizon of interest, and $\mathcal{I}\{j,k\}$ is the class-team assignments indicator function equal to 1 (0) when class k requests can be handled by agent team j (otherwise). The vectors $\mathbf{C}$ and $\mathbf{P}$ are the decision variables we seek to obtain, with all other variables as input parameters. Note that the objective function in Equation (1.1) is a formal representation of the profit-performance model described in the previous section.

The optimization (OPT-SDP) is a general formulation that captures the trade-off between the revenues of request delivery and the costs of staffing and SLA violation. Due to factors such as the highly competitive nature of the business and customer satisfaction, however, SDPs are often driven to deliver above the strict contractual terms of the SLAs. Metrics such as time to resolve high-impact customer requests (e.g., customer network outage) are monitored as a QoS metric irrespective of the target time specified in the SLA. To address these practical scenarios, one can use large values for the contractual penalties $\mathcal{P}_{j,k}$ in the objective function in Equation (1.1) to discourage solutions that allow violation of these targets over any given stationary interval as well as across a sequence of stationary intervals.

1.4 Solution Approach

Our practical solution methodology to address IT services delivery decision management is based on a two-stage approach that develops and combines a stochastic analytical methodology and a simulation-based methodology into a unified solution framework. As previously explained, a primary reason for adopting such an approach is in response to the real-world business requirement that the solution framework must be both accurate and efficient.

The first stage of our approach consists of an approximate stochastic model and analysis of the IT services delivery decision-making processes and a nearly optimal solution of a version of the corresponding profit-based optimization. We derive solutions of the approximate stochastic model of each stationary interval with respect to the profit-based objective. Then, we exploit general classes of advanced algorithms and results for various types of mathematical programming problems to determine a nearly optimal solution in an efficient manner that supports approximate scenario and what-if analysis. These algorithms and results include the best-known interior-point methods, convex programming methods, and network flow model algorithms. The resulting nearly optimal solution also renders an important starting point for simulation-based optimization, which is performed as part of the second stage of our approach.

In the second stage of our solution approach, we develop a discrete-event simulation that models the detailed characteristics of the arrivals of requests within the different request classes and the detailed characteristics of the resolution of these requests within the SDE. We develop this discrete-event simulation model to analyze in detail the performance of the existing system by computing SL attainment given current business decisions, which includes staffing of each of the teams of agents, shift schedules, and contractual SLAs. This simulation model can also be used to perform detailed scenario and what-if analysis to explore the impact of various changes in the current team operations, such as modified staffing decisions, changes in shift schedules, changes in workload, or changes in request handling times. A simulation-based optimization methodology is developed on top of this model to search among alternative solutions for an optimal solution, including support for determining the required staffing decisions for each team of agents as well as the optimal shift schedule. We exploit advanced methods from the area of simulation-based optimization to develop our methodology for guided and intelligent searching among simulation models and solutions to maximize the profit-based objective.

1.4.1 Stochastic Analytical Methodology

1.4.1.1 Stochastic Analysis

In general, our stochastic analysis is based on strong approximations (Chen and Yao 2001) of the IT services delivery model defined in Section 1.3.1. We decided upon this approach because the alternative analytical methodology of product–form networks imposes strong restrictive assumptions that do not hold in practice for the system environments of interest, resulting in very poor approximations for this class of systems. Similarly, the alternative analytical methodology of heavy-traffic stochastic-process limits has well-known accuracy problems for the types of priority queues comprising the IT services delivery model. Our approach based on strong approximations is therefore more general than product–form solutions and more accurate than heavy-traffic stochastic-process limits for the IT services delivery model.

Many different staffing options are available to the SDP to realize any capacity levels C_j, since it is often the case that the skill sets of different agent teams intersect. In addition, there are secondary business tasks that agents can perform outside of handling requests. We therefore consider a particular capacity simplification in our analytical methodology where C_j is interpreted as a capacity scaling variable for the processing rate of a single-server queueing system for each team j. Namely, the sequence of r.v.s for the handling times of class k requests by team j is given by $S_{j,k}^{(1)}/C_j, S_{j,k}^{(2)}/C_j, \ldots$. It is important to note that this capacity approximation is known to be exact in the limit of heavy traffic (Whitt 2002), and our experience with such approximations demonstrates the benefits of this approach in practice for our decision-making purposes.

To simplify the exposition of our stochastic analysis, we shall henceforth assume that the arrival processes $A_k(t)$ are independent Poisson processes and that $L = K$. Given our modeling assumption of probabilistic assignment decisions, the per-class arrival processes to each team j queueing system remains Poisson, that is, $A_{j,k}(t)$ are independent Poisson processes with rates $\lambda_{j,k} = P_{j,k}\lambda_k$. Therefore, the model of Section 1.3.1 for team j reduces to a multiclass fixed-priority M/G/1 preemptive-resume queueing system with capacity scaling C_j, and our strong approximation approach recovers the exact solutions for response and waiting times in expectation. It is important to note, however, that the general case for our model is addressed within our stochastic analytical methodology through analogous strong approximation results in a similar fashion; refer to, for example, Chen and Yao (2001). To further simplify the presentation of our analysis, we shall henceforth focus on performance metrics and SLAs related to $T_{j,k}^{(n)}$, noting that $W_{j,k}^{(n)}$ can be addressed in an analogous manner. Our stochastic analysis methodology is therefore based on exploiting various stationary response time results.

Specifically, the first two moments of the response time distribution for class k requests at the team j queueing system with capacity C_j are given by

$$\mathbb{E}[T_{j,k}] = \frac{\sum_{k'=1}^{k} \lambda_{j,k'} s_{j,k'}^{(2)}}{2(1-\rho_{j,k-1}^{+})(1-\rho_{j,k}^{+})} + \frac{s_{j,k}}{1-\rho_{j,k-1}^{+}} \tag{1.6}$$

$$\mathbb{E}[T_{j,k}^2] = \frac{\sum_{k'=1}^{k} \lambda_{j,k'} s_{j,k'}^{(3)}}{3(1-\rho_{j,k-1}^{+})^2(1-\rho_{j,k}^{+})} + \frac{s_{j,k}^{(2)}}{(1-\rho_{j,k-1}^{+})^2}$$
$$+ \left(\frac{\sum_{k'=1}^{k-1} \lambda_{j,k'} s_{j,k'}^{(2)}}{(1-\rho_{j,k-1}^{+})^2} + \frac{\sum_{k'=1}^{k} \lambda_{j,k'} s_{j,k'}^{(2)}}{(1-\rho_{j,k-1}^{+})(1-\rho_{j,k}^{+})} \right) \mathbb{E}[T_{j,k}], \tag{1.7}$$

where $s_{j,k}$, $s_{j,k}^{(2)}$, $s_{j,k}^{(3)}$ are the first three moments for the handling times of class k requests by team j, $\rho_{j,k} \equiv \lambda_{j,k}/(C_j \mu_{j,k})$ denotes the corresponding traffic intensity, and $\rho_{j,k}^{+} \equiv \sum_{k'=1}^{k} \rho_{j,k'}$; see, for example, Takagi (1991). Moreover, it can be shown that the asymptotic tail behavior of the class k response time distribution at the team j queueing system with capacity C_j is given by

$$\log \mathbb{P}[T_{j,1} > z] \sim -\Lambda_{j,1}^{*} z, \quad \text{as } z \to \infty \tag{1.8}$$

$$\Lambda_{j,1}^{*} = \sup\{\theta : \mathcal{M}_{A_{j,1}}(-\theta)\mathcal{M}_{S_{j,1}}(\theta) \leq 1\} \tag{1.9}$$

$$\log \mathbb{P}[T_{j,k} > z] \sim -\Lambda_{j,k}^{*} z, \quad \text{as } z \to \infty \tag{1.10}$$

$$\Lambda_{j,k}^{*} = \sup_{\theta \in [0, \hat{\Lambda}_{j,k}^{*}]} \{\theta - \Lambda_{j,k}(\theta)\} \tag{1.11}$$

$$\Lambda_{j,k}(\theta) = -\mathcal{M}_{\hat{A}_{j,k}}^{-1}(1/\mathcal{M}_{S_{j,k}}(\theta)) \tag{1.12}$$

$$\hat{\Lambda}_{j,k}^{*} = \sup\{\theta : \mathcal{M}_{\hat{A}_{j,k-1}}(-\theta)\mathcal{M}_{S_{j,k-1}}(\theta) \leq 1\} \tag{1.13}$$

for $k = 2, \ldots, K$, where $\hat{A}_{j,k}$ ($\hat{S}_{j,k}$) are generic r.v.s for the aggregated interarrival (handling) times of requests from classes $1, 2, \cdots, k$, $\mathcal{M}_X(\theta) = \mathbb{E}[e^{\theta X}]$ denotes the moment generating function of a r.v. X, and $f(n) \sim g(n)$ denotes that $\lim_{n \to \infty} f(n)/g(n) = 1$ for any functions f and g; refer to, for example, Abate et al. (1995) and Lu and Squillante (2005).

The representation of $I\{S_k(T_{j,k}^{(n)}) > Z_k, \mathcal{A}_k\}$ within our stochastic analysis methodology depends upon the details of the SLA for class k requests. When $S_k(T_{j,k}^{(n)})$ is related to the first or second moment of class k response times, then Equation (1.6) or Equation (1.7) can be used directly. On the other hand, when $I\{S_k(T_{j,k}^{(n)}) > Z_k, \mathcal{A}_k\}$ relates to the response time tail distribution for class k requests, we seek to obtain accurate approximations of $\mathbb{P}[S_k(T_{j,k}^{(n)}) > Z_k]$ for class k requests handled by agent team j in order to formulate and solve the corresponding optimization problem in the next section. When Z_k is sufficiently large, we can exploit the large-deviations decay rate results in Equations (1.8) through (1.13) for class k requests handled by team j. Conversely, when Z_k is not sufficiently large, we shall consider an exponential form for the tail distribution of class k requests handled by agent team j. More precisely, we assume there exist variables $\gamma_{j,k}$ and $\theta_{j,k}$ such that

$$\mathbb{P}[T_{j,k} > Z_k] \simeq \gamma_{j,k} e^{-\theta_{j,k} Z_k} \tag{1.14}$$

These approximations are justified by various results in the queueing literature, including the exponential response time distribution in M/M/1 queues where $\mathbb{P}[T > Z] = e^{-(C\mu - \lambda)Z}$ (see for example, Kleinrock 1975), the heavy-traffic approximation of an exponential waiting time distribution in a GI/G/1 queue due to Kingman (refer to Kleinrock 1976, for example), and the large aforementioned deviations results for single-server queueing systems.

To support efficient optimization algorithms in the next section, we need to appropriately choose the variables $\theta_{j,k}$ and $\gamma_{j,k}$. Although different schemes can be envisioned, we consider fitting the two variables with the first two moments of the response time distribution. Since $\mathbb{E}[X^m] = m \int_0^\infty x^{m-1} \mathbb{P}[X > x] dx$ for any nonnegative r.v. X, it then follows from Equation (1.14) that

$$\mathbb{E}[T_{j,k}] = \gamma_{j,k} / \theta_{j,k} \quad \text{and} \quad \mathbb{E}[T_{j,k}^2] = 2\gamma_{j,k} / \theta_{j,k}^2 \tag{1.15}$$

and thus

$$\theta_{j,k} = \frac{2\mathbb{E}[T_{j,k}]}{\mathbb{E}[T_{j,k}^2]} \quad \text{and} \quad \gamma_{j,k} = \frac{2\mathbb{E}[T_{j,k}]^2}{\mathbb{E}[T_{j,k}^2]}, \tag{1.16}$$

where $\mathbb{E}[T_{j,k}]$ and $\mathbb{E}[T_{j,k}^2]$ are provided in Equations (1.6) and (1.7). Although the exact per-class response time tail distributions in preemptive priority

M/G/1 queues can be obtained by numerically inverting the corresponding Laplace transforms (refer to Takagi 1991, for example), such an approach typically will not be as useful for optimization within our stochastic analytical methodology whose goal is to efficiently obtain sufficiently accurate approximations and nearly optimal solutions. Indeed, under the aforementioned stochastic analysis approach, the objective functions can be evaluated analytically (instead of numerically) and they tend to be (reasonably close to) concave in the rate variables $\lambda_{j,k}$ more often than in general.

1.4.1.2 Optimization Analysis

The optimization analysis of our analytical methodology consists of combining general classes of the mathematical programming methods with the stochastic analysis of the previous section. We first determine the minimum capacities required to ensure that each of the agent team queueing systems is stable. This must hold for any feasible solution of the stochastic optimization problem (OPT-SDP) in Section 1.3.2 and requires that $\rho_{j,K}^+ < 1$ for all $j = 1, \ldots, J$.

The next step of our general approach consists of determining fundamental properties of this mathematical programming problem and then leveraging the most efficient and effective solution methods that exploit these properties. Since such properties and methods directly depend upon the specific problem formulation of interest, we shall discuss as representative examples some of the most common cases encountered in practice. For notational convenience, and without loss of generality, we shall replace the decision variables $P_{j,k}$ with $\lambda_{j,k}$, recalling that $\lambda_{j,k} = P_{j,k}\lambda_k$ and defining $\Lambda_k \equiv (\lambda_{j,k})$ and $\Lambda \equiv (\Lambda_k)$.

To elucidate the exposition, let us consider a general formulation where $I\{\mathcal{S}_k(T_{j,k}^{(n)}) > \mathcal{Z}_k, \mathcal{A}_k\}$ is related to the response time tail distribution, noting that our solution approach holds more generally upon substituting appropriate combinations of Equations (1.6) through (1.16) in a similar manner. Specifically, the SLT for class k is $\mathcal{Z}_k$, the corresponding PA is $\mathcal{A}_k$, and the SDP does not incur any penalties provided $\mathbb{P}[T_{j,k} \leq \mathcal{Z}_k] \geq \mathcal{A}_k$. Namely, monetary penalties will be applied whenever $\mathbb{P}[T_{j,k} > \mathcal{Z}_k] > 1 - \mathcal{A}_k \equiv \alpha_k$. Recalling our initial assumption of linear revenue, penalty, and cost functions, we then have the following general formulation of the optimization problem (OPT-SDP):

$$\max_{C,\Lambda} \sum_{j=1}^{J} \sum_{k=1}^{K} \mathbb{E}\left[N_{j,k}(T)\right]\mathcal{R}_{j,k} - \mathbb{E}\left[N_{j,k}(T)\right]\mathcal{P}_{j,k}\left(\mathbb{P}\left[T_{j,k} > \mathcal{Z}_k\right] - \alpha_k\right)^+ - \mathcal{C}_{j,k}C_j \quad (1.17)$$

$$\text{s.t.} \quad \sum_{j=1}^{J} \lambda_{j,k} = \lambda_k, \quad k = 1,\ldots,K \quad (1.18)$$

$$\lambda_{j,k} = 0, \quad \text{if } \mathcal{I}\{j,k\} = 0, \quad k = 1,\ldots,K \quad (1.19)$$

$$\lambda_{j,k} \geq 0, \quad \text{if } \mathcal{I}\{j,k\} = 1, \quad k = 1,\ldots,K, \quad (1.20)$$

where $(x)^+ \equiv \max\{x,0\}$. The next step is to deal with the probabilities $\mathbb{P}[T_{j,k} > Z_k]$ based on the stochastic analysis of the previous section. In particular, when the value of Z_k is relatively small, we substitute Equations (1.14), (1.16), (1.6), and (1.7) into the general formulation. Otherwise, when the value of Z_k is sufficiently large, we instead substitute Equations (1.8) through (1.13) in an analogous manner.

The best mathematical programming methods to solve (OPT-SDP) will depend upon the fundamental properties of the specific problem instance of interest. When the objective and constraint functions are convex in $\mathbf{C}$, Λ, we exploit properties of these objective and constraint functions together with the methods of convex programming to efficiently obtain the unique optimal solution of (OPT-SDP). The interested reader is referred to Boyd and Vandenberghe (2004) for additional technical details. More generally, when the objective or constraint functions are not known to be convex, we instead exploit advanced nonlinear programming methods to efficiently obtain a (local) solution of the optimization problem (OPT-SDP). In particular, we leverage some of the best-known nonlinear programming algorithms available, such as the interior-point algorithms due to Wachter and Biegler (2006), together with some of the best-known implementations of these methods, such as the open-source software package (Ipopt 2007). Our experience demonstrates that these classes of mathematical programming algorithms are very efficient for large-scale instances of (OPT-SDP) and can effectively support an interactive decision-making framework for IT SDEs. We further note that this mathematical programming solution approach applies more generally to problem instances with nonlinear revenue, penalty, and cost functions.

In some formulations encountered in practice, the staffing capacity levels $\mathbf{C}$ are given and the mathematical programming problem reduces to determining the optimal assignments Λ that maximize expected profit. Such formulations fall within the general class of resource allocation problems and various advanced algorithms exist to solve these problems, where we exploit the strict priority ordering and preemptive-resume scheduling to obtain a separable resource allocation problem. More precisely, we can consider class 1 in isolation across all agent team queueing systems since lower priority classes do not interfere with class 1 requests under a fixed-priority preemptive-resume discipline. The objective function in Equation (1.17) is then replaced by

$$\max_{\Lambda_k} \sum_{j=1}^{J} \mathbb{E}[N_{j,k}(T)]\mathcal{R}_{j,k} - \mathbb{E}[N_{j,k}(T)]\mathcal{P}_{j,k}\left(\mathbb{P}[T_{j,k} > Z_k] - \alpha_k\right)^+ - \mathcal{C}_{j,k}C_j \quad (1.21)$$

with k set to 1. When this objective function is convex in Λ_k, we exploit network flow model algorithms (among others) for separable and convex resource allocation problems to efficiently obtain the unique optimal solution of (OPT-SDP) for class 1 requests. Next, we proceed in a recursive manner for subsequent service classes $k = 2, ..., K$ such that, upon solving the resource allocation problem for higher priority classes to determine $\Lambda_1^*, ..., \Lambda_{k-1}^*$ and fixing these variables accordingly, network flow model (and other) algorithms for separable (and convex) resource allocation problems are exploited to efficiently obtain the (unique) optimal solution Λ_k^* of (OPT-SDP) with objective function in Equation (1.21) for class k requests. The interested reader is referred to Ibaraki and Katoh (1988) for additional technical details on algorithmic approaches to solve resource allocation problems. We note that this solution approach applies more generally to problem instances with nonlinear revenue, penalty, and cost functions. Even more generally, when $\mathbf{C}$ is also a decision variable, this general algorithmic approach for resource allocation problems of our stochastic analytical methodology can be leveraged together with a gradient descent algorithm on $\mathbf{C}$.

1.4.2 Simulation-Optimization Methodology

As the basis for the second stage of our general approach, a stochastic discrete-event simulation capability is developed and used to model the IT SDE over the entire nonstationary long-term time horizon consisting of a collection of intervals of stationary system behaviors as described in Section 1.3. This simulation model provides a detailed and accurate representation of the real-world IT SDE in which the arrival and handling time processes of the workload for each class vary over time (from one stationary interval to the next stationary interval) to reflect the various temporal and seasonal patterns found in practice. The simulation model can also support intervals within which stationarity is not necessarily reached. This is in contrast to our stochastic analytical methodology in which each (stationary) interval is modeled separately and then the respective solutions of these models are combined. The stochastic simulation model is used to determine SL attainment under any given staffing level for each of the teams of agents and any given work shift assignment.

A stochastic simulation-optimization framework is then implemented on top of this discrete-event simulation model in which a control module is used to identify the sequence of simulations that will be run and to decide the procedure halting criteria, with the overall goal of determining the optimal staffing levels for each agent team, the policy for assigning requests to agent teams, and the optimal work shift schedule over the entire time horizon. In real-world problem instances, complete enumeration of all possible solutions is computationally prohibitive and highly impractical. On the other hand, ad hoc selection of scenarios for evaluation will not guarantee the same solution quality with improved computational effort. The field of *simulation optimization* focuses on the development of more intelligent procedures for guiding the selection of scenarios for evaluation. The simulation-optimization methodology consists of exploiting advanced approaches within this field developed on top of the stochastic discrete-event simulation model of the IT SDE.

The simulation-optimization control module is based on a combination of metaheuristics, that is, scatter search and tabu search, which are the most commonly available methods in simulation software packages supporting simulation-based optimization. Specifically, at each step of the procedure, an evaluation is made of the simulation results for the current solution together with a comparison against the sequence of previous solutions evaluated via simulation, and then the optimization control module suggests another simulation scenario for evaluation. Scatter search is used to guide this step-by-step procedure, initialized with a diverse set of starting solutions. This initial set of solutions attempts to strike a balance between the diversity and quality of the solutions set, where the former is evaluated using a diversification measure, whereas the latter is evaluated using the objective function. The set of solutions generated during this diversification generation phase is typically large.

In the next phase of the procedure, the initial set of solutions is transformed in an effort to improve the quality or feasibility of the set of solutions. Quality is evaluated using the objective function, whereas feasibility is evaluated using a measure of constraint violation. The input to this improvement phase is a single solution and the output is a solution that attempts to enhance the quality or feasibility of the solution being considered, where different approaches toward improvement can be adopted. Local search is one class of such approaches that explore the *neighborhood* of the solution that is input to this phase in order to improve the solution being considered. If no further improvement to the solution is observed, then the search terminates. Tabu search is used in our simulation-optimization methodology for the local (neighborhood) search of the control module. The fields of stochastic simulation and simulation optimization are active areas of ongoing research. We refer the interested reader to Asmussen and Glynn (2007) for general details and to Glover et al. (1999) and Glover and Laguna (1977) for details on the combination of scatter and tabu search approaches used in our simulation-optimization methodology.

1.5 Numerical Experiments

In this section we report on a set of numerical experiments conducted to evaluate the benefits of our two-stage approach over a more traditional approach based solely on simulation optimization. Although a vast number of numerical experiments was performed, in this section we present only a representative sample of these results. The data set for our numerical experiments, including arrival rates, request handling times, agent team sizes and shifts, and types of requests routed to the teams, is based upon data from our motivating global IT SDP. The numerical experiments are motivated by agent teams that we observed during extensive experience in implementing data-driven optimal staffing solutions at the motivating global IT SDP. In cases where data confidentiality required that exact data values could not be shared, the exact values were altered to ensure that the spirit of the situation was retained.

1.5.1   Canonical Business Scenarios

For ease of presentation, due to the complexity of the real-world environment, we focus first on a subset of real examples. Specifically, we first describe a set of numerical experiments that consider a planning horizon over which the arrival rates of service requests are stationary. The results observed for this subset of the real-world examples provide guidance and insight into the performance of the two-stage approach in real-world scenarios. We begin with a base scenario, demonstrating the team operations for a representative agent team at the motivating IT SDP. We then systematically vary parameters in this base scenario to measure how the performance of the proposed two-stage approach may be impacted by the specification of different system input parameters. The input parameters for the base scenario are provided in Table 1.1; Table 1.2 provides the modified parameters for the remaining scenarios.

TABLE 1.1

Base Scenario Parameters

	Incident Class 1	Incident Class 2	Incident Class 3	Ad Hoc Requests
Avg. weekly volume	212.0	738.0	2240.0	916.0
Avg. handling time (min)	52.1	38.7	15.3	34.5
Std. dev. handling time	38.6	58.5	16.0	60.5
Target time (min)	240.0	480.0	1440.0	7200.0
Percent attainment	95%	95%	95%	95%

TABLE 1.2

Adjusted Parameters for Each Numerical Experiment

Request Class	Scenario 1: Avg. Weekly Volume	Scenario 2: Avg. Handling Time (min)	Scenario 3: Std. Dev. Handling Time	Scenario 4: 90% Target Attainment
Incident class 1	424	104.2	77.4	90
Incident class 2	1476	77.4	117.0	90
Incident class 3	448	30.6	32.0	90
Ad-hoc requests	1833	69.0	121.0	90

Our base scenario considers data from a dispatcher who dispatches requests to a single agent team supporting a single customer account. The dispatcher is dedicated to this single agent team for a number of reasons. First, the volume of requests for this single customer is high. Second, the range of skills required to support these customer requests (breadth of knowledge) is narrow. Finally, the account is governed by strict privacy regulations and is thus limited by the set of accounts with which it can be "bundled" should the SDP wish to gain efficiencies and have the agent team support additional accounts.

The dispatcher receives four classes of requests that are tagged with one of four levels of urgency. The first three request classes are incident tickets, and the fourth request class represents ad-hoc requests. Requests in higher-urgency request classes can preempt requests in lower-urgency request classes if these requests arrive and find all agents busy handling requests in lower-urgency request classes. The requests differ in their arrival rate distributions, handling time distributions, and target response times. The lowest urgency incident tickets have the highest volume (but least stringent SLTs) of all incident tickets. Ad-hoc requests are the lowest urgency request class but are most varied in nature, as is reflected by the high standard deviation in handling time relative to their mean handling time. As indicated earlier, Table 1.1 provides basic input parameters describing the arriving workload to this agent team for the different request classes including mean weekly volume of workload for each of the four request classes, mean and standard deviation of handling time, contractual target system time, and contractual PA.

We now describe the results of our attempts to identify the required staffing levels using both a traditional simulation-optimization procedure and our proposed two-stage framework. We first utilize the simulation-optimization framework to search for the minimal required capacity. The simulation-optimization procedure identifies a required capacity of 17 agents after a total processing time of 261 minutes. In comparison, our analytical methodology identifies the required capacity of 17 agents within 1 second. The second stage of our approach confirms the required capacity of 17 agents within 15 minutes of processing time, yielding a total savings in processing time of 246 minutes.

Our experience with agent teams at our motivating IT SDP revealed significant variation between the different agent teams across a number of dimensions, including customer contractual requirements, types of requests routed to the agent team, arrival rate patterns, request handling times, and number of customers supported. We now describe a number of alternative scenarios, where system parameters are varied from the base scenario, to measure the impact on benefits gained by our proposed two-stage approach.

The volume of requests handled by any agent team and the agent utilization varies significantly across agent teams as a result of factors such as restrictions on account bundling, agent attrition, seasonality in end-customer processes, and natural variability. Scenario 1 considers the impact of increases in workload, by doubling the per hour arrival rate for each request class keeping all other model parameters fixed, to measure the impact on efficiency gained from applying the two-stage approach. The simulation-optimization procedure identifies the required capacity of 36 agents after a total processing time of 381 minutes. In comparison, the analytical methodology identifies a required capacity of 37 agents within 1 second. (In this case, the capacity suggested by the analytical methodology is higher by one agent.) Then, as part of the second stage of our approach, this suggested required capacity is used as a starting point for the simulation-optimization procedure. With this starting point, the total processing time required to identify the optimal staffing level of 36 agents is 21 minutes, yielding a total savings of 360 minutes.

We now consider the impact of the parameters of the request handling time distribution on required staffing and on the efficacy of our proposed method. The base scenario proposes parameters for the handling time distributions for each of the types of requests handled by the agent team. However, extensive analysis of request handling times across a large number of agent teams and request classes at our motivating IT SDP revealed significant differences in the distribution of handling times for the same service request class across different agent teams. Table 1.3 provides handling time statistics for one request class across a sample of agent teams.

These differences in handling times may significantly impact an agent team's operational efficiency. One explanation for the differences is varied levels of agent training. Other explanations include technology differences across SDLs. Additionally, differences in customer expectations regarding required agent response to requests can impact the request handling times. The SDP can monitor the handling times observed over time in the different agent teams to identify opportunities for information exchange and improved agent training, identify areas for increased automation, and understand where customer demands may be impacting overall service request handling times.

Considering these common differences in service handling times across the agent teams, in scenario 2 we double the mean handling time (as compared with the base scenario) for all service classes. The simulation-optimization identifies a required capacity of 30 agents with a total processing time of 182 minutes. Our analytical methodology identifies the required capacity of 30

TABLE 1.3

Statistics on Handling Times for Single Request Class across Different Agent Teams

Agent Team	No. Agents	No. Requests/ Day/Shift	Mean	Median	Std. Deviation
AT-1	33	160	14.16	7.7	12.50
AT-2	15	123	14.56	11.9	11.21
AT-3	21	43	12.11	8.5	12.61
AT-4	21	66	9.76	6.8	8.89

agents within 1 second, with the second stage of our approach confirming the required capacity in 14 minutes of processing time. This results in a total of 166 minutes of savings in processing time.

In scenario 3, we double the standard deviation of handling time for each service request class handled by the agent team (as compared with the base scenario). The simulation-optimization procedure identifies the required capacity of 18 agents within 108 minutes. In this case, our analytical methodology identifies a required capacity one lower than the number of required agents, identifying a required capacity of 17 agents within 1 second. Then the second stage of our approach, initialized with the capacity level suggested by the analytical methodology, identifies the required capacity of 18 agents within 18 minutes, yielding a total of 90 minutes savings in processing time.

As discussed in the previous sections, the terms of the contractual SLA typically vary for each of the different customers supported by the SDP. For a single customer there could be dozens of SLAs in place; an SDP typically has hundreds of SLAs (with different contractual terms) in place at any point in time. Scenario 4 considers a simple variation on the base scenario where the contractual PA is reduced from 95% to 90% across all request classes handled by the agent team. The simulation-optimization procedure identifies a reduction in the required capacity, where the required capacity is reduced to 16 agents. This capacity level is identified after 146 minutes of processing time. In comparison, our analytical methodology identifies the required capacity level of 16 agents within 1 second. The second stage of our approach is initialized with the output of the analytical methodology and identifies the optimal required staffing level within 13 minutes, yielding a total savings of 133 minutes in processing time.

1.5.2 Summary and Real-World Business Scenario

We now summarize the results of the base scenario and scenarios 1 to 4. In each case we compare the processing time required to generate results using the simulation-optimization procedure alone against the total processing time using our two-stage approach (analytical and simulation-optimization methodologies), by taking the ratio of these two processing times in order to characterize the savings derived by adopting our two-stage approach.

Table 1.4 summarizes the results of these numerical experiments, where the table contains the following information for each experiment. Column I provides the required capacity level as indicated by the simulation-optimization procedure and column II provides total required processing time by the simulation-optimization procedure. Column III provides the required capacity level as indicated by our analytical methodology. Column IV indicates the time required to obtain the required capacity level using our two-stage approach (i.e., after initializing the simulation-optimization procedure with the results of the analytical methodology). Finally, column V indicates the ratio of the processing time using the simulation-optimization procedure

TABLE 1.4

Results of Numerical Experiments

Numerical Experiment	(I) Req'd Capacity	(II) Sim-Opt Runtime (min)	(III) Analytical Capacity	(IV) Two-Stage Runtime (min)	(V) Factor of Time Savings
Base case	17	261	17	15	17.5
Scenario 1	36	381	37	21	14.5
Scenario 2	30	182	30	14	13.0
Scenario 3	18	108	17	18	6.0
Scenario 4	16	146	16	13	11.2

with the reduced time achieved by using our two-stage approach. Of the five numerical experiments reported in the table, the solution output by the analytical methodology is exact in three cases, the solution output by the analytical methodology is one higher than the optimal capacity in one case, and the solution output by the analytical methodology is one lower than the optimal capacity in one case.

Most important to note is the total time savings obtained by the introduction of our proposed two-stage approach. The mean processing time to obtain recommended capacity levels using the purely simulation-optimization procedure is 200 minutes. After introducing the analytical methodology and using the results of this first stage of our approach to initialize the second simulation-optimization stage for each of the experiments, the mean processing time drops to 16 minutes, yielding a factor of 12.4 reduction in processing time, on average, across the reported cases. The impact is significant considering the large number of agent teams involved and the dynamic nature of the business, necessitating frequent rebalancing of agent team capacity levels.

As discussed, the base scenario and scenarios 1 to 4 represent a subset of a real planning problem. The benefits achieved when the methodology is applied to more complex real-world problems is significantly greater, given the large number of possible solutions evaluated by the simulation-optimization procedure before the optimal solution is identified. We considered a number of numerical experiments, incorporating the complexities of the real-world business environment that include, for example, nonstationary workload, overlapping and complicated agent shift structures, and multiple agent teams. We present limited results to highlight some additional complexities observed when applying our two-stage approach to determine capacity levels in a dynamic SDE.

Scenario 5 considers an agent team servicing three types of requests: console alerts, change tickets, and ad-hoc requests. Each request is tagged with one of five levels of urgency. Each of the types of requests is subject to contractual SLAs that specify both SLT times and PA for responding to

TABLE 1.5

Scenario 5: Details of Requests and Service Level Targets

	Alerts	Changes	Ad Hoc
Avg. weekly volume	6784.0	403.0	940.0
Avg. arrivals per hour	40.4	2.4	5.6
Avg. handling time (min)	5.5	2.4	20.3
Std. dev. handling time	3.5	1.8	18.5
Target time (min)	10.0	30.0	60.0
Percent attainment	99%	95%	95%

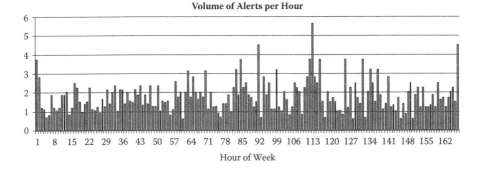

FIGURE 1.1

Scenario 5: Volume of alerts per hour of the week.

the requests in each of the associated service request classes. Table 1.5 provides the details regarding the requests and SLTs for the work arriving to this agent team.

The arrival rate of requests to this agent team is not stationary. In fact, it varies quite significantly from hour to hour throughout the week. Figure 1.1 depicts the volume of alerts per hour for each hour of the week for one of the levels of urgency. The x-axis denotes the hour of the week and the y-axis denotes the total volume of alerts.

The agent team provides 24-hour support. Agents are organized into multiple 12-hour shifts, enabling the 24-hour support. The shifts are "paired" such that, for example, agents in shift 1 work from 7 a.m. until 7 p.m. followed by agents in shift 2 who work from 7 p.m. to 7 a.m. Shifts follow a pattern of three or two days off followed by three or two 12-hour working days. Figure 1.2 provides an image of the shift schedule for this agent team.

Management in this agent team specified that a minimum of three agents is required during any hour of the day due to physical constraints on the locations of the systems that are used to monitor for alerts. Fewer than three agents would not provide sufficient staff to ensure that the agents could reach the different systems in a timely manner.

Shift 1	Mon	Tue	Wed	Thu	Fri	Sat	Sun	Mon	Tue	Wed	Thu	Fri	Sat	Sun	Mon	Tue	Wed	Thu	Fri	Sat	Sun	Mon	Tue	Wed	Thu	Fri	Sat	Sun
Start	19	19			7	7	7			19	19				7	7			19	19	19			7	7			
Stop	7	7			19	19	19			7	7				19	19			7	7	7			19	19			

Shift 2	Mon	Tue	Wed	Thu	Fri	Sat	Sun	Mon	Tue	Wed	Thu	Fri	Sat	Sun	Mon	Tue	Wed	Thu	Fri	Sat	Sun	Mon	Tue	Wed	Thu	Fri	Sat	Sun
Start	7	7			19	19	19			7	7				19	19			7	7	7			19	19			
Stop	19	19			7	7	7			19	19				7	7			19	19	19			7	7			

Shift 3	Mon	Tue	Wed	Thu	Fri	Sat	Sun	Mon	Tue	Wed	Thu	Fri	Sat	Sun	Mon	Tue	Wed	Thu	Fri	Sat	Sun	Mon	Tue	Wed	Thu	Fri	Sat	Sun
Start			19	19				7	7			19	19	19			7	7				19	19			7	7	7
Stop			7	7				19	19			7	7	7			19	19				7	7			19	19	19

Shift 4	Mon	Tue	Wed	Thu	Fri	Sat	Sun	Mon	Tue	Wed	Thu	Fri	Sat	Sun	Mon	Tue	Wed	Thu	Fri	Sat	Sun	Mon	Tue	Wed	Thu	Fri	Sat	Sun
Start			7	7				19	19			7	7	7			19	19				7	7			19	19	19
Stop			19	19				7	7			19	19	19			7	7				19	19			7	7	7

FIGURE 1.2
Shift schedule for scenario 5.

We use our two-stage approach to determine the capacity levels for each shift in this agent team. The procedure recommends optimal required capacity of 16 agents per shift. The purely simulation-optimization procedure took over 3 days of processing time to identify the required capacity levels as compared with our proposed two-stage approach which took tens of minutes to identify the required capacity levels.

In scenario 6, we compare the system described in scenario 5 with an identical system except the arrival rate is stationary and set equal to the mean arrival rate of scenario 5. Here, the required capacity is eight agents per shift. Exploring this significant difference in required capacity between scenarios 5 and 6 we note the high volume of alerts, the tight SL target time, and the restrictive PA. The combination of these factors leads to a required high capacity to meet the spikes in arrivals of alert requests due to the tight SLT times. Correspondingly, we find low average utilization in the system with this highly nonstationary pattern of request arrivals; average utilization per shift is 52.2% across all shifts. Similar to the results of scenario 5, our two-stage approach identifies the required capacity in many orders of magnitude less time than that needed by the purely simulation-optimization approach.

1.6 Conclusion

We conclude with a discussion of our experience with the motivating real-world IT SDE and the significant benefits to the business in practice of our two-stage stochastic decision-making solution framework. Our motivation for exploring the benefits of such a two-stage approach was driven by our experience with a massive deployment of a purely simulation-optimization approach to support optimal staffing decisions at a large IT SDP. The realities of the actual SDE resulted in complicated possible assignments of the arriving requests to the different agent teams and a large number of agent team-staffing level combinations. Thus, the search space was huge and the simulation-optimization procedure often took many hours or even a few days to determine the optimal staffing levels. The problem was exacerbated by cases where, after completion of the simulation-optimization exercise, the agent teams provided input data changes. Sometimes these changes were due to the dynamic nature of the work environment (e.g., changes in supported customers) or errors in the data provided. In all cases, the time required for simulation-optimization needed to be reinvested. We realized that this approach, where the time required to produce solutions is on the order of many hours or a few days, could not be sustained on an ongoing basis to support decision making for hundreds of globally located agent teams. A more advanced dynamic approach was required that could quickly produce optimal solutions based upon the provided data.

In Sections 1.3 and 1.4 we presented our two-stage stochastic decision-making solution framework to address these critical business issues in practice. Although the presentation focuses on certain instances of the real-world IT SDE, our solution framework can be applied more generally. As a specific example, for certain types of service requests, agents are required to handle additional work items immediately upon completing each request of this type. Similarly, the agents may also need to perform other business tasks outside the actual handling of service requests, which is typically handled in between service request completions. We include these effects in our system model by allowing "server vacations" within the context of the queueing system for each agent team. In the case of Poisson arrival processes, we exploit the corresponding results for a multiclass fixed-priority M/G/1 preemptive-resume queueing system with server vacations (see for example, Takagi 1991). More generally, we extend our stochastic analysis by developing strong approximations of the corresponding queueing system for each agent team (refer for example, to Chen and Yao 2001). Then the remainder of our general solution approach described in Sections 1.3.2 and 1.4 is applied in a similar manner to this extended stochastic analysis to obtain the desired results.

Other variations of our decision-making solution framework encountered in practice include service request priorities as decision variables of the stochastic optimization. In this case, as well as similar cases encountered in practice (e.g., various alternative forms for the SLAs), we adjust and extend the stochastic analysis and optimization analysis of our analytical methodology as appropriate, and then apply our two-stage decision-making framework in the manner described herein. When this concerns adding the priority structure as a decision variable, we simply adjust our optimization analysis to explore different priority orderings of the various service request classes. In most instances encountered in practice, one is typically not interested in investigating all possible priority structures and there can be a significant reduction in the priority orderings considered. However, even when such reductions are not possible, the critical point is that our stochastic analytical methodology supports investigating the entire design space for a wide range of complex IT SDE issues in a way that reduces, by many orders of magnitude, the amount of time and resources required under traditionally deployed methods. This in turn makes it possible for us to leverage our simulation-based optimization in a more surgical manner to obtain globally optimal solutions for these complex issues in small neighborhoods of the performance–profit space with the greatest importance or sensitivity. Such an approach can further support higher level IT SDM and decision-making capabilities, such as optimal pricing decisions, optimization of contractual agreements with customers, and so on.

We conclude this chapter by briefly discussing some additional examples where our two-stage solution framework is supporting higher-level decision making for real-world IT SDEs in practice. One example concerns the impact of hiring new agents and losing existing agents through attrition, each representing a challenging and costly problem faced by SDPs in all industries. Through attrition, an SDP may lose highly qualified and fully cross-trained agents and be required to hire additional agents with limited knowledge and skill sets, which can in turn require additional expenses to train and educate new hires, and additional time for new agents to ramp up and become upskilled or cross-skilled. To understand the impact of such a scenario on an SDP, our solution framework can be used to determine the number and type of additional agents that must be hired to fill gaps resulting from the attrition of fully cross-trained agents, and determine the optimal dispatching of service requests among teams comprised of fully cross-trained and more seasoned agents and those teams comprised of newly hired agents with limited skills and experience (and thus able to only handle a subset of service request types). Furthermore, our solution framework can be used by an SDP to investigate the optimal mix of, and optimal assignments among, more costly teams whose agents are fully cross-trained/seasoned and less costly teams whose agents have limited skills/experience under various work shifts and workloads. Finally, our decision-making framework can be used to address a challenge often faced by the agent teams of an IT service delivery organization in which the contractual terms of the SLAs are modified without full consideration of the operational impact including the ability of the agent teams to meet the new terms of the agreements. In such cases, our solution methodology is used to consider how changes in the terms of the SLAs (e.g., changes to the contractual PA for each request type) may impact the required staffing levels.

References

Abate, Joseph, Gagan L. Choudhury, David M. Lucantoni, and Ward Whitt. Asymptotic analysis of tail probabilities based on the computation of moments. *Annals of Applied Probability*, 5:983–1007, 1995.

Aksin, Zeynep, Mor Armony, and Vijay Mehrotra. The modern call center: A multidisciplinary perspective on operations management research. *Production and Operations Management*, 16:665–688, 2007.

Asmussen, Soren, and Peter W. Glynn. *Stochastic Simulation: Algorithms and Analysis*. Springer, 2007.

Atlason, Jalaus, Marina A. Epelman, and Shane G. Henderson. Optimizing call center staffing using simulation and analytic center cutting-plane methods. *Management Science*, 54:295–309, 2008.

Balseiro, Santiago, Awi Federgruen, and Assaf Zeevi. Integrated planning models for capacity planning and call routing in call center networks with general SLAs. In preparation.

Boyd, Stephen, and Lieven Vandenberghe. *Convex Optimization*. Cambridge University Press, 2004.

Cezik, Mehmet Tolga, and Pierre L'Ecuyer. Staffing multiskill call centers via linear programming and simulation. *Management Science*, 54:310–323, 2008.

Chen, Hong, and David D. Yao. *Fundamentals of Queueing Networks: Performance, Asymptotics, and Optimization*. Springer-Verlag, 2001.

Dieker, Antonius B., Soumyadip Ghosh, and Mark S. Squillante. Capacity management in stochastic networks. In preparation.

Feldman, Zohar, and Avishai Mandelbaum. Using simulation based stochastic approximation to optimize staffing of systems with skills based routing. In B. Johansson, S. Jain, J. Montoya-Torres, J. Hugan, and E. Yucesan, editors, *Proceedings of the 2010 Winter Simulation Conference*, pages 3307–3317, Baltimore, MD, 2010. The Society for Computer Simulation International.

Gans, Noah, Ger Koole, and Avishai Mandelbaum. Telephone call centers: Tutorial, review, and research prospects. *Management Science*, 5:79–141, 2003.

Glover, Fred, James Kelly, and Manuel Laguna. New advances for wedding optimization and simulation. In P. A. Farrington, H. B. Nembhard, D. T. Sturrock, and G. W. Evans, editors, *Proceedings of the 1999 Winter Simulation Conference*, pages 255–260, Phoenix, AZ, 1999. The Society for Computer Simulation International.

Glover, Fred, and Manuel Laguna. Heuristics for integer programming using surrogate constraints. *Decision Sciences*, 8:156–166, 1977.

Gurvich, Itay, James Luedtke, and Tolga Tezcan. Staffing call centers with uncertain demand forecasts: A chance-constrained optimization approach. *Management Science*, 56:1093–1115, 2010.

Gurvich, Itay, and Ward Whitt. Queue-and-idleness-ratio controls in many-server service systems. *Mathematics of Operation Research*, 34:363–396, 2009.

Gurvich, Itay, and Ward Whitt. Service-level differentiation in many-server service systems: A solution based on fixed-queue-ratio routing. *Operations Research*, 29:567–588, 2007b.

Ibaraki, T., and N. Katoh. *Resource Allocation Problems: Algorithmic Approaches*. The MIT Press, Cambridge, Massachusetts, 1988.

Ipopt. Project description. 2007. http://www.coin-or.org/projects/Ipopt.xml.

Kingman, J. F. C. The effect of queue discipline on waiting time variance. In *Proceedings of the Cambridge Philosophical Society*, 58:163–164, 1962.

Kleinrock, Leonard. *Queueing Systems Volume I: Theory*. John Wiley and Sons, 1975.

Kleinrock, Leonard. *Queueing Systems Volume II: Computer Applications*. John Wiley and Sons, 1976.

Lu, Yingdong, and Mark S. Squillante. Dynamic scheduling to optimize sojourn time moments and tail asymptotics in queueing systems. Technical report, IBM Research Division, November 2005.

Takagi, Hideaki. *Queueing Analysis: A Foundation of Performance Evaluation*, Volume 1: Vacation and Priority Systems, Part 1. North Holland, Amsterdam, 1991.

Wachter, Andreas, and Lorenz T. Biegler. On the implementation of an interior-point filter line-search algorithm for large-scale nonlinear programming. *Mathematical Programming*, 106(1):25–57, 2006.

Whitt, Ward. *Stochastic-Process Limits*. Springer-Verlag, New York, 2002.

Woods, Rose A. Industry output and employment projections to 2018. *Monthly Labor Review*, 132(11):53–81, 2009. Bureau of Labor Statistics, U.S. Department of Labor.

2

Simulation-Based Studies on Dynamic Sourcing of Customers under Random Supply Disruptions

Sanad H. Liaquat, Pragalbh Srivastava, and Lian Qi

CONTENTS

2.1 Introduction

Dynamic sourcing is a kind of flexible sourcing strategy in which a customer can be temporarily served by other source(s) when its assigned source is disrupted or out of stock. It is similar to contingent sourcing ("a tactic in which the firm turns to a backup supplier in the event of a failure at its normal supplier" as defined in Tomlin, 2009) that has been studied in many research works but is mainly under multiple customer–multiple source settings, whereas existing research on contingent sourcing focuses on single customer–multiple source problems.

We use a simple example in Figure 2.1 to illustrate the basic idea of dynamic sourcing and its difference from the commonly used fixed assignment strategy. Each customer is initially assigned to and served by one source

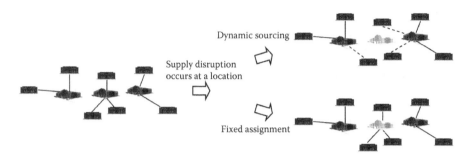

FIGURE 2.1
Dynamic sourcing strategy versus fixed assignment strategy.

of supply, as shown in the left part of Figure 2.1. However, when any of the sources is out of stock or disrupted (e.g., the second source in Figure 2.1), by following the dynamic sourcing strategy, all customers assigned to it will be temporarily served by other available sources until the disrupted source of supply is back to the normal operations. In the upper-right part of Figure 2.1, we use dash lines to represent the temporary service relationships between customers and their temporary sources of supply. However, if a fixed assignment strategy is followed, then no customer will temporarily get services from any other source of supply, even though its own source is disrupted. Please refer to the lower-right part of Figure 2.1 for this case.

In practice, the dynamic sourcing strategy can sometimes enhance customer satisfaction and service level compared to the fixed assignment strategy by temporarily assigning a customer to other sources in case its assigned source is not available. We have already seen implementations of dynamic sourcing in industries. For instance, the largest distributor of soft contact lenses in the United States, ABB Concise, is offering the dynamic sourcing option to its customers. In case the assigned distribution center is out of stock, a customer is allowed to get services from alternative locations, with a certain surcharge. Interested readers can refer to the company's Web site for details: www.abbconcise.com/Customer_Service/Contact_Lens_Shipping. aspx.

On the other hand, however, as we are going to point out in the "Literature Review" (Section 2.2) of this chapter, the current research on dynamic sourcing is still far from sufficient. Most of the related research works are either under simple settings with single retailer and multiple (usually two) sources (contingent sourcing) or studying dynamic sourcing while ignoring some important supply chain management issues, such as the inventory management decisions. To the best of our knowledge, only Snyder, Daskin, and Teo (2007) design a comprehensive scenario-based integrated supply chain design model that considers dynamic sourcing while making facility location, customer assignment (determine which customer should be served by which distribution center), and inventory management decisions. However, the number of scenarios greatly limits the size of the problem they can study and solve.

In this chapter, we study supply chains with one supplier, several retailers, and customers. The supplier and retailers might be subject to random disruptions. Traditional research in the literature on integrated supply chain design under supply uncertainties, such as Qi and Shen (2007), and Qi, Shen, and Snyder (2010), usually assumes that the assignments of customers to retailers are fixed, and cannot be temporarily adjusted even when some of the retailers are disrupted. On the contrary, we now allow dynamic sourcing to investigate if cost savings can be achieved by applying dynamic sourcing for the assignments of customers to retailers.

Specifically, we conduct the following two simulation-based studies. In the first study, we consider a supply chain with one supplier and multiple retailers and customers. The supplier is perfectly reliable, but all retailers are subject to random disruptions. We assume that retailers are using the inventory policy proposed by Qi, Shen, and Snyder (2009) in managing their on-hand inventory and making replenishment decisions. By assuming that the locations of the supplier, retailers, and customers, and the fixed and unit inventory and transportation costs are given, we investigate the advantages of using the dynamic sourcing strategy to assign customers to retailers in case of disruptions at any retailers. To do so, we compare the total costs corresponding to the dynamic sourcing strategy with that of the fixed assignment strategy. Though dynamic sourcing is subject to additional reassignment charges and higher inventory and transportation costs, it can reduce a shortage penalty, and hence might bring the total cost down. In our simulation experiments, the optimal customer assignment decisions for the fixed assignment case and the optimal assignments of customers to available (undisrupted) retailers for the dynamic sourcing case are determined using the algorithm designed by Daskin, Coullard, and Shen (2002). In addition, we also want to point out that although we fix all the location and cost information in each experiment, we use a number of input instances in our simulation experiments to make our conclusions applicable for various practical problems.

The second experiment examines a supply chain with one supplier, two retailers, and several customers. Both the supplier and one of the retailers are subject to random disruptions, but this unreliable retailer has lower operating cost. The other retailer is perfectly reliable but corresponds to higher operating costs. If the unreliable retailer is disrupted, all customers served by this unreliable retailer can be temporarily served by the reliable retailer (dynamic sourcing) but will be charged reassignment fees. In addition, during the disruptions at the supplier, lateral transshipments between two retailers are allowed. By assuming that the locations of the supplier, retailers, and customers, and the fixed and unit inventory and transportation costs are given, we evaluate if cost savings can be achieved by incorporating the perfectly reliable retailer into this supply chain where dynamic sourcing and transshipment are allowed. Our evaluation is carried out by comparing the total cost incurred in the above supply chain to that of another supply chain with a very similar structure, but with two unreliable retailers.

The main contributions of this chapter are

- We study supply chain design problems that consider dynamic sourcing while making customer assignment and inventory management decisions.
- We design a novel method (in Section 2.6) to integrate optimal supply chain design algorithms coded in C++ into the Arena® simulation models so that real-time optimal supply chain design decisions are made during the simulations whenever it is necessary.
- We demonstrate using simulation models that significant cost savings can often be achieved by using dynamic sourcing, and propose the conditions under which such cost savings are pronounced.
- Our simulation models also show that cost savings may not often be realized by keeping a perfectly reliable retailer in the supply chain network when dynamic sourcing and transshipment are allowed, considering the dynamic sourcing and the transshipment features of the supply chain, and the high operating costs associated with the perfectly reliable retailers.

The rest of this chapter is organized as follows. Section 2.2 reviews some related research on contingent sourcing and integrated supply chain design/facility location problems with supply disruptions. We conduct two simulation-based studies in Sections 2.3 and 2.4 to evaluate the cost savings that can be achieved by using dynamic sourcing, as well as the value of perfectly reliable facilities in a supply chain with dynamic sourcing. Finally, in Section 2.5 we conclude the chapter and discuss some extensions. Our simulation model setups and its integration with the optimal supply chain design algorithm coded in C++ are introduced in the Appendix.

2.2 Literature Review

The dynamic sourcing studied in this chapter is related to the research on contingent sourcing, hence we first review studies on contingent sourcing in this section. Babich (2006) considers a problem with two suppliers that have different lead times. The supplier with shorter lead time can be used as a contingent source since the firm can first order from the supplier with the longer lead time and partially observe its delivery before it determines the order size from the supplier with shorter lead time. This order deferral option provides the firm some tractability when it responds to the uncertainty associated with the supplier with longer lead time.

Tomlin (2006) works on a problem that consists of one firm and two suppliers. The firm can place orders with either a cheap but unreliable supplier or an expensive but reliable supplier. The reliable supplier has limited capacity but can offer capacity flexibility that goes beyond its regular capacity, at an additional charge, in the case if the cheap supplier is disrupted. This flexible capacity serves as a contingent source in a certain sense. In that paper, Tomlin (2006) evaluates different risk-mitigation strategies, including inventory mitigation, sourcing mitigation, and contingent sourcing, under different problem settings determined by the combinations of the unreliable supplier's percentage of uptime and expected disruption length.

Chopra, Reinhardt, and Mohan (2007) study a one-period problem in which one supplier is cheaper but is subject to both yield uncertainty and random disruptions, and the other is more expensive but is perfectly reliable. The retailer can pay the reliable supplier a premium to use it as a contingent source. At the beginning of the period, the retailer orders from the unreliable supplier and observes the amount of the received products. In the case of any shortage, the retailer is allowed to order up to I units of products from the reliable supplier with the premium it pays. The authors formulate the optimal replenishment quantity from the unreliable supplier and the optimal reservation quantity from the reliable supplier. They conclude that the retailer should optimally order more from the unreliable supplier if most of the supply risk comes from yield uncertainty; on the other hand, it should order more from the reliable supplier if most of the supply risk is caused by supply disruptions. Based on the work of Chopra et al. (2007), Schmitt and Snyder (2012) consider a multiple period problem.

Tomlin (2009) compares disruption mitigation strategies when the attributes of the suppliers, products, and firm vary. He shows that contingent sourcing becomes attractive when the probability of supply failure increases or the decision maker is risk averse. Similar to Chopra, et al. (2007), Tomlin (2009) only considers one-period problems when he studies contingent sourcing.

The aforementioned research only considers problems with a single retailer and multiple facilities. On the contrary, we investigate problems with multiple retailers and multiple facilities in this chapter. This chapter is also closely related to the literature on integrated supply chain design/facility location with disruptions, in which the assignment of retailers to distribution centers (DCs) is one of the main issues to be determined.

Snyder and Daskin (2005) examine facility location problems that consider the random disruptions at some DCs. Their models are based on two classical facility location models and incorporate a given probability to model facility disruptions. During the disruption at a DC, customers assigned to this DC are allowed to be reassigned to alternate DCs. They use their models to minimize the weighted sum of the nominal cost (which is incurred when no disruptions occur) and the expected transportation cost that is used for retailer reassignment during disruptions. They do not consider inventory costs.

The uniform failure probability assumption in Snyder and Daskin (2005) is relaxed in two recent papers, in which the failure probabilities are allowed to be facility specific. Specifically, Shen, Zhan, and Zhang (2010) apply a two-stage stochastic program as well as a nonlinear integer program to formulate the resulting problem. They further propose several effective heuristics for the problem. They show that these heuristics can bring near-optimal solutions. The authors also design an approximation algorithm, which is with a constant worst-case bound, for a special case with constant facility failure probability. Instead of using the nonlinear formulation as in Shen et al. (2010), Cui, Ouyang, and Shen (2010) design a compact linear mixed integer program (MIP) formulation and a continuum approximation (CA) formulation for the problem. They propose a custom-designed Lagrangian relaxation (LR) solution algorithm for the MIP. In addition, the CA model can predict the total systemwide cost without the detailed information about facility locations and customer assignments. This provides a fast heuristic to find near-optimal solutions. According to their computational experiments, the LR algorithm they propose is effective for midsized reliable uncapacitated fixed charge location problems, and the CA solutions are close to the optimal in most of the test instances. When the problem size is going large, the CA method is a good alternative to the LR algorithm, since it can avoid prohibitively long running times.

Different from the aforementioned papers that only consider the facility location and retailer assignment decisions, Snyder, Daskin, and Teo (2007), and Qi, Shen, and Snyder (2010) further incorporate the inventory decisions at open DCs into the problems. Snyder, Daskin, and Teo (2007) design a scenario-based integrated supply chain design model in which they use different scenarios to describe the potential uncertainties. They then design solution algorithms to minimize the expected total cost of the system across all scenarios. However, a limitation of using a scenario-based model is that the number of scenarios has a big influence on the problem size. Qi, Shen, and Snyder (2010) consider disruptions at both the supplier and DCs, whereas the earlier literature only considers disruptions at DCs. However, dynamic sourcing of retailers, the temporary reassignment of retailers to DCs when some of the DCs are disrupted, is not incorporated in this paper. Instead, the authors assume that the assignments of retailers to DCs are fixed and cannot be temporarily adjusted even when some of the DCs are disrupted.

Some other related research works on integrated supply chain design/ facility location with disruptions are Church and Scaparra (2007), Berman, Krass, and Menezes (2007), and Scaparra and Church (2008). Interested readers can refer to Snyder and Daskin (2007), in which the authors compare reliable facility location models under a variety of risk measures and operating strategies. Snyder, Scaparra, Daskin, and Church (2006) provide another detailed literature review for supply chain design models with disruptions.

Finally, we mention Qi and Shen (2007), who study a supply chain design problem for a three-echelon supply chain that considers yield uncertainty/ product defects using ideas from the random yield literature. However, supply disruptions are not considered, and dynamic sourcing of retailers is not allowed in this paper.

2.3 Simulation-Based Study 1: Evaluation of Dynamic Sourcing

2.3.1 Experiment Design

In this experiment we study the advantages of using dynamic sourcing over fixed sourcing by comparing the total costs incurred by these two strategies. The supply chain network we investigate in our scenarios in this section has five customers/demand points and three retailers. We assume single sourcing between customers and retailers. In other words, one customer can only be served by one retailer at a time. In addition, the arrivals of customer demand are assumed to follow Poisson processes with known rates.

To serve the customers, all retailers in our scenarios are getting replenishments from a single supplier. The supplier is always available to serve the retailers. However, we consider random disruptions at retailers, which can be caused by natural phenomenon or disasters such as tornados, earthquakes, and hurricanes. When a retailer is available and running, it is said to be in an ON state; and when it is disrupted and during the recovery, it is in an OFF state. No retailer can serve customers when it is in an OFF state, and unserved customer demand is assumed to be lost with a lost-sales penalty, π, for each unit of lost sales. We assume that the retailers in our scenarios are using the inventory policy investigated in Qi et al. (2009), which is a zero-inventory-ordering policy, to manage their inventory. The durations of the ON and OFF cycles of a retailer are assumed to follow independently and identically distributed (iid) exponential distributions, with rates α and β, respectively (so we call α the disruption rate of the retailer and β its recovery rate). Figure 2.2 illustrates the ON and OFF states of a retailer, as well as the distributions of their durations. Since we use α and β, respectively, to represent the rate parameters of the distributions of the ON and OFF durations, the scale parameters of these distributions are $1/\alpha$ and $1/\beta$, respectively, as shown in Figure 2.2.

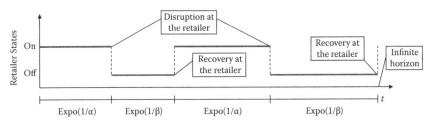

FIGURE 2.2
Retailer states.

Following the results in Qi et al. (2009), when the supplier is perfectly reliable but the retailer is subject to random disruptions, the optimal replenishment quantity from the supplier to this retailer each time is

$$\hat{Q} \equiv \sqrt{\frac{2FD}{\alpha a + h}} \tag{2.1}$$

where

F = Fixed ordering cost from the retailer to the supplier
a = Unit ordering cost from the retailer to the supplier
h = Yearly holding cost at the retailer
D = Expected annual demand rate (units/year)

For the supply chain network constructed earlier, we design two simulation models in Arena, one with fixed assignments of customers to retailers and the other follows dynamic sourcing.

2.3.1.1 Fixed Assignment Model

In the fixed assignment model, the assignment decisions of customers to retailers are made using the algorithm proposed in Daskin et al. (2002) and are fixed after that. If a disaster disrupts the retailer that is serving a customer, any demand from this customer to this retailer will not be served and is lost until the retailer recovers from its disruption.

The total costs we consider in this model include: fixed and unit ordering costs, inventory holding cost and the lost-sales penalty at retailers, and the transportation costs from retailers to customers.

2.3.1.2 Dynamic Assignment Model

The dynamic assignment model does not have customers assigned to retailers permanently. As soon as a retailer gets disrupted, the dynamic assignment takes place and the customers that were being served by this disrupted retailer are reassigned to other available retailers. The reassignment decisions are made using an algorithm proposed in Daskin et al. (2002) with the difference that only the available retailers are taken into consideration in the current algorithm. We design a novel method to first code the reassignment algorithm using C++ and compile it into a dynamic-link library (DLL) file, and then call this DLL file from the Arena simulation model to make real-time optimal customer reassignments in Arena. (Refer to the Appendix for details about the development of our simulation models in Arena and how we call the DLL file coded in C++ from Arena to make the optimal reassignment decisions.) This protocol guarantees no customer demand will be stocked out. Thus, no lost-sales cost will be incurred in this model if at least one retailer is ON.

However, whenever customers who were served by a retailer that is disrupted and under recovery are reassigned, a reassignment cost will be charged (as what ABB Concise does). This cost corresponds to those additional administrative expenses needed to process the customer reassignments.

The main reason for us to call the DLL file coded in C++ from Arena to make the reassignment decision is that there is no closed-form solution for the reassignment problem, which can only be solved numerically. This is different from the way we used to handle the inventory decisions in our simulation models. Since the inventory decision at a retailer can be written in closed-form as shown in Equation (2.1), we can directly incorporate the inventory decision into the Arena models as parameter values. However, as demonstrated in Daskin et al. (2002) and many other follow-up research work, such as Shen and Qi (2007), the reassignment problem can be solved very quickly in C++, even for large-scale problems. Therefore, calling the DLL file from Arena for the reassignment decision will not influence the efficiency of the simulation studies, and we can apply our simulation model for big realistic problems with many retailers and customers, as long as the capability of the Arena software allows.

The total costs we consider in the dynamic sourcing model include: fixed and unit ordering costs, inventory holding cost, lost-sales penalty (in the case if all retailers are disrupted at the same time), reassignment costs at retailers, and the transportation costs from retailers to customers.

For both models we constructed in Sections 2.3.1.1 and 2.3.1.2, we run Arena simulations on the following 50 data sets for 100 replications per model per data set over 365 days.

Fixed ordering cost from a retailer to the supplier, $F = 100$

Unit ordering cost from a retailer to the supplier, $a = 1$

Yearly holding cost at the retailer, $h = 0.2$

Expected annual demand rate, $D = 365$

Per unit lost-sales penalty, $\pi = 4$

Cost per reassignment of a customer: 10 or 20

Disruption rate of retailers α (times/year): 365/1000, 365/500, 365/200, 365/100, or 365/50

Recovery rate of retailers β (times/year): 365/50, 365/30, 365/20, 365/10, or 365/5

2.3.2 Result Analysis

We compare the total costs associated with the fixed assignment model and the dynamic assignment version, and report experimental results in Table 2.1 and Table 2.2 for comparisons when the reassignment cost (RAC) is equal to 10 and 20, respectively.

REMARK

The basic time unit associated with the rate parameters α and β defined at the beginning of Section 2.3 (refer to Figure 2.2) is a year, whereas in our Arena model, we use day as the basic time unit. Therefore, in Tables 2.1, 2.2, and 2.3, the scale parameters in our simulation for the exponential distributions of the ON and OFF durations of retailers are defined to be $\alpha' = 365/\alpha$ and $\beta' = 365/\beta$. ∎

TABLE 2.1

Comparison between Fixed and Dynamic Assignment Models When RAC = 10

Set Index	$\alpha' = 365/\alpha$	$\beta' = 365/\beta$	Estimated Mean Difference	Standard Deviation	0.950 C.I. Half-Width
1	1000	50	50.40	7.64	1.52
2	1000	30	49.50	7.36	1.46
3	1000	20	48.60	6.96	1.38
4	1000	10	47.60	6.61	1.31
5	1000	5	46.80	6.57	1.30
6	500	50	48.40	8.33	1.65
7	500	30	45.90	7.49	1.49
8	500	20	44.00	7.25	1.44
9	500	10	42.40	7.05	1.40
10	500	5	41.80	6.88	1.36
11	200	50	48.20	8.74	1.73
12	200	30	45.00	8.59	1.70
13	200	20	43.20	8.22	1.63
14	200	10	41.50	6.78	1.35
15	200	5	40.30	6.22	1.23
16	100	50	48.60	8.86	1.76
17	100	30	45.80	8.25	1.64
18	100	20	40.70	7.47	1.48
19	100	10	40.70	7.18	1.42
20	100	5	39.10	7.33	1.45
21	50	50	53.40	8.98	1.78
22	50	30	49.80	7.96	1.58
23	50	20	47.20	8.65	1.72
24	50	10	45.10	8.35	1.66
25	50	5	43.70	8.73	1.73

For each set of α and β, Table 2.1 and Table 2.2 show the corresponding mean, standard deviation, and 95% confidence interval of the difference between the costs of the fixed and dynamic assignment models, when RAC equals 10 and 20, respectively. By observing these two tables, we can see that

TABLE 2.2

Comparison between Fixed and Dynamic Assignment Models When RAC = 20

Set Index	$\alpha' = 365/\alpha$	$\beta' = 365/\beta$	Estimated Mean Difference	Standard Deviation	0.950 C.I. Half-Width
1	1000	50	49.70	7.57	1.50
2	1000	30	49.60	7.55	1.50
3	1000	20	47.90	6.88	1.37
4	1000	10	46.90	6.52	1.29
5	1000	5	46.10	6.46	1.28
6	500	50	47.50	8.23	1.63
7	500	30	45.00	7.36	1.46
8	500	20	43.10	7.11	1.41
9	500	10	41.50	6.90	1.37
10	500	5	40.90	6.74	1.34
11	200	50	47.00	8.63	1.71
12	200	30	43.70	8.41	1.67
13	200	20	41.90	8.04	1.60
14	200	10	40.00	6.59	1.31
15	200	5	38.80	6.03	1.20
16	100	50	47.10	8.76	1.74
17	100	30	44.10	8.05	1.60
18	100	20	38.80	7.18	1.42
19	100	10	38.60	6.92	1.37
20	100	5	36.90	7.03	1.40
21	50	50	51.30	8.89	1.76
22	50	30	47.20	7.69	1.53
23	50	20	44.20	8.38	1.66
24	50	10	41.50	8.01	1.59
25	50	5	39.80	8.19	1.63

- Significant cost savings can usually be achieved by doing dynamic sourcing, since Table 2.1 and Table 2.2 show that the 95% confidence intervals of the differences between the costs of fixed and dynamic assignment solutions are always in the positive zone.
- When the recovery processes at retailers become slower (i.e., when recovery rate β is getting small or β' in the tables becomes big), the advantage of doing dynamic sourcing is more pronounced, evidenced by the increasing of the average differences with the increasing of β' in both tables.
- When the reassignment cost is low, the cost saving of dynamic sourcing is high, which is shown in Table 2.1 and Table 2.2. For given α and β, the average difference in Table 2.2 is usually smaller than that in Table 2.1.

2.4 Simulation-Based Study 2: The Advantage of Incorporating a Perfectly Reliable DC

2.4.1 Experiment Design

Following the conclusions we made in Section 2.3 that significant cost savings can usually be achieved by dynamic sourcing, in this experiment, we further study if it often brings us benefits by using perfectly reliable retailers in a supply chain in which dynamic sourcing is allowed. Considering the higher operating costs at perfectly reliable retailers, such benefits might be trivial when dynamic sourcing is allowed, since dynamic sourcing can also significantly reduce lost-sales and save lost-sales penalty, which is the main saving of using perfectly reliable suppliers. To demonstrate this conjecture, in this section, we compare the total costs incurred in two supply chain networks, each of which consists of one unreliable supplier, two retailers (served by the same supplier), and five customers. The main difference between these two supply chain networks is that both retailers in the first network are unreliable, whereas the second network has one perfectly reliable retailer and one unreliable retailer subject to random disruptions. When any unreliable retailer is disrupted, we use dynamic sourcing to temporarily reassign the customers that were served by the disrupted retailer to the other (available) retailer.

As in our simulation studies in Section 2.3, we assume that one customer can only be served by one retailer at a time, and the arrivals of customer demand follow Poisson processes with known rates. In addition, retailers in the supply chain networks are also using the zero-inventory-ordering inventory policy studied in Qi et al. (2009) to manage their inventory, and the durations of the ON and OFF cycles in a retailer are exponentially distributed, with rates α and β, respectively (refer to the beginning of Section 2.3.1 and Figure 2.2 for details). As for the unreliable supplier, we similarly assume its ON and OFF cycles follow exponential distributions with rates λ and ψ, respectively, and all the aforementioned random ON and OFF cycles are independently and identically distributed.

If the unreliable supplier is disrupted at any given time point but any retailer needs replenishment, we allow transshipment between two retailers. In other words, if a customer demand arrives at a retailer, but this retailer does not have any inventory on hand due to the delay shipment from the supplier, the retailer can temporarily get support from the other retailer with the quantity needed. We consider the unit cost for each product transshipped.

Since all retailers are assumed to use the inventory policy studied in Qi et al. (2009) to make the replenishment decision, by it, we know that for any unreliable retailer, its replenishment quantity from the unreliable supplier is given by

$$\hat{Q} \equiv D \times \dfrac{-\bar{A} + \sqrt{\bar{A}^2 + \dfrac{2\alpha(\bar{A}+\bar{B})\left(\dfrac{\alpha F \bar{B}}{D} + \bar{A}(\pi - a)\right)}{\alpha a + h}}}{\alpha(\bar{A}+\bar{B})}$$

where

$$\bar{A} = \frac{\lambda(\alpha+\beta)}{\beta\psi(\alpha+\lambda+\psi)}$$

$$\bar{B} = \frac{1}{\alpha} + \frac{1}{\beta}$$

and for the reliable retailer (if there is any), its replenishment quantity each time should be

$$\frac{-\dfrac{\lambda h D}{\psi+\lambda} + \sqrt{\left(\dfrac{\lambda h D}{\psi+\lambda}\right)^2 + 2h\psi\left[\psi F D + \dfrac{\lambda D^2}{\psi+\lambda}(\pi - a)\right]}}{h\psi}$$

Parameters F, a, h, and π are defined in Section 2.3.

We build simulation models in Arena for both supply chain networks introduced earlier and run simulations on the following 32 data sets for 100 replications per model per data set over 365 days. In each model, we compute the total costs that include fixed and unit ordering costs, inventory holding cost, reassignment cost, transshipment cost, and the lost-sales penalty at retailers (caused by the random disruptions at retailers and the supplier), and the transportation costs from retailers to customers.

Fixed ordering cost from a retailer to the supplier, $F = 100$
Unit ordering cost from a retailer to the supplier, $a = 1$
Yearly holding cost at any unreliable retailer, $h = 0.2$
Yearly holding cost at any reliable retailer, $h_r = 0.3$ or 0.5
Expected annual demand rate, $D = 365$
Per unit lost-sales penalty, $\pi = 4$
Cost per reassignment: 10
Unit transshipment cost: 0.1 or 0.2
Disruption rate of any unreliable retailer α (times/year): $365/200$ or $365/50$
Recovery rate of retailers β (times/year): $365/10$
Disruption rate of the unreliable supplier λ (times/year): $365/500$ or $365/200$
Recovery rate of the unreliable supplier ψ (times/year): $365/10$ or $365/5$

TABLE 2.3

Comparison of Models with and without the Reliable Retailer

Set Index	h_r	Unit Transshipment Cost	$\lambda' = 365/\lambda$	$\psi' = 365/\psi$	$\alpha' = 365/\alpha$	Estimated Mean Difference	Standard Deviation	0.950 C.I. Half-Width
1	0.3	0.1	500	10	200	−287	131	26
2	0.3	0.1	500	10	50	−78	33	6
3	0.3	0.1	500	5	200	−1090	454	90
4	0.3	0.1	500	5	50	−261	115	23
5	0.3	0.1	200	10	200	−130	54	11
6	0.3	0.1	200	10	50	−34	15	3
7	0.3	0.1	200	5	200	−461	203	40
8	0.3	0.1	200	5	50	−104	43	8
9	0.3	0.2	500	10	200	−287	131	26
10	0.3	0.2	500	10	50	−78	33	6
11	0.3	0.2	500	5	200	−1090	454	90
12	0.3	0.2	500	5	50	−261	115	23
13	0.3	0.2	200	10	200	−130	54	11
14	0.3	0.2	200	10	50	−34	15	3
15	0.3	0.2	200	5	200	−461	203	40
16	0.3	0.2	200	5	50	−104	43	8
17	0.5	0.1	500	10	200	−288	128	25
18	0.5	0.1	500	10	50	−80	32	6
19	0.5	0.1	500	5	200	−1090	443	88
20	0.5	0.1	500	5	50	−264	115	23
21	0.5	0.1	200	10	200	−131	53	11
22	0.5	0.1	200	10	50	−35	15	3
23	0.5	0.1	200	5	200	−462	198	39
24	0.5	0.1	200	5	50	−106	42	8
25	0.5	0.2	500	10	200	−288	128	25
26	0.5	0.2	500	10	50	−80	32	6
27	0.5	0.2	500	5	200	−1090	443	88
28	0.5	0.2	500	5	50	−264	115	23
29	0.5	0.2	200	10	200	−131	53	11
30	0.5	0.2	200	10	50	−35	15	3
31	0.5	0.2	200	5	200	−462	198	39
32	0.5	0.2	200	5	50	−106	42	8

2.4.2 Result Analysis

In Table 2.3, we compare the total costs associated with the models we construct in Section 2.4.1. Both retailers in the first model are subject to random disruptions (Model 2.1), whereas one of the retailers in the second model is perfectly reliable (Model 2.2). For each given data set, we calculate the average value, standard deviation, and 95% confidence interval of the difference between the total costs of Models 2.1 and 2.2, and report them in the last three columns.

The same as the Remark mentioned earlier, the basic time unit associated with the rate parameters λ and ψ is year, whereas in our Arena model, we use day as the basic time unit. Therefore, in Table 2.3, the scale parameters in our simulation for the exponential distributions of the ON and OFF durations of the supplier are defined to be $\lambda' = 365/\lambda$ and $\psi' = 365/\psi$.

By observing Table 2.3, one can summarize that

- Using perfectly reliable retailers usually may not bring significant cost savings to supply chain networks considering the high inventory holding cost associated with reliable retailers, as we can see from Table 2.3 that most 95% confidence intervals of the differences between the costs of Model 2.1 (without any perfectly reliable retailer) and Model 2.2 (with one perfectly reliable retailer) are in the negative zone.

    An intuitive explanation of this observation is that since we allow dynamic sourcing in our simulation models, this can help the supply chain reduce the chances of lost sales and hence reduce the lost-sales penalty. This is also the primary advantage of using perfectly reliable retailers. Now considering the higher operating costs associated with perfectly reliable retailers, using dynamic sourcing only might be a better choice, which, itself, can reduce the lost-sales penalty already.

- Perfectly reliable retailers might be gradually getting favored when

    - The inventory holding costs associated with perfectly reliable retailers are lower.

    - Unreliable retailers are often disrupted, since we can see from Table 2.3 that the differences of the costs of Model 2.1 and Model 2.2 are usually getting larger when α' decreases (or α increases).

    - The supplier is often unavailable, evidenced by the fact that the differences of the costs of Model 2.1 and Model 2.2 increase when λ' decreases (or λ increases) or ψ' increases (or ψ decreases).

The first two observations conform to our intuition. As for the last observation, it can be explained as follows: When the supplier is often unavailable, having a reliable retailer in the supply chain can help us keep more inventory at the retailer tier, and therefore maintain a certain customer service level at retailers (by doing dynamic sourcing or transshipment) even when the supplier is not available. This can greatly save the lost-sales penalty caused by shortages incurred otherwise and hence save the total costs.

2.5 Conclusions and Extensions

In this chapter, we conduct simulation-based studies for supply chains that include one supplier and multiple retailers and customers. Considering the random disruptions at the supplier and retailers, we investigate the advantages of using the dynamic sourcing strategy to assign customers to retailers in case of any disruptions at retailers or the supplier as well as the conditions under which such kind of advantages are significant. In addition, our simulations show that it may not often bring cost savings by keeping perfectly reliable retailers in a supply chain when dynamic sourcing and transshipment are allowed, considering the higher operating costs associated with them. However, we provide some cases, by evaluating our simulation results, in which using perfectly reliable retailers might be getting preferred.

In our current simulation studies, we assume that the locations of all retailers are given, based on which we determine the assignment of customers to retailers and the inventory management decisions at retailers. This assumption can be easily relaxed so that our study can be extended to an integrated supply chain design problem that simultaneously considers facility location, retailer assignments, and inventory management decisions. To do so, we will need to apply the Lagrangian relaxation based solution technique proposed in Daskin et al. (2002) and Shen and Qi (2007) to determine the locations of DCs.

In addition, we assume in our studies that retailers are using the inventory policy proposed in Qi et al. (2009) in managing their on-hand inventory and making replenishment decisions. This policy assumes that the inventory at retailers is to be destroyed upon any disruptions. As pointed out by Sargut and Qi (2012), a more realistic inventory policy is to imply that the disruptions at a retailer do not destroy the inventory at this retailer but simply prevent this retailer from serving its customers during the recovery periods. Based on our discussions in this chapter, to use the policy proposed in Sargut and Qi (2012) to replace the inventory policy we are using in this chapter, the model setups and experiments will not be greatly influenced except that we need to apply the inventory model formulated in Sargut and Qi (2012) instead when we make the replenishment decisions at retailers.

Appendix

We use Rockwell Arena to create models for simulating the behaviors of the supply chains we consider in this chapter. In this appendix, we focus on the introduction of how we realize the interactions between the Arena model and the DLL file coded in C++, so that we can make the real-time optimal customer assignment decisions by calling the DLL file from Arena whenever

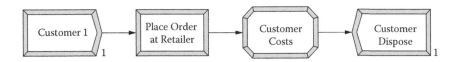

FIGURE 2.3
Arena template for customers (for the first simulation study; refer to Section 2.3).

it is necessary. The model setups introduced in the following are for the first simulation study (introduced in Section 2.3), and those for the second simulation study (Section 2.4) are pretty similar and are omitted in this section due to space limitations. Interested readers may refer to Arena-related books and materials to get further understanding of the Arena model setups.

Figures 2.3, 2.4, and 2.5 show the templates we set up in Arena for customers, retailers, and the supplier. When a simulation starts, the model will first access a local text file to get the parameter values, such as the numbers of customers and retailers, and various costs at retailers. It will then generate and set up a complete simulation model for the supply chain that consists of multiple customers and retailers based on the templates and parameter values.

Upon the arrival of a customer order generated by the "Customer 1" module in Figure 2.3, the "Place Order at Retailer" signal module in the next will call a prewritten VBA (Visual Basic for Applications, which is equipped in Arena) user function to see which retailer will be used to serve this order. This user function contains customer ID as its argument. It is the key connection between the Arena model and the DLL file, where the optimal assignment decisions are made. Please refer to Figure 2.6, which also illustrates the discussions in the next paragraph.

Whenever this user function is called (e.g., an order from customer 2 arrives), it will first check if there has been any change with the retailers' availability statuses since it was called last time. If one or several retailers' availability statuses are changed, then the user function will call the DLL file, and pass two pieces of information: an array of retailers' ON/OFF (0: OFF; 1: ON) statuses and the total number of retailers. This DLL file, coded in C++, then reads a static text file to collect all the cost information and makes the optimal decision on the assignments of customers to available retailers using the algorithm proposed by Daskin, Coullard, and Shen (2002). After the user function receives the return from the DLL file, which is an array showing the optimal assignments of customers to available retailers, it returns which retailer should be used to serve the order from the current customer (in our example, customer 2) to the Arena model. On the other hand, if no retailer's availability status has been changed since the user function was called last time, then the user function simply returns the ID of the retailer who should be serving this customer according to the return of the DLL's last run.

The "Place Order at Retailer" signal module in Figure 2.3 then sends a signal (representing an order) to the retailer the customer is assigned to based on the return of the user function. At the retailer side (Figure 2.4), this signal

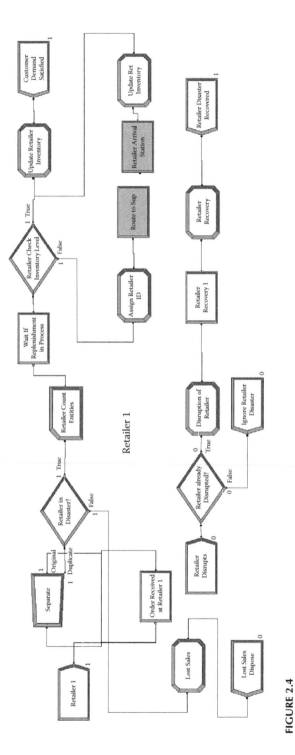

FIGURE 2.4
Arena template for retailers (for the first simulation study; refer to Section 2.3).

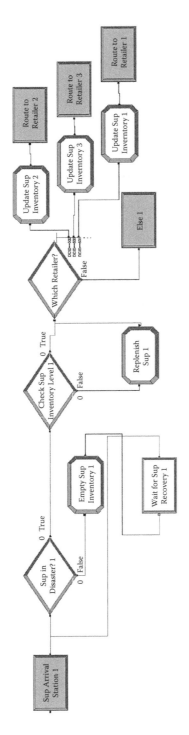

FIGURE 2.5
Arena model of the supplier (for the first simulation study; refer to Section 2.3).

FIGURE 2.6
Integration of Arena models and the DLL file coded in C++.

is received at the "Order Received at Retailer 1" module. A check is made at the "Retailer in Disaster?" module to see if the retailer is in OFF state. (Note that although in our dynamic sourcing case, customer demand is always being assigned to retailers with ON state if there is any retailer available, we still need this check in our template for retailers for two reasons: (1) this template is also used for the model for the fixed assignment case, in which a customer's assigned retailer might be temporarily OFF; and (2) in a dynamic sourcing case, it is possible that all retailers are OFF, when we still have to assign a customer demand to a retailer in order to collect the lost-sales penalty into the total cost.) If it is in OFF state, the demand entity is disposed and the lost-sales cost is added to the total cost. If the retailer is in ON state, the demand entity is routed through the model until it is satisfied.

Furthermore, disruptions at a retailer are generated by the "Retailer Disrupts" module in Figure 2.4 and the whole disruption handling process including disruption and recovery is simulated using the flow in the lower-right part of Figure 2.4.

References

Babich, V. 2006. Vulnerable options in supply chains: Effects of supplier competition. *Naval Research Logistics* 53: 656–673.

Berman, O., D. Krass, and M.B.C. Menezes. 2007. Facility reliability issues in network p-median problems: Strategic centralization and co-location effects. *Operations Research* 55: 332–350.

Chopra, S., G. Reinhardt, and U. Mohan. 2007. The importance of decoupling recurrent and disruption risks in a supply chain. *Naval Research Logistics* 54: 544–555.

Church, R. L., and M. P. Scaparra. 2007. Protecting critical assets: The r-interdiction median problem with fortification. *Geographical Analysis* 39: 129–146.

Cui, T., Y. Ouyang, and Z. M. Shen. 2010. Reliable facility location design under the risk of disruptions. *Operations Research* 58: 998–1011.

Daskin, M. S., C. Coullard, and Z. J. Shen. 2002. An inventory-location model: Formulation, solution algorithm and computational results. *Annals of Operations Research* 110: 83–106.

Qi, L., and Z. M. Shen. 2007. A supply chain design model with unreliable supply. *Naval Research Logistics* 54: 829–844.

Qi, L., Z. M. Shen, and L. Snyder. 2009. A continuous-review inventory model with disruptions at both supplier and retailer. *Production and Operations Management* 18: 516–532.

Qi, L., Z. M. Shen, and L. Snyder. 2010. The effect of supply uncertainty on supply chain design decisions. *Transportation Science* 44: 274–289.

Sargut, Z., and L. Qi. 2012. Analysis of a two-party supply chain with random disruptions. *Operations Research Letters* 40: 114–122.

Scaparra, M. P., and R. L. Church. 2008. A bilevel mixed-integer program for critical infrastructure protection planning. *Computers and Operations Research* 35: 1905–1923.

Schmitt, A. J., and L.V. Snyder. 2012. Infinite-horizon models for inventory control under yield uncertainty and disruptions. *Computers and Operations Research* 39: 850–862.

Shen, Z. M., and L. Qi. 2007. Incorporating inventory and routing costs in strategic location models. *European Journal of Operational Research* 179: 372–389.

Shen, Z .M., R. Zhan, and J. Zhang. 2010. The reliable facility location problem: Formulations, heuristics, and approximation algorithms. *INFORMS Journal on Computing*.

Snyder, L., and M. S. Daskin. 2005. Reliability models for facility location: The expected failure cost case. *Transportation Science* 39: 400–416.

Snyder, L., and M. S. Daskin. 2007. Models for reliable supply chain network design. In *Critical Infrastructure: Reliability and Vulnerability*, ed. A. T. Murray and T. H. Grubesic, Chapter 13, 257–289. Springer-Verlag, Berlin.

Snyder, L., M. S. Daskin, and C. P. Teo. 2007. The stochastic location model with risk pooling. *European Journal of Operational Research* 179: 1221–1238.

Snyder, L., M. P. Scaparra, M. S. Daskin, and R. L. Church. 2006. Planning for disruptions in supply chain networks. In *TutORials in Operations Research*, ed. M. P. Johnson, B. Norman, and N. Secomandi, Chapter 9, 234–257. INFORMS, Baltimore, MD.

Tomlin, B. T. 2006. On the value of mitigation and contingency strategies for managing supply-chain disruption risks. *Management Science* 52: 639–657.

Tomlin, B. T. 2009. Disruption-management strategies for short life-cycle products. *Naval Research Logistics* 56: 318–347.

3

Customer Perceptions of Quality and Service System Design

George N. Kenyon and Kabir C. Sen

CONTENTS

3.1 Introduction

Quality is a broad and pervasive philosophy that can significantly affect a firm's competitiveness (Powell 1995; Sousa and Voss 2002). The academic literature contains an extensive discussion of various quality management systems and philosophies (Crosby 1979, 1984, 1996; Deming 1986; Garvin 1987; Juran 1988), together with the factors and practices driving these systems (Flynn, Schroeder, and Sakakibara 1994; Hackman and Wageman 1995; Ahire, Golhar, and Walter 1996). Since the inception of quality as a management philosophy, numerous companies have adopted various quality control systems such as total quality management (TQM), six-sigma, and just-in-time (JIT) with the aim of improving quality that results in reduction of costs and increased revenues. However, as reported in *The Economist* magazine, in a survey conducted by Arthur D. Little, only a third of the respondents believed that their quality management efforts were having an impact on their competitiveness ("The Cracks in Quality" 1992).

The design of quality systems for service industries is particularly difficult because of the property of "intangibility" associated with services. Thus, the customer cannot easily touch or measure the level of services he or she receives. Given this distinction, it follows that the success of any service offering is strongly related to the firm's ability to control customer's expectations and their perceptions of the quality of the service. In order to do this, it is essential that service system designers understand the attributes that define service quality as well as the underpinnings of the process through which consumers create their perceptions about the service received. An appreciation of the relationship between these two factors is important as many tangible goods are now marketed with complementary services. In many cases, it is the service rather than the tangible part of the offering that provides an opportunity for the firm to separate itself from the competition and gain a sustainable competitive advantage.

During the last two decades, numerous advancements in information technology have provided service providers with the tools to garner and analyze a wide array of information about consumers. This has given service providers an opportunity to change the philosophy of the overall firm to a new service dominant logic away from the old goods-based logic (Vargo and Lusch 2004; Lusch, Vargo, and Wessels 2008). As opposed to the mere exchange of goods, this new paradigm stresses the importance of the use of applied knowledge and skills for the benefit of the consumer. The proposed service design recognizes this potential, but also notes the nuances in the kinds of knowledge and the challenges facing the modern service provider.

In this chapter, we draw on the existing literature to derive a prescriptive model for determining the important factors that influence consumer perceptions of service. This model is the basis for designing a system that consistently delivers quality to a firm's customers. The rest of the chapter is organized as follows: Section 3.2 discusses the various dimensions of service quality. In Section 3.3, we develop a framework for evaluating how customers are likely to evaluate the quality of services they purchase. This is followed by Section 3.4, which develops the key elements of the service design. The section contains important do's and don'ts for the practical design of efficient service systems. Here, we underline the potential for a service provider to gain a competitive edge over the competition by identifying the key dimensions that should play a part in the service design. The importance of these dimensions in service design is discussed in detail in Section 3.5. We conclude the chapter in Section 3.6 with a discussion of the important implications for service designers and suggestions for future research.

3.2 The Dimensions of Service Quality

Zeithaml, Bitner, and Gremler (2006) distinguish between overall customer satisfaction and service quality. According to them, service quality is one of the components of customer satisfaction. Customer satisfaction is also affected by product quality (i.e., the quality of the tangible part of the firm's offering), the price as well as situational and personal factors. From the service design perspective, excellence in service quality should help distinguish the firm's overall market offering. Among the first researchers to evaluate the different dimensions of service quality were Parasuraman, Zeithaml, and Berry (1985). They devised a survey, titled SERVQUAL, which can be used to gauge the quality in different service environments. The SERVQUAL scale concentrates on evaluating 21 service attributes, which are classified under five broad quality dimensions. These dimensions are:

1. *Reliability*—This denotes the degree of consistency with which the expected level of service is delivered to the customer. Thus, reliability implies that different interactions with the customer invariably lead to the same desirable result.

2. *Responsiveness*—While reliability encapsulates the consistency of the service offering, the service provider must also have the ability to respond to the customer's needs. Thus, responsiveness denotes the firm's degree of flexibility and the ability to design the service offering to a customer's special needs (Zeithaml et al. 2006).

3. *Assurance*—The ability of the service provider to convey trust and confidence in customers is denoted as assurance. The degree with which the service employee can convey these qualities at the points of contact with customers is critical to service quality. Both proper training and effective promotional campaigns help develop this dimension.

4. *Empathy*—The service provider's willingness to see the service interaction from the eyes of the customer and to design a service that caters to each individual's needs are encapsulated by the dimension of empathy. Often, "empathy" depends on the intuitive knowledge that some firms have about their customer's actual needs and wants.

5. *Tangibles*—While services are essentially intangible, many services are accompanied by tangible cues. These cues, such as the physical state of the equipment at the service site, together with its cleanliness and design, influence how the consumer perceives the service.

These dimensions are important factors to consider in the design of service systems. However, there are large variances in service situations across different sectors. In order to understand the rudiments of the ideal service design for each firm in a given situation, we must have a proper under-standing of how consumers form expectations of the service they get from a provider as well as how they form their opinions of the actual service they receive. Thus, the model of consumer expectations and perceptions of ser-vice should be the cornerstone of the design of the service system. It should help identify the salient attributes that must be present in the service design. At the same time, these five service quality dimensions should pervade all elements of this system.

The design of an efficient service system is particularly important given the gradual transition from a goods dominant driven culture to a service dominant driven culture. In a goods dominant culture, the primary focus is on the efficient production of units of output (Lusch et al. 2008). In contrast, a service dominant culture emphasizes the efficient application of knowledge and skills for the benefit of another party. Lusch, Vargo, and O'Brien (2007) note that all sections and levels of an organization must be involved in the creation of value. Thus, customers are also cocreators of the value creation process.

Chesbrough and Spohrer (2006) recognize the importance of collabora-tive knowledge sharing between the provider and the customer in service industries. At the outset, both parties lack complete information about the other's capabilities or preferences. The collaboration between the two par-ties can be challenging given the nature of the knowledge that has to be shared. Advances in information technology provide both the provider and customer collaborative opportunities that were not available a few decades ago. However, the latest technical tools cannot overcome barriers to infor-mation sharing that arise from the very nature of the knowledge that is to be shared.

Chesbrough and Spohrer (2006) distinguish between codified (also termed as explicit) and tacit knowledge. Whereas codified knowledge can be trans-mitted in a formal, systematic manner, tacit knowledge is difficult to transfer between two parties. For example, setting up a home theater system with the help of diagrams and detailed instructions is likely to be classified as codified knowledge. On the other hand, as Chesbrough and Spohrer note, the ability to ride a bicycle is an example of tacit knowledge. This distinc-tion between codified and tacit knowledge is similar to the classification of specific and general knowledge made earlier by Jensen and Meckling (1992). Whereas general knowledge is inexpensive to transmit between two par-ties, the sharing of specific knowledge is prohibitively costly. The challenge for modern service systems is to make the collaboration between all parties involved in the transaction efficient, after taking into account the nuances between the different kinds of knowledge that is to be shared. However, the starting point for the design is an evaluation of the underpinnings of the customer's expectations and perceptions of the service. This is discussed in the next section.

3.3 A Model for Understanding Consumer Expectations and Perceptions

Economists (Nelson 1970; Darby and Karni 1973) identify three distinct properties that are present in different offerings in the marketplace. These are search qualities (factors that can be determined before the actual purchase), experience qualities (factors that can only be evaluated after the actual purchase of the product), and credence qualities (attributes that the consumer might find difficult to judge even after product purchase). Previous researchers (Zeithaml 1981) argue that most service offerings, as opposed to physical goods, are relatively higher in experience and credence qualities and lower in search qualities. However, the design of services must take into account all three qualities. One way to integrate all three qualities for services is to introduce the dimension of time as the customer interacts with the provider's service offering. Gronroos (1990) suggests that an important point in time when a customer gets a chance to judge service quality is during the encounter with the service provider. This could be a phone conversation or an actual visit to the provider's outlet. This is denoted in the literature as the "moment of truth." However, in all probability, the prospective customer hears of the service provider's offerings before the actual interaction through word of mouth or through the provider's promotional activities. Thus, the model for understanding consumers' expectations and perceptions must start at the initial point of awareness of the provider's service offering. We therefore divide our model into different stages, based on the dimension of time.

3.3.1 Precontact Awareness Stage

At the outset, the prospective consumer has a chance of evaluating certain aspects of the provider's service. This occurs even before a visit to the service center or a purchase from the provider's Web site. From the promotional literature or information on the Web site, the prospective customer can gather information on the service provider's locations, hours of operation, prices of various items, description of the retail outlet, warranties of services rendered, and so on. As this information is available before the consumer purchases the service, these attributes are examples of search qualities. During the awareness stage, the consumer might get the requisite information and be motivated to visit the service provider's retail store or make a purchase at the Web site. However, three different types of errors might occur right at the outset, during the availability of search properties.

First, the consumer might not receive information on attributes that are important to him or her. This is the error of "saliency" and is similar to Gap 1 of Zeithaml, Parasuraman, and Berry (1990). Here, a gap arises because of the difference between management's perceptions of customer expectations and actual consumer expectations. To prevent this type of error, the information given in promotional literature or Web sites must be about attributes that consumers care about.

Second, expectations might be raised so high that the actual delivery of the service cannot live up to the promises implied in the promotional literature or Web site. This leads to a difference between customer expectations and perceptions, which is likely to lead to overall customer dissatisfaction. This is the error of "overpromising" and is similar to the customer gap described by Zeithaml et al. (2006). Exaggerating service capabilities or stating vague promises in management communications at the awareness stage is one of the root causes of Gap 4. This gap is the difference between actual service performance and the provider's communications about the service to the outside world (Zeithaml et al. 1990). Thus, all external communications about the service's attributes and capabilities must be carefully designed.

The third error that can arise during the awareness stage is the customer's lack of interest in the provider's service offering, after his or her initial evaluation of the promotional literature or other information on the provider's Web site. This lack of interest is likely to lead to lower traffic to the service provider's store or Web site. The underlying reason behind the lack of interest is that information about the service conveys inadequacy in either the service's "points of parity" or "points of differences" (Keller, Sternthal, and Tybout 2002). Whereas points of parity ensure that the service is on par with other competitive products, points of differences are required to position the brand ahead of the competition. The prospective consumer might find that the information from the search qualities suggests that the provider lags behind the competition on important attributes or is not offering better value than his or her regular service provider. In these cases, the transition to the next step, which could be a visit to the provider's store or a purchase on its Web site, is not likely to occur.

Those errors that occur during the awareness stage are components of the search properties of services. They can be classified into "capability" or "engagement" factors. Capability factors encapsulate the impressions that the consumer gets from the search properties about what the service provider is capable of delivering. Attestations from previous customers, certificates, and awards won by the provider present evidence about the provider's capabilities. Engagement factors are information about the various transaction costs involved in engaging in an interaction with the provider. These include prices, the location, hours of operation, and delivery charges. For example, a customer might decide not to go to a restaurant based on the prices posted on an online menu. Thus, right at the outset, the service design must carefully think about the information that is included in the search properties as it is instrumental in creating the "customer gap" (Zeithaml et al. 2006).

3.3.2 First and Subsequent Contacts

Based on the information gathered under search qualities, the consumer might go on to the next stage, which is visiting the actual store or using the Web or the phone to place an order. These actual points of contact with the service provider are the "moments of truth" alluded to by Gronroos (1990). However, in order to have a better understanding of the service design, it is imperative to separate the first contact from subsequent ones. If the distribution of a potential stream of moments of truth abruptly ends with the first contact, it is essential to understand what led to the service failure. Thus, our discussion of these contact points is separated into two sections.

3.3.2.1 First Moment of Truth

The first interaction with the service provider, either at the store or through a phone call or a visit to the Web site, is often the most important as in many cases the first impression can be the last impression. Thus, in some cases, the first contact is the only point in the theoretical distribution of all moments of truth. As customers do not opt to evaluate the service provider over a period of time, it is important to find out why the first trial of the provider's services causes the customer not to come back for subsequent visits. Sometimes, the failure to revisit the service can be because of circumstances beyond the provider's control. For example, the customer might move to another residence, where there is no convenient access to the provider's service. Alternatively, the consumer's impression of the service might result in the customer opting not to try the service for a second time. This loss of consumer confidence falls under the "assurance" dimension of service quality, identified by Parasuraman et al. (1985). Both personal human interactions with the service staff as well as nonpersonal factors can lead to a lack of assurance on the part of the consumer. An inattentive medical attendant is an example of the former, while a glitch in the software that results in duplicate bills from the doctor's office for the same visit is a nonpersonal factor. As the consumer gets evidence of these properties on his or her first purchase or encounter, all of these can be classified under experience qualities. However, there are subtle differences between the properties of these interactions.

The factors that trigger the consumer opting not to revisit can result from either the failure to meet his or her basic needs or an unsatisfactory encounter with the service staff. Examples of the former include smaller-than-expected portions at a restaurant or incorrect tax returns filed because of faulty use of tax software. The basic needs of the consumer are not met in these cases. Rude and unfriendly behavior from service staff are examples of unsatisfactory service encounters. Although the consumer's basic needs might be met, the interaction itself is unsatisfactory under these circumstances.

The customer might also decide not to come back after a first visit for reasons other than the actual encounter with service personnel. There are two separate factors that can account for this. First, the consumer might not feel assured about the provider's capabilities based on the tangible cues that he or she observes. For example, dilapidated equipment in a medical center will not inspire confidence in a patient, even if the medical staff is polite and friendly. These are examples of defects in the provider's servicescape (Bitner 1992). Second, even if the provider's facilities have the appropriate equipment, the consumer might not feel comfortable about the environment. The proximity of the service outlet to crime-ridden areas or the type of fellow consumers that frequent the store are environmental factors that might prompt the consumer not to return. Thus, other than the inability to meet basic needs and unsatisfactory personal interactions, both tangible cues and environmental reasons are experience qualities that can deter the customer from coming back after the first visit.

3.3.2.2 Subsequent Moments of Truth

After the first encounter with the service provider, the consumer might decide to frequent the service outlet for future visits. The degree of patronage is likely to depend on the consistency of the service offering and the emergence of new points of differences compared with the competition. The different moments of truth can be visualized as a band of performance. Consistent performances by the provider on every visit will lead to a narrow band of moments of truth. This should result in the "reliability" dimension, identified by Parasuraman et al. (1985). In this case, provided the provider's service has not been overtaken by competitive firms that have created new points of differences, the customer will continue to patronize the service.

Figure 3.1 illustrates the development of this band of interactions (or moments of truth denoted as MT) with the service provider over time. The band should have a positive slope over time in order to stay ahead of other companies who attempt to bridge the competitive gap established by the provider. Also, the band contains spikes in performance that might occur on isolated service occasions that are not part of the norm.

Under normal conditions, moments of truth will fall within a predictable band of performance (Scenario 1). Over time, each time that the customer experiences the service and the performance level is consistent with pervious experiences, the customer's expectations will force a narrowing of the expectable band of performance. However, there will be cases where moments of truth will not fall into a predictable narrow band of performance. Sometimes, there are spikes in service, when a particular moment of truth falls outside the normal band of performance. These spikes are ultra moments of truth, that is, a significant departure from the usual interaction with the service provider. Ultra moments of truth can be either a traumatic or a delightful

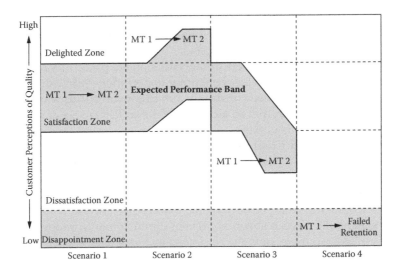

FIGURE 3.1
Customer expectation shifts between moments of truth. Note: MTs are moments of truths. MT 1 is followed by MT 2.

event, as perceived by the consumer. In the case of traumatic moments of truth (Scenario 3), some part of the interaction goes awry. This can sometimes arise because of a breakdown in the provider's operations. For example, moviegoers might see a sudden interruption in a film's screening because of a malfunction in the movie projector. When a traumatic event occurs, it is important that the service provider make every effort to attenuate the consumer's negative experience, for example, by giving free tickets for a future movie of their choice to every person in the theater. Alternatively, the traumatic occasion might result because of circumstances that are beyond the provider's control. For example, diners with a reservation at a popular restaurant could be delayed by an unforeseen traffic jam. The restaurant management should make every effort to seat the dining party even though almost every table is taken. The responses to these traumatic moments of truth by the provider can often partially obliterate a consumer's bad memories. The provider's extra effort might instill a seed of loyalty in the consumer because of a sense of gratitude for services rendered under difficult circumstances.

Ultra moments of truth can also lead to a pleasant surprise on the part of the customer leading to delight (Scenario 2). These spikes in performance often arise when a provider provides an unusual benefit to the consumer that the latter did not expect. Often, these can result from purely human endeavors that are dependent on the provider's staff anticipating or remembering a customer's special preferences. For example, hotel management might have a frequent patron's favorite dinner ready for him although he missed his flight and is late for the traditional dinner times at the hotel. Alternatively, the service provider might use technology to reward loyal customers on a particular occasion, either based on some personal data or on the frequency of their visits. These are usually based on the innovative use of information technology. Thus, either human endeavors or technology-based techniques can lead to positive spikes in performance that create pleasant surprises for the consumer. When this occurs, consumer expectations often shift higher. As a result, the provider must constantly aim to convert the upward spike(s) into a regular occurrence, that is, incorporate the spike(s) in Scenario 2 into a future band of moments of truth.

In contrast, Scenario 3 represents a situation where the provider's service is below the consumer's expectations based on previous expectations. For example, an ATM machine could be out of cash and disappoint the customer who always frequented it before. Too many traumatic moments of truth can lead the customer to stop patronizing the bank. Alternatively, a one-off traumatic moment of truth can abruptly prevent the customer from revisiting the store again. For example, damage to clothes by a dry cleaning service could stop the customer from revisiting the store. In these cases, the customer moves to Scenario 4, the zone of disappointment. Here, the task of inviting the customer to frequent the provider's service in the future becomes almost impossible and in all likelihood, the store will lose the customer's future patronage.

The nature of the experience qualities during the moments of truth shapes future consumer perceptions of the service. Here, two properties of service encounters play important roles. First, narrow bands of performance indicate consistent experience qualities. This leads to the desired "reliability" quality described by Parasuraman et al. (1985). Second, the characteristics of the spikes, which can be measured as the ratio of positive to negative spikes, indicate the "responsiveness" and "empathy" quality of services (Parasuraman et al., 1985). These two different experience characteristics lead the customer to form strong opinions, either positive or negative, about the nature of the service. The feelings about the service are particularly strong if the spikes relate to saliency, that is, service attributes that are important to the consumer. Over time, if future encounters with the service provider bolster their existing views, customers form strong beliefs about the service provider. These beliefs about the service are an example of credence qualities. These credence qualities can be about three different properties of the service provider. For example, the consumer might feel that the equipment related to the service is modern compared to the competition. This is an example of a tangible factor.

Second, the consumer might believe that the provider's service locations are one of the few within the industry that are open around the clock. This is an example of a convenience factor. Third, the consumer might feel confident that her current hairstylist never fails to give her a haircut that makes her look her best. This is an example of a capability factor. All three factors are examples of credence qualities that develop over time. It is important that the provider is knowledgeable about both the positive and negative credence

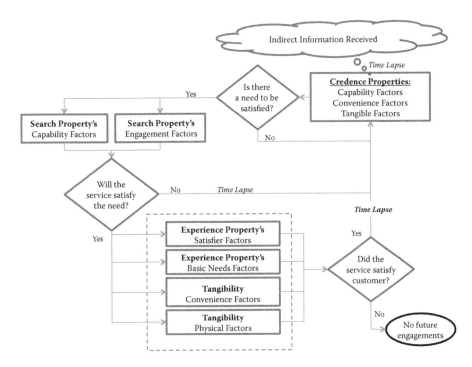

FIGURE 3.2
Creating customer perceptions of quality.

qualities that consumers believe about the service. Also, the provider must attempt to understand the exact nature of both the search and experience qualities that led to the credence qualities developing over a span of time. Figure 3.2 illustrates the model that shows how consumer perceptions about a service build up over time.

3.4 The Framework for Service Design

In order to design a successful new service, or the improvement of an existing service, the provider has to attempt to gain and maintain a sustainable competitive advantage. This is similar to Herzberg's "two-factor" theory. Here, Herzberg (1973) distinguishes between motivators that give positive satisfaction from factors that do not. Perception properties also have factors that are the basic requirements for consideration of the service and other factors that are associated with deriving satisfaction from the service experience. With credence properties, if the customer does not believe that the service provider is capable of providing a satisfying service experience, he or she will not even worry about whether the service is convenient or if the tangibles are fulfilling. The same logic applies with search properties. If the customer does not find validation of the service provider's capabilities, he or she will not even consider engaging the service in the future. Furthermore, if the customer engages in the service experience and the provider does not meet his or her basic needs, the probability of satisfying the customer is greatly reduced.

Based on the model described in Section 3.3, the design of service systems has four stages. These are: (1) the gathering of information; (2) the translation of information into policies; (3) the implementation of policies; and finally (4) the continuous monitoring of consumer perceptions and the resultant changes and alterations in policies and practice. Although each service sector has its own unique challenges, the framework of the design should avoid particular pitfalls that are likely to creep in. Keeping these do's and don'ts in mind, we discuss some salient aspects of the design process with reference to the model for each of the four stages.

3.4.1 Getting Information from the Consumer

The process of gathering information from existing and potential consumers should be preceded by careful thought as to what the provider needs to know about being successful in the service industry. The first step in this process is to strive to include all the salient attributes that consumers use in choosing a provider. Also, every effort must be made to include all alternative competitive providers that the consumer also considers. The digital age has seen the proliferation of information about various types of inter-competitive forces offering different alternatives for the same basic need. For example, a simple dinner can involve choosing between different restaurants, various food delivery services, microwave choices, and so forth. Here, the provider must be careful to avoid concentrating only on types of food services that are similar to its own. Thus, it is essential that the "content space" of information that has to be garnered is carefully determined before the actual collection process.

The second step in the process is to strive to get the unadulterated consumer view about the actual service. Several types of errors can contaminate the information-gathering process. For example, in an effort to lower costs, some providers opt to make the service personnel who interact with the consumer administer the survey. This raises a moral hazard problem, with the actual questioner more intent on getting a better rating rather than collecting the consumer's true opinion about the service encounter. Thus, the collection process must be designed to prevent exogenous factors from contaminating the veracity of the information gathered.

The third point to remember in the information gathering process is to focus on the collection of the total distribution of all moments of truth. For example, consumers are more likely to be willing to report only on ultra moments of truth, rather than on service interactions that fall into the band of reliability. The provider must make every effort to garner opinions on both types of interactions. While consumer experiences about ultra moments of truth are likely to provide clues to future remedial action, feedback on "typical" satisfactory interactions are also instrumental in providing information on any possible advantages that the provider has over the competition.

It is important to note that reactions to a service can be markedly different among different segments of customers (Lin 2010). Thus, all consumer segments should be covered in customer surveys. Overall, consumer information has to be analyzed with the aim of discerning the linkage between the three properties of perception—search, experience, and credence—that influence the service selection and patronage process. Thus, the questionnaire or interviews should focus on the attributes that are the underpinnings of the linkages between the three qualities. The collection of information relevant to this should reduce Gap 1, the difference between management knowledge about consumer expectations and actual consumer expectations (Zeithaml et al. 1990).

3.4.2 Formulation of Policies

Based on the information gathered from consumers, the provider should formulate a set of standards that employees should follow in their interactions with customers. The standards should be set in quantifiable form for each of the typical interactions that the consumer is expected to encounter at the service location. The mapping of consumer expectations to the formulated standards and policies involves a series of studies that measure how actual policies translate into final results that the consumer expects from a provider. Here again, it is important to understand the linkages between the three properties of perception—search, experience, and credence—and focus on the aspects of each that are salient to the customer. Standards that arise from an understanding of the network of linkages between the three properties of perception applicable for a particular service provider will result in a reduction in Gap 2 (Zeithaml et al. 1990). This is the difference between management's perceptions of customer expectations and the final design of service design and standards. Basing a provider's policies on competitive standards might not always result in a reduction in Gap 2, as consumer expectations for each provider can be different.

During policy formulation, it is important to distinguish between the actual service encounters that involve the customer from those incidences and events that do not. Mapping the service delivery system using a service blueprint format, as represented in Figure 3.3, can be very useful. The service blueprint is a flowchart of the service delivery process that had been divided into three zones. Each zone defines a level of interaction between the customer and the service delivery system. Zone 1 defines the activities that the customer engages in prior to his or her entry into the service system. Zone 2

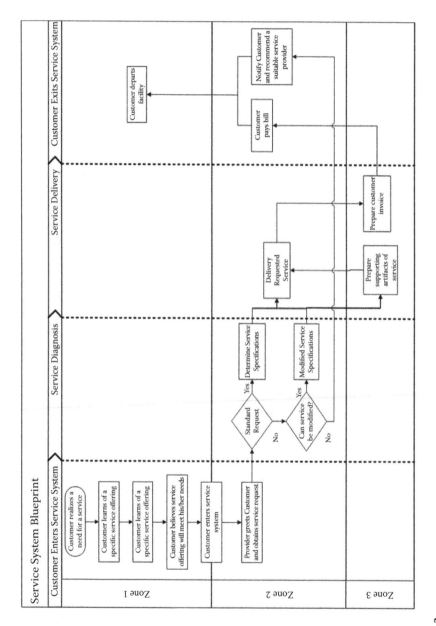

FIGURE 3.3
The service system blueprint.

defines the service-based activities that the customer is directly involved with. Zone 2 is frequently referred to as "front-end" or "front-office" operations. Zone 3 defines the service-based activities that are performed without direct customer interaction. Zone 3 is often referred to as "back-end" or "back-office" operations. An example of these divisions can be seen in bank operations. A customer receives a large sum of money in Zone 1. The customer then enters the bank and informs the teller that he wishes to deposit the money into his account. The teller receives the money, counts it, and then executes a deposit transaction in Zone 2. The transaction is received by a clerk in the bank's back office where the transaction is posted to the customer's account, which is part of Zone 3. Upon receiving notification that the transaction has been correctly posted, the teller hands the customer a receipt of the transaction within Zone 2. Finally, after completing all his transactions, the customer exits the bank in Zone 1.

An applied example of using the service blueprint in policy formulation is the incidence of accidents for a line of taxi cabs serving a metropolitan area. This might be related to the number of hours that each taxi operator has to drive without a break. A prudent firm would want to improve customer perceptions of the company's safety record by reducing the number of accidents. Thus, by monitoring the operating hours for each driver, the taxi service could work to reduce the incidence of accidents affecting the customer. However, all front-end and back-end operations may not be related. For example, mistakes happening at the operating theater of a medical clinic could occur solely because of wrong decisions made at the patient's bedside. Thus, the provider must carefully formulate policies for both front-end and back-end operations and recognize the links between the two types when they do exist. Some of the linkages between the two types of operations might already be required by regulatory agencies. For example, restaurant workers in many counties in the United States are mandated to wear gloves when working on food items, as unhygienic handling of these products are related to health hazards for the customer. Thus, service providers not only have to recognize the major linkages between these two types of operations but also formulate policies for both kinds so that they do more than merely follow procedures that are set by governmental or civic agencies.

A second dimension in policy formulation is the distinction between a "normal" service encounter from traumatic moments of truth. Here, two separate types of guidelines come into play. For example, in the case of the movie theater with a malfunctioning projector, there should be guidelines about the number of operational hours after which each projector has to be serviced and inspected. This is an example of a back-end operation. However, a second set of guidelines should also be set defining the expectable parameters for employee interventions to traumatic moments of truth. Thus, theater employees will be able to react in a timely and appropriate fashion in the event of a projector breakdown to mitigate customer dissatisfaction. This is an important component of front-end operations.

Whereas the first set of guidelines are preventive measures, the second set attempts to reduce the damage caused by unexpected events. Often, the second set of guidelines should be in place to cater to the "perishability" aspect of services. As services cannot be stored, providers have to deal with cases where demand does not match supply (Zeithaml et al. 2006). Thus, employees must follow established guidelines to react to situations where an unanticipated consumer demand or a breakdown in supply has the potential to lead to traumatic moments of truth.

3.4.3 Implementation of Policies

Gap 3, as postulated by Zeithaml et al. (1990), refers to situations where the actual performance by the provider's employees does not conform to established service standards. To reduce Gap 3, the provider must ensure that employees have the proper training so as to conform to the guidelines. Given the quality of perishability, in many cases, employees should be versatile enough to respond to situations where mismatched supply and demand give rise to the need for possessing a wide array of job skills. In addition, in order to be competitive, the provider must ensure that service employees are supported by the appropriate technology. This aspect is important, as committed employees might not be able to match a competitor's responses to a customer, if they are at a disadvantage because of inferior technological support.

Two components of service design come into play at the policy implementation stage. The first is the knowledge component where advances in information technology provide more opportunities for the service provider to satisfy his or her customer. The second is the choice of the mix between centralization and decentralization for the whole chain. Both these dimensions will be discussed in greater detail in Section 3.5 in our analysis of the critical elements of service design.

3.4.4 Monitoring Consumer Perceptions and Actual Performance

There are many opportunities for service providers to duplicate and improve upon the unpatented features of a competitor's service offering. Thus, the provider must continuously monitor customer perceptions about both the provider and its competition. Feedback from consumers, the adjustment to standards and policies, and the monitoring of performance must be managed as a continuous process. One way to ensure that the feedback loop is working properly is to periodically have new customers evaluate the quality of the service with regard to experiences with other providers.

During the monitoring stage, the provider should identify the underlying reasons behind the ultra moments of truth that result in customer delight. For example, diners at a particular restaurant location within a service chain might like the extra special attention given to them by the wait staff. The

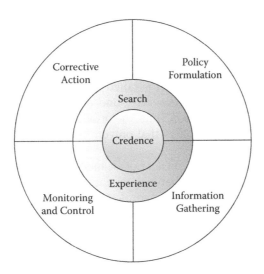

FIGURE 3.4
Service system design versus consumer perceptions.

provider should try to ascertain the practice followed by the restaurant management that led to this customer experience. For example, the provider could find that management at that location made two members of the waitstaff share the responsibilities for each table. Thus, if a waiter is unavailable, a patron has the option of asking another waitstaff for service. Given this scenario, the provider should attempt to implement a similar policy at other locations within the chain.

Overall, the service provider must strive to learn what promises (during the search stage) and interactions (during the experience stage) make consumers hold certain beliefs about a service (during the credence stage). A deeper understanding of this linkage will make the provider more knowledgeable about the competitive forces at play and the requisite course of actions that are required to succeed in its service sector. The complete service design incorporating all four stages is presented in Figure 3.4.

3.5 Striving for the "Sweet Spot" in Service Design

Although the different stages of service design were described earlier, it is important to consider the various dimensions that come into play for the actual design. A better understanding of these underlying factors provides important insights for the service provider. The first dimension relates to the knowledge component. Here, advances in information technology, specifically in the gathering, organization, and analysis of data, provide opportunities for a more scientific basis of bridging the gap between the service provider and the customer (Chesbrough and Spohrer 2006). However, it is important to note that certain types of knowledge cannot easily be transmitted or stored (Jensen and Meckling 1992). Thus, "on the spot" observations over many years can give astute store managers unique insights into both back-end and front-end operations. This is often based on detailed knowledge of work staff, equipment, and individual customers served; and often cannot be easily codified (Chesbrough and Spohrer 2006). Thus, the provider has to recognize the limitations of information technology that prevent service design from becoming an exact science.

Given the potential for the existence of tacit knowledge in downstream operations, an overzealous adherence to standards established by a central authority might sometime result in employees not responding to the specialized needs of customers or operating at subpar efficiency. These incidents will negatively affect consumer perceptions of the service provider on the responsiveness and empathy dimensions, identified by Parasuraman et al. (1985). In order to avoid these situations, a second dimension for service design has to be introduced. While the first dimension relates to the efficient use of information on both front-end and back-end operations, based on the best available technical tools, the second relates to areas where there are limits to codified knowledge. Here, in contrast to depending on a centralized system of codified knowledge, a better alternative might be to allow the tacit knowledge of the downstream, on-the-spot manager to play a part. Thus, a change in the organizational format from a centralized to a decentralized structure might be the corrective action needed in this situation.

The dimension of centralized–decentralized organizational format becomes an important second factor. The change in format becomes particularly important in unusual situations where established procedures are not appropriate. For example, a hardware store chain might have a policy of selling on strictly cash or credit. However, after a hurricane, the chain might allow store managers in affected areas to go against established policy and accept checks. This is because credit card transactions become impossible during long periods of communications breakdown. Although this temporary change in policy might give rise to some bad checks, the goodwill gained within the community is likely to outweigh the additional financial costs involved. Thus, the second component touches on the issue of a shift in the centralization–decentralization mix in particular situations.

A possible alternative to centralized control is to allow for selective decentralization in cases where the bundle of codified knowledge normally available within the chain is suddenly not effective in providing value to the customer. Notwithstanding the temporary shift within the centralization–decentralization dimension, the chain should continue to use available information technology to build a database of incidents related to extraordinary situations, together with an inventory of effective actions that were taken as a response. By consulting this database in the future, the service provider will reduce the dependence on the "gut feeling" of downstream managers. For example, by following the delivery and reception records of all consumers, post offices, and their own warehouses, a DVD mailing service can make an informed guess about the source of a problem if an individual customer has returned a DVD that has not been received. However, if the provider decides to have a more decentralized structure, he or she must also decide on the exact strata of employees who should be empowered to make decisions in a given situation. Finally, each of these decisions should be analyzed after the event, to ensure that individual managers, perhaps inadvertently, have not indulged in unequal treatment of the service provider's customers. In sum, the effective use of information technology should help build a base of knowledge that incorporates most situations facing the service provider into a scientific paradigm.

Over time, the service provider can design operational policies as a near science. Thus, over the long run, the service is converted from its dependence on individual, uncorrelated decisions based on the tacit knowledge of individual managers into a science based on codified knowledge; thus improving the consistency of service delivery. Nevertheless, as unanticipated situations arise in future service encounters, the provider must always be prepared to move the point of centralization or decentralization in situations where previously designed policies are no longer applicable. Campbell and Frei (2011) note the importance of local knowledge in service operations. Thus, an adherence to a fixed policy where all decisions are based on a centralized bank of codified knowledge is perhaps an unachievable goal for the provider.

Campbell and Frei (2011) also note that there is a wide variation in customer sensitivity to service times. The variance between different customer segments can be particularly significant across national outlets within the same service chain. Thus, the service chain must avoid adopting a "boiler plate" approach for all of its outlets. The difference between generalizing and discriminating across different outlets or situations is the third dimension in the service design.

Careful analysis of information collected from consumers can help the provider get a better understanding of nuances in customer preferences and expectations across different locations and times. Generalized solutions are applicable for uniform preferences. However, the provider must not hesitate to provide alternatives when customer preferences are markedly different. For example, a nationwide coffee chain might offer alternative flavors in areas where there are differences in taste. The change between generalization and discrimination is an important one to consider for the service provider. Sometimes a shift toward discrimination is accompanied by a greater dependence on a decentralized structure that allows downstream managers greater leeway to cater to variances in local tastes.

The fourth and perhaps most important dimension within service design is the role of value creation. Previous researchers (Vargo and Lusch 2004;

Maglio and Spohrer 2008) have noted the potential for services to cocreate value for the benefit of all parties involved in an interaction. Unfortunately, the provider can make the correct decisions on all of the first three dimensions in order to create value for everyone involved in the distribution chain, except for the consumer. In contrast to this approach, a bank might decide not to enter the residential mortgage business in a region, as it might feel that current terms and conditions reduce the chances of individual customers paying off their loans and owning their homes during their lifetimes. As the consumer is an essential part of the service interaction, the creation of value for both the customer and the provider is a critical component that should be considered in service design.

The four dimensions to be considered in service design relate to the nuances in the nature of knowledge involved in a service transaction, the degree of centralization–decentralization allowed within the service provider's organization, the mix of generalization–discrimination that the provider allows throughout the chain, and finally the role of value for all parties involved in the transaction (particularly the customer). A movement toward the correct mix on each of these dimensions will ultimately hit the "sweet spot" of service design, that is, the point at which the provider delivers value to the customer in the most efficient manner possible. However, it is important to note that the sweet spot area could vary between different customers. For example, Lin (2010) finds subtle differences in attitudes in mobile banking between potential and repeat customers. Moreover, the sweet spot can change over time in response to changes in the competitive landscape and customer preferences. Finally, advances in information technology are likely to change the optimal point within the centralization–decentralization continuum. Thus, the provider must always strive to discern the sweet spot by continuously examining the point on each dimension where it should position itself. Figure 3.5 illustrates the concept of the sweet spot in service design.

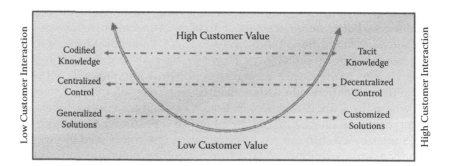

FIGURE 3.5
Customer value creation model.

3.6 Conclusions

In this chapter, we present a model that builds on existing research to demonstrate how different stages and choices in the service system design process influence the final consumer's perception of service quality. The service designs should be built around a continuous loop of improvement based on information gathered from consumers, the formulation of policies based on this information, and the continuous monitoring of actual performance. The resultant service should minimize the gaps based on the SERVQUAL scale and establish the underlying five dimensions of service quality (Parasuraman et al. 1985). Future research, based on consumer surveys, can further explore how consumer beliefs (credence qualities) arise from the prior stages, where search and experience qualities come into play.

The suggested framework for the service design is a prescriptive plan for building a system that seeks to continuously improve a provider's service quality. We identify four dimensions on which the service provider must carefully decide to position itself in order to reach the sweet spot for providing the best value to the customer. Future researchers can examine the differential impact of various elements of the model in context of the final design across different service sectors. In some sectors, such as health services, the tangible elements might be critical in determining quality, as part of the servicescape (Bitner 1992). In other sectors, such as financial and legal services, personal interaction might be the most important element within the final design. Overall, the suggested design framework contains procedural details that can be applied to almost all sectors. Practitioners can therefore use the suggested design framework in different situations. The important element in the design process is to garner accurate consumer feedback about the service and use this to build a system that maintains a sustainable competitive edge for the provider. This will result in the buildup of positive credence qualities based on the research and experience qualities that have been carefully formulated at earlier stages.

Acknowledgments

This research was supported in part by the William and Katherine Fouts Scholar in Business Award.

References

Ahire, S. L., Golhar, D. Y., Walter, M. A., 1996. Development and validation of TQM implementation constructs. *Decision Sciences* 27(1): 23–56.

Bitner, M. J., 1992. Servicescapes: The impact of physical surroundings on customers and employees. *Journal of Marketing* 56(April): 57–71.

Campbell, D., Frei, F., 2011. Market heterogeneity and local capacity decision in services. *Manufacturing and Service Operations Management* 12(1): 2–19.

Chesbrough, H., Spohrer, J., 2006. A research manifesto for services science. *Communications of the ACM* 49(7): 35–40.

The cracks in quality. 1992, April 18. *The Economist*, 323 (7755): 67–68.

Crosby, P. B., 1979. *Quality Is Free*, New York, NY: McGraw-Hill.

Crosby, P. B., 1984. *Quality Without Tears.* New York, NY: McGraw-Hill.

Crosby, P. B., 1996. *Quality Is Still Free.* New York, NY: McGraw-Hill.

Darby, M., Karni, E., 1973. Free competition and the optimal amount of fraud. *Journal of Law and Economics* 16 (April): 67–86.

Deming, W. E., 1986. *Out of the Crisis*, Massachusetts Institute of Technology, Center for Advanced Engineering Studies, Cambridge, MA.

Flynn, B. B., Schroeder, R. G., Sakakibara, S., 1994. A framework for quality management research and an associated measurement instrument. *Journal of Operations Management* 11(4): 339–366.

Garvin, D.A., 1987. Competing on the eight dimensions of quality. *Harvard Business Review* 65(6): 101–109.

Gronroos, C., 1990. *Service Management and Marketing: Managing the Moments of Truth in Service Competition.* Lexington, MA: Lexington Books.

Hackman, J., Wageman, R., 1995. Total quality management: Empirical, conceptual, and practical issues. *Administrative Science Quarterly* 40: 309–342.

Herzberg, F., 1973. *Work and the Nature of Man.* New York, NY: Mentor Book, New America Library.

Jensen, M. C., Meckling, W. H., 1992. Specific and general knowledge and organizational structure. In *Contract Economics,* Lars Werin and Hans Wijkander, eds., Oxford: Blackwell, pp. 251–274. Reprinted in *Journal of Applied Corporate Finance* (Fall 1995).

Juran, J. M., 1988. *Juran on Planning for Quality.* New York, NY: McMillan.

Keller, K. L., Sternthal, B., Tybout, A., 2002. Three questions you need to ask about your brand. *Harvard Business Review* 80(9): 80–89.

Lin, H.-F., 2010. An empirical investigation of mobile banking adoption: The effect of innovation attributes and knowledge-based trust. *International Journal of Information Management* 31: 252–260.

Lusch, R. F., Vargo, S. L., O'Brien, M., 2007. Competing through service: Insights from service-dominant logic. *Journal of Retailing* 83(1): 5–18.

Lusch, R. F., Vargo, S. L., Wessels, G., 2008. Toward a conceptual foundation for service science: Contributions from service-dominant logic. *IBM Systems Journal* 47(1): 5–14.

Maglio P., Spohrer, J., 2008. Fundamentals of service science. *Journal of the Academic Marketing Science* 36: 18–20.

Nelson, P., 1970. Information and consumer behavior. *Journal of Political Economy* 78(2): 311–329.

Parasuraman, A., Zeithaml, V. A., Berry, L. L., 1985. A conceptual model of service quality and its implications for future research. *Journal of Marketing* 49(4): 41–50.

Powell, T., 1995. Total quality management as competitive advantage: A review and empirical study. *Strategic Management Journal* 16: 15–37.

Sousa, R., Voss, C. A., 2002. Quality management re-visited: A reflective review and agenda for future research. *Journal of Operations Management* 20: 91–109.

Vargo, S. L., Lusch, R. F., 2004. Evolving to a new dominant logic for marketing. *Journal of Marketing* 68 (January): 1–17.

Zeithaml, V. A., 1981. How consumer evaluation processes differ between goods and services. In *Marketing of Services,* J. H. Donnelly and W. R. George, eds. Chicago, IL: American Marketing Association Proceedings Series.

Zeithaml, V. A., Bitner, M. J., Gremler, D., 2006. *Services Marketing: Integrating Customer Focus Across the Firm*, 4th ed. New York, NY: McGraw-Hill.

Zeithaml, V. A., Parasuraman, A., and Berry, L.L., 1990. *Delivering Quality Service: Balancing Customer Perceptions and Expectations*. New York, NY: The Free Press.

4

Data Mining and Quality in Service Industry: Review and Some Applications

Teresa Oliveira, Amílcar Oliveira, and Alejandra Pérez-Bonilla

CONTENTS

4.1 Introduction to Data Mining: Literature Review

Over the last three decades, increasingly large amounts of data have been stored electronically and this volume will continue to increase in the future. Branco (2010) mentions that the idea behind data mining (DM) appears when the automatic production of data leads to the accumulation of large volumes of information, which is conclusively established in the 1990s. Chatfield (1997) argues that the term first appears in a book on econometric analysis dated 1978 and written by E. E. Leamer. However, there are indications that the term has already been used by statisticians in the mid-1960s (Kish 1998). Frawley et al. (1992) present a data mining definition: "Data Mining is the nontrivial extract of implicit, previously unknown, and potentially useful information from the data." Data mining is part of the knowledge discovery process and a current challenge offering a new way to look at data as an extension of exploratory data analysis. It is a process for exploring large data sets in the search for consistent and useful patterns of behavior leading to the detection of associations between variables and to the distinguishing of new data sets. Data mining is then the process of discovering meaningful new correlations, patterns, and trends by sifting through vast amounts of data using statistical and mathematical techniques. It uses machine learning and statistical and visualization techniques to discover and present knowledge in an easily comprehensible way. As this knowledge is captured, this can be a key factor to gaining a competitive advantage over competitors in an industry and to improve the quality of the interaction between the organization and their customers. To several authors, data mining is known as an essential step in the process of knowledge discovery in databases (KDD), which can be represented by the following steps:

1. *Cleaning the data*—A stage where noise and inconsistent data are removed.
2. *Data integration*—A stage where different data sources can be combined, producing a single data repository.
3. *Selection*—A stage where user attributes of interest are selected.
4. *Data transformation*—A phase where data is processed into a format suitable for application of mining algorithms (e.g., through operations aggregation).
5. *Prospecting*—An essential step consisting of the application of intelligent techniques in order to extract patterns of interest.
6. *Evaluation or postprocessing*—A stage where interesting patterns are identified according to some criterion of the user.
7. *Display of results*—A stage of knowledge representation techniques, showing how to present the extracted knowledge to the user.

Antunes (2010) refers to the Web site www.kdnuggets.com, which centralizes information on data mining, and where one can see that the consumption sectors, banking, and communications represent about 60% of the applications of techniques of data mining.

Aldana (2000) emphasizes the difference between DM and KDD. The author states that KDD concerns itself with knowledge discovery processes applied to databases; it deals with ready data, available in all domains of science and in applied domains such as marketing, planning, and control. KDD refers to the overall process of discovering useful knowledge from data, whereas data mining refers to the application of algorithms for extracting patterns and associations from the data.

Besides the literature review and an introduction to the links between data mining and most frequent statistical techniques, the goal of this chapter is to review an application of data mining in the service industry.

4.2 Data Mining and Statistics

DM profits from tools in several major fields, but statistics plays a key role since DM can be used with the three main objectives: explanation, confirmatory, and exploration.

However, there are some main differences between statistics and data mining pointed out in the paper by Branco (2010):

1. The most striking difference is the magnitude and complexity of the data in DM.

2. Cleaning, compression, and data transformation operations are essential initial processes in DM but are not generally required in statistical methods.

3. Traditional statistics are generally not directly applicable to the treatment of major data sets that DM can analyze.

4. While statistics can be viewed as a process of analyzing relations, DM is a process of discovering relationships.

5. The statistical approach is confirmatory in nature while the DM follows an exploratory approach.

Branco (2010) refers to several authors who gave precious contributions in this area, including Friedman (1998), Hand (1998, 1999a, 1999b), Kuonen (2004), and Zhao and Luan (2006).

The major statistical techniques used in data mining are as follows:

Regression—Data mining within a regression framework will rely on regression analysis, broadly defined, so that there is no necessary commitment *a priori* to any particular function of the predictors. Based on a numerical data set, a mathematical formula that fits the data is developed. Advanced techniques, such as multiple regression, allow the use of more than one input variable and allow for the fitting of more complex models, such as a quadratic equation.

Time series analysis—Construction of models that detail the behavior of the data through mathematical methods. These models look for identifying temporal patterns for characterization and prediction of time series events.

Classification—In data mining this technique is applied to the classification of customers, for example, to classify them according to their social level.

Clustering—In this technique data are grouped according to their classification and group belonging, and these groups are built based on the similarity between the elements of this group.

Recognition patterns—Data items are present together with some probability. For example, a supermarket cart: through it one can extract information for the provision of supermarket products to please consumers by placing products purchased in conjunction with another.

4.3 Data Mining Major Algorithms

Wu et al. (2008) make an effort to identify some of the most influential algorithms that have been widely used in the data mining community. The most important are described in the sections that follow.

4.3.1 Classification

- C4.5 is an extension of Quinlan's earlier ID3 algorithm. The decision trees generated by C4.5 can be used for classification, and for this reason, C4.5 is often referred to as a statistical classifier. Details on this algorithm can be seen in Quinlan (1993, 1996) and in Kotsiantis (2007).

- Classification and regression tree (CART) analysis is a nonparametric decision tree learning technique that produces either classification or regression trees, depending on whether the dependent variable is categorical or numeric, respectively. We refer to Loh (2008) and to Wu and Kumar (2009) for further developments.

- *k*-Nearest neighbor (*k*NN) is a method for classifying objects based on closest training examples in the feature space. This is a nonparametric method, which estimates the value of the probability density function (pdf) or directly the posterior probability that an element x belongs to class Cj from the information provided by the set of prototypes. The book by Wu and Kumar (2009) and the paper by Shakhnarovish et al. (2005) are recommended references on this issue.

- Naive Bayes is a probabilistic classifier based on Bayes theorem. In simple terms, a naive Bayes classifier assumes that the absence or presence of a particular feature of a class is unrelated to the absence or presence of any other feature, as we can see, for example, in Frank et al. (2000).

4.3.2 Statistical Learning

- Support vector machines (SVMs)—given a set of training examples (pattern) can label the classes and train an SVM to build a model that predicts the kind of a new sample. For further details on this algorithm we refer to Press et al. (2007).
- Expectation–maximization (EM) algorithm is a parameter estimation method that falls into the general framework of maximum likelihood estimation and is applied in cases where part of the data can be considered to be incomplete. Many references can be found under this issue, such as Hastie et al. (2001) and more recently Gupta and Chen (2010).

4.3.3 Association Analysis

- An *a priori* algorithm is used to find association rules in a data set. This algorithm is based on prior knowledge of frequent sets, and this serves to reduce the search space and increase efficiency. Agrawal and Srikant (1994) consider the problem of discovering association rules between items in a large database of sales transactions and they present *a priori* algorithms for solving this problem.
- Frequent-pattern tree (FP-tree) is an efficient algorithm based in a compact tree structure for finding frequent patterns in transaction databases, as shown in Han et al. (2004).

4.3.4 Link Mining

- PageRank is a search-ranking algorithm using hyperlinks on the Web. This algorithm is described in the paper by Page et al. (1999) available at http://ilpubs.stanford.edu:8090/422/1/1999-66.pdf, as well as in the paper by Richardson and Domingos (2002) available at http://research.microsoft.com/en-us/um/people/mattri/papers/nips2002/qd-pagerank.pdf.

- Hyperlink-induced topic search (HITS) is an algorithm designed to assess and classify the importance of a Web site. Li et al. (2002) present two ways to improve the precision of HITS-based algorithms on Web documents as well as list interesting references on this issue.

4.3.5 Clustering

- k-Means is a simple iterative method to partition a given data set into a user-specified number of clusters, k. It is also referred to as Lloyd's algorithm, particularly in the computer science community, since it was introduced in Lloyd (1957), being published later in Lloyd (1982). Many references can be found on developing this issue (e.g., Mahajan et al. 2009).

- BIRCH (balanced iterative reducing and clustering using hierarchies) clustering algorithm is an unsupervised algorithm used to perform hierarchical clustering over particularly large data sets. Zhang (1997) demonstrates that this algorithm is especially suitable for very large databases. BIRCH incrementally and dynamically clusters incoming multidimensional metric data points to try to produce the best quality clustering with the available resources, such as available memory and time constraints.

4.3.6 Bagging and Boosting

- AdaBoost (adaptive boosting) is a machine-learning algorithm and is a statistical classifier using a set of weak classifiers to form a strong classifier, which is the one who gives the final answer of the algorithm. AdaBoost is one of the most important ensemble methods, formulated by Freund and Schapire (1997).

4.3.7 Sequential Patterns

- Generalized sequential patterns (GSP) is an algorithm proposed to solve the mining of sequential patterns with time constraints, such as time gaps and sliding time windows. A description of the GSP algorithm can be found, for example, in Pujari (2001). In the work by Hirate and Yamana (2006), the authors refer to the importance of identification of item intervals of sequential patterns extracted by sequential pattern mining, presenting an application using Japanese data on earthquakes.

- PrefixSpan (mining sequential patterns efficiently by prefix-projected pattern growth) is a sequential pattern mining method, described in Pei et al. (2001).

4.3.8 Integrated Mining

- Classification based on association rules (CBA) is a method proposed to integrate classification and association rule mining. The main algorithm is described in Liu et al. (1998).

4.4 Software on Data Mining

Several packages have been developed to consider DM applications (Oliveira and Oliveira 2010). The R GNU project in statistics is very well known as open-source software, freely available, and supported by an international team of specialists in the areas of statistics, computational statistics, and computer science. Nowadays R is also increasingly being recognized as providing a powerful platform for DM. The Rattle package (http://rattle.togaware.com) in R is dedicated to data mining and it is widely used in industry by consultants and for teaching. Other nonopen sources also present packages considering DM developments and applications. We emphasize that in SPSS there is the package Clementine devoted to DM in use in the social sciences (http://credit-new. com/software/1681-spss-clementine-v12.html). The Oracle Database includes the package Oracle Data Mining, with several models included, which we will detail. Also in StatSoft-Statistica, there is an introduction to DM application. A 2009 survey by KDnuggets about analytic tools used in data mining can be found at www.kdnuggets.com/polls/2010/data-mining-analytics-tools.html.

4.5 Service Industry and Challenges

With the technological advances of the late 20th century, it was feared by many that computers and other new technologies would tend to replace people and destroy jobs. However, these new features, products, and services have also created jobs that were not predicted or previously available, most of them concerning service industry occupations.

According to the United States Standard Industry Classification System, there are two major sectors for jobs: the goods-producing sector and the service-producing sector. Goods-producing industries are those that create some kind of tangible object, like construction, manufacturing, mining, and agriculture. The service-producing sector, or service industry, rather than tangible objects, creates services such as banking, insurance, communications, wholesale, engineering, software development, medicine, educational, and government services. Among others, service industry companies are involved in retail, transport, distribution, hotels and tourism, cleaning, leisure and entertainment, and law and consultancy. Currently, it is very well known that service industry plays a key role in the international economy, considering the underlying social, cultural, and economic benefits. Economic development in countries is reflected in the inverse proportion to its potential focus respectively in goods-producing and service-producing sectors. Castro et al. (2011) present current and future trends on innovation in service industries. The purpose of the authors is to compare the behavior of service companies and manufacturing companies in technological, organizational, and commercial innovation, and they point out that future tendency should be determined by the use of models to reflect all types of innovation with no sector discrimination. Parellada et al. (2011) present an overview of the service industries' future (priorities: linking past and future) and the authors emphasize the role of industry services on the growth of the international economy. They refer to the ease with which services can be implemented in certain activities with notably less capital than that required for most industries as well as the impact on the job market, which are factors that have brought about belief and reliance on this sector to provide the recovery from any international economic crisis.

4.6 Data Mining and Quality in Service Industry

Data mining is ready for application in industry services and in the business community, since it is supported by mature technologies such as massive data collection, powerful multiprocessor computers, and data mining algorithms (Aldana 2000). The purpose of service industry collections is to provide information on the size, structure, and nature of the industry under study. Data collections can have origins in several areas, for example, the information on business income and expenses and employment status. Seifert (2004) notes that DM is used for a variety of purposes in private and public services and that industries such as banking, insurance, pharmacy, medicine, and retailing commonly use this technique to reduce costs, enhance research, and increase sales. Some examples are credit scoring in banking and the prediction of effectiveness of a medicine. Zwikael (2007) focuses his paper on quality management and points to it as a key process in the service industries. Zwikael presents a benchmarking research study aimed at improving project planning capabilities in the service industry. It was concluded that managers in the service sector would benefit from acquiring proper knowledge and techniques relating to quality management in the planning phase of projects. Quality should be present in every project from the beginning and particularly in the data acquisition. In their paper Farzi and Dastjerdi (2010) introduce a method for measuring data quality based on a DM algorithm. This algorithm has three steps that calculate the data transaction quality. Ograjenšek (2002) emphasizes that to provide a satisfactory definition of services, some authors like Mudie and Cottam (1993), Hope and Mühlemann (1997), and Kasper et al. (1999) characterize them with these important features: intangibility, inseparability, variability, and perishability. Murdick et al. (1990) define quality of a service as the degree to which the bundle of services attributes as a whole satisfies the customer; it involves a comparison between customer expectations (needs, wishes) and the reality (what is really offered). Whenever a person pays for a service or product, the main aim is getting quality. There are two particular reasons why customers must be given quality service:

1. Industry is very competitive and customers have a huge variety of alternatives.
2. Most customers do not complain about meeting problems, they simply take their business elsewhere.

In a quality service one faces a special challenge since there is the need to meet the customer requests while keeping the process economically competitive. To ensure quality assurance of the companies there are shadow shopping companies who act as quality control laboratories. They develop experiments, draw inference, and provide solutions for improvement. In the case of the service industry the quality assurance service is provided by the various surveys and inspections that are carried out from time to time.

Another survey by KDnuggets about fields where the experts applied data mining can be seen at www.kdnuggets.com/polls/2010/analytics-data-mining-industries-applications.html. We see that the biggest increases from 2009 to 2010 were for the following fields: manufacturing (141.9%), education (124.1%), entertainment/music (93.3%), government/military (56.5%). The biggest declines were for social policy/survey analysis (−44.8%), credit scoring (−48.8%), travel/hospitality (−49.7%), security/anti-terrorism (−62.4%).

4.7 Application of Data Mining to Service Industry: Example of Clustering

Based on the examples mentioned earlier we will now outline a specific case study where data mining techniques and knowledge discovery process have been applied to the service industry. An example using automatic classification and knowledge discovery in data mining is the methodology proposed in Pérez-Bonilla (2010). The application focuses on a wastewater treatment plant (WWTP), because this is one of the domains where conventional approaches work worse. The methodology benefits from the hierarchical structure of the target classification to induce concepts iterating with binary divisions from dendrogram (or hierarchical tree, see Figure 4.3), so that, based on the variables that describe the objects belonging to a certain domain, the specifics of each class can be found, thus contributing to the automatic interpretation of the conceptual description of clustering. Industrial wastewater treatment covers the mechanisms and processes used to treat waters that have been contaminated in some way by anthropogenic industrial or commercial activities, prior to its release into the environment or its reuse. Most industries produce some wet waste although recent trends in the developed world have been to minimize such production or recycle such waste within the production process. However, many industries remain dependent on processes that produce wastewater. First, clustering is a well-known technique used in knowledge discovery processes (www.kdnuggets.com/polls/2006/data_mining_methods.htm). This section presents the methodology of conceptual characterization by embedded conditioning (CCEC), aimed at the automatic generation of conceptual descriptions of classifications that can support later decision making, as well as its application to the interpretation of previously identified classes characterizing the different situations on a WWTP. The particularity of the method is that it provides an interpretation of a partition previously obtained on an ill-structured domain, starting from a hierarchical clustering. This section addresses the usefulness of CCEC for building domain theories as models supporting later decision making.

In automatic classification, where classes composing a certain domain are to be discovered, one of the most important required processes and one of the less standardized, is the interpretation of the classes (Gordon 1994), closely related to validation and critical for the later usefulness of the discovered knowledge. The interpretation of the classes, so important to understanding the meaning of the obtained classification as well as the structure of the domain, used to be done in an artistic-like way. But this process becomes more complicated as the number of classes grows. This chapter is involved with the automatic generation of useful interpretations of classes, so that decisions about the action associated to a new object can be modeled. The aim is to develop, in the long term, a decision support system.

4.7.1 Case Study

The case study in this chapter was a pilot plant, which is located in the Domale-Kamnik wastewater treatment plant in Slovenia. A scheme of the pilot plant with sensors and actuators is shown in Figure 4.1. In the pilot plant the moving bed biofilm reactor (MBBR) technology is tested for the purpose of upgrading the whole plant for nitrification and denitrification. The pilot plant with the volume of 1125 m^3 consists of two anoxic and two aerobic tanks that are filled with the plastic carriers on which the biomass develops, and a fifth tank, which is a dead zone without plastic carriers and

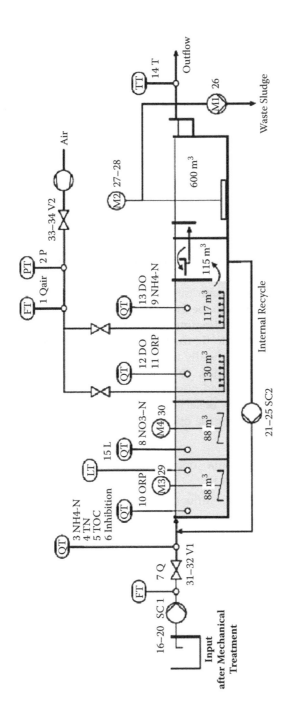

FIGURE 4.1
MBBR (moving bed biofilm reactor) pilot plant with sensors and actuators.

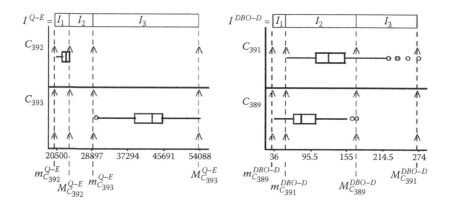

FIGURE 4.2

Boxplot-based discretization for two classes partition.

a settler. The total airflow to both aerobic tanks can be manipulated online so that oxygen concentration in the first aerobic tank is controlled at the desired value. The wastewater rich with nitrate is recycled with the constant flow rate from the fifth tank back to the first tank. The influent to the pilot plant is wastewater after mechanical treatment, which is pumped to the pilot plant. The inflow is kept constant to fix the hydraulic retention time. The influent flow rate can be adjusted manually to observe the plant performance at different hydraulic retention times. The database that was used in this study consisted of 365 daily averaged observations that were taken from June 1, 2005, to May 31, 2006. Every observation includes measurements of the 16 variables that are relevant for the operation of the pilot plant. The variables that were measured were:

- NH4-influent—Ammonia concentration at the influent of the pilot plant (pp) (3 in Figure 4.1)
- Q-influent—Wastewater influent flow rate of the pp (7 in Figure 4.1)
- TN-influent—Concentration of the total nitrogen at the influent of the pp (4 in Figure 4.1)
- TOC-influent—Total organic carbon concentration at the influent of the pp (5 in Figure 4.1)
- Nitritox-influent—Measurement of the inhibition at the influent of the pp (6 in Figure 4.1)
- h-wastewater—Height of the wastewater in the tank (NO in Figure 4.1)
- O2-1 aerobic—Dissolved oxygen concentration in the first aerobic tank (third tank) (12 in Figure 4.1)
- Valve-air—Openness of the air valve (between 0% and 100%), highly related with Q-air (V2 in Figure 4.1)

- Q-air—Total airflow that is dosed in both aerobic tanks (1 in Figure 4.1)
- NH4-2 aerobic—Ammonia concentration in the second aerobic tank (9 in Figure 4.1)
- O2-2 aerobic—Dissolved oxygen concentration in the second aerobic tank (fourth tank) (13 in Figure 4.1)
- TN-effluent—Concentration of the total nitrogen at the effluent of the pp (NO in Figure 4.1)
- Temp-wastewater—Temperature of the wastewater (14 in Figure 4.1)
- TOC-effluent—Total organic carbon concentration at the effluent of the pp (NO in Figure 4.1)
- Freq-rec—Frequency of the internal recycle flow rate meter (NO in Figure 4.1)
- FR1-DOTOK-20s (Hz)—Frequency of the motor that pumps the wastewater into the pilot plant; this frequency is highly correlated with Q-influent

4.7.2 The Methodology

The methodology proposed tries to approximate in a formal model the natural process that follows an expert in its phase of interpretation of results by making an iterative approximation based on a hierarchical clustering. This methodology:

- Provides a systematizing of the process of interpretation of classes from a hierarchical cluster and represents a significant advance to the current situation in which the interpretation is done manually, and more or less crafted.
- Likewise, it helps to systematize and objectify the mechanisms of interpretation used by human experts.
- The results generated by the methodology allow the expert to more easily understand the main characteristics of the classification obtained by generating explicit knowledge directly from the classes.
- While the methodology proposed is general, application focuses on WWTP because this is one of the domains where conventional approaches work worse and belong to the lines under research developed in the group.

From a theoretical point of view, the main focus of this methodology has been to present a hybrid methodology that combines tools and techniques of statistics and artificial intelligence in a cooperative way, using a transversal and multidisciplinary approach combining elements of the induction of concepts from artificial intelligence, propositional logic, and probability theory. Thus, this methodology contributes to the generic design of KDD, which should include modules that support the definition of the problem (including knowledge), data collection, cleaning and preprocessing, data reduction, selection of data mining technique, interpretation, and production of *a posteriori* discovered knowledge. It also contributes to objective procedures for the validation of results, as the fact that clustering has a clear interpretation is related to the usefulness of a classification (currently used as a criterion for validation) and usable to decide whether it is correct, evaluating the usefulness requires *a posteriori* mechanism of understanding the meaning of classes. The methodology of CCEC benefits from the hierarchical structure of the target classification to induce concepts iterating with binary divisions from dendrogram, so that, based on the variables that describe the objects belonging to a certain domain, the specifics of each class can be found, thus contributing to the automatic interpretation of the conceptual description of clustering.

4.7.2.1 Basic Concepts

Four main concepts of methodology CCEC are:

1. *Variable X_k is totally characterizing of class*, $C \in \mathcal{P}$, if all the values taken by X_k in class C are characteristic of C (objects of other classes take other values). As an example, see variables Q-E, which is totally characterizing the class C393 if it takes only high values. See Figure 4.2.

2. *Variable X_k is partially characterizing of $C \in \mathcal{P}$* if there is at least one characteristic value of C, although the class can share some other values with other classes. See DBO-D in Figure 4.2, which is partially characterizing the class C391 if it takes very low or high values.

3. *Boxplot-based discretization* (BbD) (Pérez-Bonilla and Gibert 2007) is an efficient way of transforming a numerical variable into a qualitative one so that the resulting qualitative variable maximizes the association with the reference partition. Basically the cut points are obtained from the minimum and maximum values that the numerical variable takes in the classes induced producing the categorical one. Identify where the subsets of intersecting classes between classes change and, given X_k, $\mathcal{P}$, it consists of the following steps:

   a. Calculate the *minimum* (m_C^k) and *maximum* (M_C^k) of X_k inside any class. Built $\mathcal{M}^k = \{m_{C_1}^k, \ldots, m_{C_e}^k, M_{C_1}^k, \ldots M_{C_e}^k\}$, where *card* ($M^k$) = 2ξ.

   b. Build the *set of cut points* $\mathcal{Z}^k$ by sorting $\mathcal{M}^k$ in increasing way into $\mathcal{Z}^k = \{z_i^k : i = 1 : 2\xi\}$. At every z_i^k the intersection between classes changes arity.

TABLE 4.1

Summary Statistics for Q-E versus P_2 (Left) and $\mathcal{M}^{Q\text{-}E}$, Set of Extreme Values of Q-E$|C \in P_2$ and $\mathcal{Z}^{Q\text{-}E}$, Corresponding Ascendant Sorting (Right)

		Class	**C393**	**C392**	$\mathcal{M}^{Q\text{-}E}$	$\mathcal{Z}^{Q\text{-}E}$
Var.		$N = 396$	$n_c = 390$	$n_c = 6$		
Q-E	$\bar{X}$		42,112.9453		29,920	20,500
	S		4,559.2437	1,168.8481	20,500	23,662.9004
	Min		29,920	20,500	54,088.6016	29,920
	Max		54,088.6016	23,662.9004	23,662.9004	54,088.6016
	N*		1	0		

c. Build the *set of intervals I^k induced by P on X_k* by defining an interval I_s^k between every pair of consecutive values of $\mathcal{Z}^k$. $\mathcal{I}^k = \{I_1^k, \ldots, I_{2\xi-1}^k\}$ is the boxplot-based discretization of X_k. The $I_s^k, s \in \{1, 2\xi - 1\}$ intervals are variable length and the intersection among classes changes from one to another. (See Figure 4.2 and Table 4.1.)

4. The methodology *boxplot-based induction rules* (BbIR) is presented in Pérez-Bonilla and Gibert (2007). It is a method for inducing probabilistic rules ($r : x_{ik} \in I_s^k \xrightarrow{psc} C, p_{sc} \in [0,1]$ is a degree of certainty of r) with a minimum number of attributes in the antecedent, based on the BbD of X_k.

   a. Use the boxplot-based discretization to build

   $$r_s : \mathcal{I}^k = \{I_1^k, I_2^k, I_3^k, \ldots, I_{2\xi-1}^k\}$$

   b. Build table $I^k \times P$ where rows are indexed by $s \in \{1, 2\xi-1\}$ and columns by $C \in \{1, \xi\}$.

   c. Build the table $I^k | P$ by dividing the cells of $I^k \times P$ by the row totals. Cells,

   $$p_{sc} = P(C | I^k = I_s^k) = P(i \in C | x_{ik} \in I_s^k) = \frac{card\{i : x_{ik} \in I_s^k \wedge i \in C\}}{card\{i \in I : x_{ik} \in I_s^k\}} = \frac{n_{sc}}{\sum_{\forall s} n_{sc}}$$

   If $\exists C : p_{sc} = 1 \ \& \ Cov(r_s) = 100\%$ then X_k is a *totally characterizing variable* of C.

   d. For every cell in table $I^k | P$ produce the rule: If $x_{ik} \in I_s^k \xrightarrow{psc} i \in C$.

Quality measurement rules:

- Support (*Sup*)—Given a rule $r : x_{ik} \in I_s^k \rightarrow C$, where $I_s^k \subset \tau_k$, the support of r is the proportion of objects in $\mathcal{I}$ that satisfy the antecedent of the rule (Liu et al. 2000).

$$Sup(r) = \frac{card\{i \in C : x_{ik} \in I_s^k\}}{n}$$

- Relative covering (*CovR*)—Given a rule, the relative covering is the proportion of class C that satisfy the rule.

$$CovR(r) = \frac{card\{i \in C : x_{ik} \in I_s^k\}}{card\{c\}} * 100$$

- Confidence (*Conf*)—Given a rule, the confidence of r is the proportion of objects in $\mathcal{I}$ that satisfy the antecedent of the rule and belong to C (Liu et al. 2000).

$$Conf(r) = \frac{card\{i \in C : x_{ik} \in I_s^k\}}{card\{x_{ik} \in I_s^k\}}$$

4.7.2.2 Methodology of Conceptual Characterization by Embedded Conditioning (CCEC)

CCEC takes advantage of the existence of τ and uses the property of any binary hierarchical structure that $\mathcal{P}_{\xi+1}$ has the same classes of $\mathcal{P}_\xi$ except one, which splits in two subclasses in $\mathcal{P}_{\xi+1}$. Binary hierarchical structure will be used by CCEC to discover particularities of the final classes step by step, also in hierarchical way. The CCEC (Pérez-Bonilla 2010) allows generation of automatic conceptual interpretations of a given partition $\mathcal{P} \in \tau$. The steps to be followed are described next. The application of CCEC to the WWTP is illustrated in Section 4.7.3:

1. Cut the tree at highest level (make $\xi = 2$ and consider $\mathcal{P}_2 = \{C_1, C_2\}$).
2. Use BbD (Pérez-Bonilla and Gibert 2007), to find (total or partial) characteristic values for numerical variables.
3. Use BbIR to induce a knowledge base describing both classes.
4. For classes in $\mathcal{P}_2$, determine concepts $A_1^{\xi, X_k} : "[X_k \in I_s^k]", A_2^{\xi, X_k} : \neg A_1^{\xi, X_k}$ associated to C_1, C_2, by taking the intervals provided by a totally characteristic variable or the partial one with greater relative covering and $p_{sc} = 1$.
5. Go down one level in the tree by making $\xi = \xi + 1$ and so considering $\mathcal{P}^{\xi+1}$. As said before $\mathcal{P}^{\xi+1}$ is *embedded* in $\mathcal{P}^\xi$ in such a way that

there is a class of $\mathcal{P}^{\xi}$ splitting in two new classes of $\mathcal{P}^{\xi+1}$, namely $C_i^{\xi+1}$, and $C_j^{\xi+1}$ and all other classes $C_q^{\xi+1}$, $q \neq i, j$, are common to both partitions and $C_q^{\xi+1} = C_q^{\xi}$ $\forall q \neq i, j$. Since in the previous step $C_i^{\xi+1} \cup C_j^{\xi+1}$ were conceptually separated from the rest, at this point it is only necessary to find the variables that separate (or distinguish) $C_i^{\xi+1}$ from $C_j^{\xi+1}$, by repeating steps 2 to 4. Suppose $B_i^{\xi+1, x_k}$ and $B_j^{\xi+1}$, the concepts induced from $C_i^{\xi+1}$ and $C_j^{\xi+1}$, in the step $\xi + 1$.

6. Integrate the extracted knowledge of the iteration $\xi + 1$ with that of the iteration ξ by determining the compound concepts finally associated to the elements of $\mathcal{P}_{\xi+1}$. The concepts for the classes of $\mathcal{P}_{\xi+1}$ will be $A_q^{\xi+1, X_k} = A_q^{\xi, X_k}$ $A_i^{\xi+1, X_k} = \neg A_q^{\xi, X_k} \wedge B_i^{\xi+1, X_k}$, and $A_j^{\xi+1, X_k} = \neg A_q^{\xi, X_k} \wedge B_j^{\xi+1, X_k}$.

7. Make $\xi = \xi + 1$, and return to step 5 repeating until $\mathcal{P}_{\xi} = \mathcal{P}$.

4.7.3 Application and Results

4.7.3.1 Hierarchical Clustering

The standard input of a clustering algorithm is a data matrix with the values of K variables $X_1 \ldots X_K$ (numerical or not) observed over a set $\mathcal{I} = \{1, \ldots, n\}$ of individuals.

$$
\chi = \begin{pmatrix}
x_{11} & x_{12} & \cdots & x_{1K-1} & x_{1K} \\
x_{21} & x_{22} & \cdots & x_{2K-1} & x_{2K} \\
\vdots & \vdots & \vdots & \vdots & \vdots \\
x_{n-11} & x_{n-12} & \cdots & x_{n-1K-1} & x_{n-1K} \\
x_{n1} & x_{n2} & \cdots & x_{nK-1} & x_{nK}
\end{pmatrix}
$$

Variables are in columns, and individuals are represented in rows. Cells contain the value (x_{ik}), taken by individual $i \in \mathcal{I}$ for variable X_k, $(k = 1{:}K)$. The set of values of X_k is named $\mathcal{D}^k = \{c_1^k, c_2^k, \ldots, c_s^k\}$ for categorical variables and $\mathcal{D}^k = r_k$ for numerical ones, being $r_k = [\min X_k, \max X_k]$ the range of X_k. A partition in ξ classes of $\mathcal{I}$ is denoted by $\mathcal{P}_{\xi} = \{C_1, \ldots, C_{\xi}\}$, and $\tau = \{\mathcal{P}_1, \mathcal{P}_2, \mathcal{P}_3, \mathcal{P}_4, \ldots, \mathcal{P}_n\}$ is an indexed hierarchy of $\mathcal{I}$.

$$
\chi_p = \begin{pmatrix}
x_{11} & x_{12} & \cdots & x_{1K-1} & x_{1K} & C(1, \mathcal{P}_{\xi}) \\
x_{21} & x_{22} & \cdots & x_{2K-1} & x_{2K} & C(2, \mathcal{P}_{\xi}) \\
\vdots & \vdots & \vdots & \vdots & \vdots & \vdots \\
x_{n-11} & x_{n-12} & \cdots & x_{n-1K-1} & x_{n-1K} & C(n-1K), \mathcal{P}_{\xi}) \\
x_{n1} & x_{n2} & \cdots & x_{nK-1} & x_{nK} & C(n, \mathcal{P})
\end{pmatrix}
$$

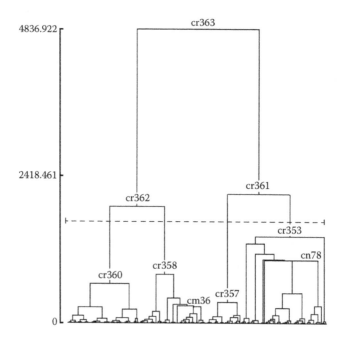

FIGURE 4.3
Hierarchical tree to case study.

Usually, τ is the result of a *hierarchical clustering* over $\mathcal{I}$, and it can be represented in a graphical way as a dendrogram (or hierarchical tree).

The dendrogram of Figure 4.3 was produced by a hierarchical clustering using Ward's method. This method involves an agglomerative clustering algorithm (mutual neighbors). It will start out at the leaves and work its way to the trunk, so to speak. It looks for groups of leaves that it forms into branches, the branches into limbs, and eventually into the trunk. Ward's method starts out with n clusters of size 1 and continues until all the observations are included into one cluster.

Finally a $P_4 = \{Cr_{353}, Cr_{357}, Cr_{358}, Cr_{360}\}$ has been obtained. Figure 4.4 contains the *class panel graph* (Gibert et al. 2005) of the 16 variables regarding the partition P_4 where the *multiple boxplot* presented by Tukey (1977) of variables for each class are displayed. As usual in hierarchical clustering, the final partition is the horizontal cut of the tree that maximizes the ratio between *heterogeneity* between classes with respect to *homogeneity* within classes, which guarantees the *distinguishability* of classes.

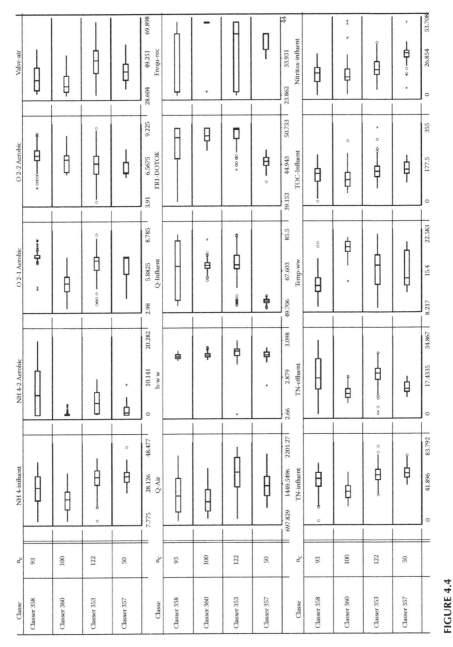

FIGURE 4.4
Class panel graph of P4.

4.7.3.2 CCEC Interpretation of Classes

The process would continue separating the Cr_{360} and Cr_{358} of the partition $\mathcal{P}_4$ which are the subdivision of Cr_{362}. Here, a Temp-ww with a relative covering of 66,67%, is chosen. Similarly, the interpretation of $\mathcal{P}_4$, which is the final partition obtained with the knowledge base (as is shown in the example to $P2_{Lj3,R2}^{EnW,G}$ in Table 4.2) with certain rules for Cr_{360} and Cr_{358} is as follows:

- Cr_{358} is such that "$x_{TN-\ influent,i} \in [28.792,83.792]$" $\wedge$ "$x_{Q-influent,i} \in [55.666,85.092]$," that is, nonlow concentration of the total nitrogen at the influent and high values of wastewater influent flow rate. Represents the plant operation under the high load. In this case influent nitrogen concentrations are high and the influent flow rate is quite high as well. Even though the oxygen concentration in the aerobic tanks is high, this cannot decrease the effluent nitrogen concentrations. It means that when the plant is overloaded and high effluent concentrations at the effluent of the plant can be expected.

- Cr_{357} is such that "$x_{TN-influent,i} \in [28.792,83.792]$" $\wedge$ "$x_{Q-influent,i} \in [49.706,55.666]$," that is, nonlow concentration of the total nitrogen at the influent and not high values of wastewater influent flow rate. Represents the situation when the influent flow rate is low, that is, when the hydraulic retention time of the plant is high. In this case we get quite low effluent nitrogen concentrations if the oxygen concentration in the aerobic tank is high enough. It means when the influent flow rate to the plant is low the effluent concentrations of the plant can be obtained at the low level if the oxygen concentration in the aerobic tanks is high.

- Cr_{358} is such that "$x_{TN-influent,i} \in [0.0,28.792]$" $\wedge$ "$x_{Temp-ww,i} \in [8.472,13.327)$," that is, low concentration of the nitrogen and low values of water's temperature. Explains the situation when the wastewater temperature is low. In this case nitrogen removal efficiency of the plant is rather low. This is so because microorganisms in the tanks do not work as intensively in cold conditions and therefore higher concentrations at the effluent of the plant can be expected.

- Cr_{360} is such that "$x_{TN-influent,i} \in [0.0,28.792]$" $\wedge$ $x_{Temp-ww,i} \in [13.327,21.896]$," that is, low concentration of the nitrogen and nonlow values of temperature. Shows the situation when the wastewater temperature is high. In warmer conditions the microorganisms in the plant work faster, so the effluent nitrogen concentrations can be low even when the oxygen concentrations in the aerobic tanks are quite low.

This set of concepts can in fact be considered as a domain model, which can support a later decision for the treatment that is applied to a new day, provided that a standard treatment is previously associated to every class by

TABLE 4.2

Summary of Knowledge Base for Cr_{361} and Cr_{361} from $P2_{Lj3,R2}^{EnW,G}$

Concep	Knowledge Base Cr_{361} (172 Days) and Cr_{362} (193 Days)	Cov	Cov R
$A_{C_{361}}^{2,NH4-influent}$	$r_{3,C_{r361}}^{NH4-influent} : x_{NH4-influent,i} \in \left(40.541, 48.477\right] \xrightarrow{1.0} C_{r361}$	5	2.91%
$A_{C_{361}}^{2,O2-1aerobic}$	$r_{3,C_{r361}}^{O2-1aerobic} : x_{O2-1aerobic,i} \in \left(8.371, 8.785\right] \xrightarrow{1.0} C_{r361}$	1	0.58%
$A_{C_{361}}^{2,O2-2aerobic}$	$r_{1,C_{r361}}^{O2-2aerobic} : x_{O2-2aerobic,i} \in \left[3.91, 4.94\right) \xrightarrow{1.0} C_{r361}$	2	1.16%
$A_{C_{361}}^{2,Value-air}$	$r_{3,C_{r361}}^{Value-air} : x_{Value-air,i} \in \left(54.777, 69.898\right] \xrightarrow{1.0} C_{r361}$	28	16.28%
$A_{C_{361}}^{2,Q-air}$	$r_{3,C_{r361}}^{Q-air} : x_{Q-air,i} \in \left(2030.77, 2201.27\right] \xrightarrow{1.0} C_{r361}$	27	15.70%
$A_{C_{361}}^{2,h-ww}$	$r_{3,C_{r361}}^{h-ww} : x_{h-ww,i} \in \left(3.058, 3.098\right] \xrightarrow{1.0} C_{r361}$	16	9.30%
$A_{C_{361}}^{2,Q-influent}$	$r_{1,C_{r361}}^{Q-influent} : x_{Q-influent,i} \in \left[49.706, 50.99\right) \xrightarrow{1.0} C_{r361}$	6	3.49%
$A_{C_{361}}^{2,Freq-rec}$	$r_{3,C_{r361}}^{Freq-rec} : x_{Freq-rec,i} \in \left(43.97, 44.0\right] \xrightarrow{1.0} C_{r361}$	3	1.74%
$A_{C_{361}}^{2,TN-influent}$	$r_{3,C_{r361}}^{TN-influent} : x_{TN-influent,i} \in \left(65.25, 83.792\right] \xrightarrow{1.0} C_{r361}$	9	5.23%
$A_{C_{361}}^{2,Temp-ww}$	$r_{3,C_{r361}}^{Temp-ww} : x_{Temp-ww,i} \in \left(21.896, 22.583\right] \xrightarrow{1.0} C_{r361}$	5	2.91%
$A_{C_{361}}^{2,TOC-influent}$	$r_{3,C_{r361}}^{TOC-influent} : x_{TOC-influent,i} \in \left(290.212, 355.0\right] \xrightarrow{1.0} C_{r361}$	20	11.63%
$A_{C_{361}}^{2,TOC-effluent}$	$r_{3,C_{r361}}^{TOC-effluent} : x_{TOC-effluent,i} \in \left(44.053, 52.57\right] \xrightarrow{1.0} C_{r361}$	10	5.81%
$A_{C_{362}}^{2,NH4-influent}$	$r_{1,C_{r362}}^{NH4-influent} : x_{NH4-influent,i} \in \left[7.775, 7.972\right) \xrightarrow{1.0} C_{r362}$	3	1.55%
$A_{C_{362}}^{2,NH4-2aerobic}$	$r_{3,C_{r362}}^{NH4-2aerobic} : x_{NH4-2aerobic,i} \in \left(9.846, 20.282\right] \xrightarrow{1.0} C_{r362}$	38	19.69%
$A_{C_{362}}^{2,O2-1aerobic}$	$r_{1,C_{r362}}^{O2-1aerobic} : x_{O2-1aerobic,i} \in \left[2.98, 3.297\right) \xrightarrow{1.0} C_{r362}$	4	2.07%
$A_{C_{362}}^{2,Value-air}$	$r_{1,C_{r362}}^{Value-air} : x_{Value-air,i} \in \left[28.604, 28.934\right) \xrightarrow{1.0} C_{r362}$	5	2.59%
$A_{C_{362}}^{2,Q-air}$	$r_{1,C_{r362}}^{Q-air} : x_{Q-air,i} \in \left[697.829, 739.819\right) \xrightarrow{1.0} C_{r362}$	2	1.04%
$A_{C_{362}}^{2,Q-influent}$	$r_{3,C_{r362}}^{Q-influent} : x_{Q-influent,i} \in \left(85.092, 85.5\right] \xrightarrow{1.0} C_{r362}$	1	0.52%
$A_{C_{362}}^{2,FR1-DOTOK}$	$r_{1,C_{r362}}^{FR1-DOTOK} : x_{FR1-DOTOK,i} \in \left[39.153, 42.276\right) \xrightarrow{1.0} C_{r362}$	5	2.59%
$A_{C_{362}}^{2,TN-influent}$	$r_{1,C_{r362}}^{TN-influent} : x_{TN-influent,i} \in \left[0.0, 28.792\right) \xrightarrow{1.0} C_{r362}$	**44**	**22.80%**
$A_{C_{361}}^{2,TN-effluent}$	$r_{3,C_{r362}}^{TN-effluent} : x_{TN-effluent,i} \in \left(28.933, 34.867\right] \xrightarrow{1.0} C_{r362}$	14	7.25%
$A_{C_{362}}^{2,TOC-influent}$	$r_{1,C_{r362}}^{TOC-influent} : x_{TOC-influent,i} \in \left[0.0, 63.22\right) \xrightarrow{1.0} C_{r362}$	20	10.36%
$A_{C_{362}}^{2,Nitritox-influent}$	$r_{1,C_{r362}}^{Nitritox-influent} : x_{Nitritox-influent,i} \in \left[0.0, 3.833\right) \xrightarrow{1.0} C_{r362}$	4	2.07%
$A_{C_{362}}^{2,TOC-effluent}$	$r_{1,C_{r362}}^{TOC-effluent} : x_{TOC-effluent,i} \in \left[0.0, 2.014\right) \xrightarrow{1.0} C_{r362}$	1	0.52%

experts. In this association the possibility of easily interpreting the classes is critical, as well as the knowledge from the experts, for easily understanding the meaning of the classes. In this sense the proposed method provides simple and short rules, which are easier to handle than those provided by other induction rules algorithms.

4.7.3.3 Discussion

Concepts associated with classes are built taking advantage of the hierarchical structure of the underlying clustering. CCEC is a quick and effective method that generates a conceptual model of the domain, which will be of great support to the later decision-making process based on a combination of boxplot-based discretization and an interactive combination of concepts upon hierarchical subdivisions of the domain. This is a preliminary proposal that has been successfully applied to real data coming from a WWTP. This proposal can be useful for the interpretation of partitions with a large number of classes. Automatic generation of interpretations is supposed to cover the important goal of KDD of describing the domain (Fayyad et al. 1996). However, in this proposal a direct connection between the generated concepts and the generation of automatic rules allows direct construction of a decision model for later class prediction. By associating an appropriate characteristic (or an appropriate standard treatment) to every class, a model for working the wastewater plant, based on a concrete day upon reduced number variables, is obtained along with an estimation of the risk associated to that decision (which is related to the certainty of the rule). Three different criteria for deciding which variable is kept at every iteration are assessed. This model can be included in an intelligent decision support system recommending decisions to be made according to certain new situations in the plant.

4.8 Considerations and Future Research

An additional key challenge for 21st century data mining companies is the creation of real-time algorithms for Web mining analysis in this age of the Internet. The problem of data analysis, whether a small set of data or an extremely large set, is a problem of statistical nature. The classical statistics were developed and prepared to study the small ensembles, and many of the methods are inappropriate for large sets. DM is not complete without the statistics and statistical techniques for improving the analysis of large data sets.

Acknowledgments

We would like to thank the staff of the Domale-Kamnik WWTP for providing us with the data from the pilot plant and the referees for their helpful comments. We also thank the Project HAROSA Knowledge Community. Research was partially sponsored by national funds through the Fundação Nacional para a Ciência e Tecnologia, Portugal–FCT under the project PEst-OE/MAT/UI0006/2011.

References

Agrawal, R., and Srikant, R. (1994). Fast algorithms for mining association rules in large databases. *Proceedings of the 20th International Conference on Very Large Data Bases*, VLDB, pp. 487–499, Santiago, Chile.

Aldana, W. A. (2000). Data mining industry: Emerging trends and new opportunities. Master thesis, Massachusetts Institute of Technology.

Antunes, M. (2010). CRM e Prospecção de Dados–ao seu serviço. *Boletim da Sociedade Portuguesa de Estatística*, Fernando Rosado, ed., Primavera de 2010, 34–39.

Branco, J. A. (2010). Estatísticos e mineiros (de dados) inseparáveis de costas voltadas? *Boletim da Sociedade Portuguesa de Estatística*, Fernando Rosado, ed., Primavera de 2010, 40–43.

Castro, L. M., Montoro-Sanchez, A., and Ortiz-De-Urbina-Criado, M. (2011). Innovation in services industries: Current and future trends. *The Service Industries Journal*, 31(1): 7–20.

Chatfield, C. (1997). Data mining. *Royal Statistical Society News*, 25(3): 1–2.

Farzi, S., and Dastjerdi, A. B. (2010). Data quality measurement using data mining. *International Journal of Computer Theory and Engineering*, 2(1): 1793–8201.

Fayyad, U., Piatetsky-Shapiro, G., and Smyth, P. (1996). From data mining to knowledge discovery: An overview. In *Advances in Knowledge Discovery and Data Mining*, Fayyad, U., Piatetsky-Shapiro, G., Smyth, P., and Uthurusamy, R., eds., pp. 1–36, AAAI Press/MIT Press, Menlo Park, CA.

Frank, E., Trigg, L., Holmes, G., and Witten, I. H. (2000). Naive Bayes for regression. *Machine Learning*, 41(1): 5–15.

Frawley, W., Piatetsky-Shapiro, G., and Mathews, C. (1992). Knowledge discovery in databases: An overview. *AI Magazine*, 213–228.

Freund, Y., and Schapire, R. E. (1997). A decision-theoretic generalization of on-line learning and an application to boosting. *Journal of Computer and System Sciences*, 55(1): 119–139.

Friedman, J. H. (1998). Data mining and statistics: What is the connection? www-stat. stanford.edu/~jhf/ftp/dm-stat.ps.

Gibert, K., Nonell, R., Velarde, J. M., and Colillas, M. M. (2005). Knowledge discovery with clustering: Impact of metrics and reporting phase by using klass. *Neural Network World*, 15(4): 319–326.

Gordon, A. D. (1994). Identifying genuine clusters in a classification. *Computational Statistics and Data Analysis,* 18: 561–581.

Gupta, M. R., and Chen, Y. (2010). Theory and use of the EM algorithm. *Foundations and Trends in Signal Processing,* 4(3): 223–296.

Han, J., Pei, J., Yin, Y., and Mao, R. (2004). Mining frequent patterns without candidate generation: A frequent-pattern tree approach. *Data Mining and Knowledge Discovery,* 8(1), 53–87.

Hand, D. J. (1998). Data mining: Statistics and more? *The American Statistician,* 52:112–118.

Hand, D. J. (1999a). Statistics and data mining: Intersecting disciplines. *SIGKDD Explorations,* 1: 16–19.

Hand, D. J. (1999b). Data mining: New challenges for statisticians. *Social Science Computer Review,* 18: 442–449.

Hastie, T., Tibshirani, R., and Friedman, J. (2001). *The Elements of Statistical Learning.* Springer, New York.

Hirate, Y., and Yamana, H. (2006). Generalized sequential pattern mining with item intervals. *Journal of Computers,* 1(3): 51–60.

Hope, C., and Mühlemann, A. (1997). *Service Operations Management: Strategy, Design and Delivery.* Prentice Hall, London.

Kasper, H., van Helsdingen, P., and de Vries, W. Jr. (1999). *Services Marketing Management: An International Perspective.* John Wiley & Sons, Chichester.

Kotsiantis, S. B. (2007). Supervised machine learning: A review of classification techniques. *Informatica* 31: 249–268.

Kuonen, D. (2004). Data mining and statistics: What is the connection? *The Data Administrative Newsletter.*

Leamer, E. E. (1978). *Specification Searches: Ad Hoc Inference with Nonexperimental Data.* John Wiley & Sons, New York.

Li, L., Shang, Y., and Zhang, W. (2002). *Improvement of HITS-based algorithms on web documents.* Proceedings of the 11th International World Wide Web Conference (WWW 2002). Honolulu, HI.

Liu, B., Hsu, W., Chen, S., and Ma, Y. (2000). Analyzing the subjective interestingness of association rules. *IEEE Intelligent Systems,* 47–55.

Liu, B., Hsu, W., and Ma, Y. (1998). *Integrating classification and association rule mining.* Proceedings of the Fourth International Conference on Knowledge Discovery and Data Mining (KDD-98, Plenary Presentation), New York.

Lloyd, S. P. (1957). *Least square quantization in PCM.* Bell Telephone Laboratories Paper.

Lloyd, S. P. (1982). Least squares quantization in PCM. *IEEE Transactions on Information Theory* 28(2): 129–137.

Loh, W.-Y. (2008). Classification and regression tree methods. In *Encyclopedia of Statistics in Quality and Reliability,* Ruggeri, F., Kenett, R., and Faltin, F., eds., pp. 315–323, Wiley, Chichester.

Mahajan, M., Nimbhorkar, P., and Varadarajan, K. (2009). The planar k-means problem is NP-hard. *Lecture Notes in Computer Science,* 5431: 274–285.

Mudie, P., and Cottam, A. (1993). *The Management and Marketing of Services.* Butterworth-Heinemann, Oxford.

Murdick, R. G., Ross, J. E., and Claggett, J. R. (1990). *Information System for Modern Management.* Prentice Hall, New Jersey.

Ograjenšek, I. (2002). Applying statistical tools to improve quality in the service sector. *Developments in Social Science Methodology* (Metodološki zvezki), 18: 239–251.

Oliveira, T. A., and Oliveira, A. (2010). *Data mining and quality in service industry: Review and some applications.* Presentation in IN3-HAROSA Workshop, Media-TIC, Barcelona, Spain, November 22–23. http://dpcs.uoc.edu/joomla/images/stories/workshop2010/Harosa_2010_Teresa_1.pdf.

Page, L., Brin, S., Motwani, R., and Winograd, T. et al. (1999). *The PageRank citation ranking: Bringing order to the web.* Available at http://ilpubs.stanford.edu:8090/422/1/1999-66.pdf.

Parellada, F. S., Soriano, D. R., and Huarng, K.-H. (2011). An overview of the service industries' future (priorities: linking past and future). *The Service Industries Journal*, 31(1): 1–6.

Pei, J., Han, J., Mortazavi-Asl, B., Pinto, H., Chen, Q., Dayal, U., and Hsu, M.-C. (2001). *PrefixSpan: Mining sequential patterns efficiently by prefix-projected pattern growth.* ICDE '01 Proceedings of the 17th International Conference on Data Engineering. IEEE Computer Society Washington, DC.

Pérez-Bonilla, A. (2010). Metodología de Caracterización Conceptual por Condicionamientos Sucesivos. Una Aplicación a Sistemas Medioambientales. PhD thesis, Universitat Politecnica de Catalunya. http://www.tesisenxarxa.net/TDX-0226110-123334/.

Pérez-Bonilla, A., and Gibert, K. (2007). Towards automatic generation of conceptual interpretation of clustering. In *Progress in pattern recognition, image analysis and application*, Volume 4756 of Lecture Notes in Computer Science, Valparaiso–Vica del Mar, L. Rueda, D. Mery, and J. Kittler, eds., Chile, pp. 653–663. Springer-Verlag, London.

Pérez-Bonilla, A., Gibert, K., and Vrecko, D. (2007). Domzale-Kamnik Wastewater Treatment Plant (Ljubljana–Slovenia). Clustering and Induction Knowledge Base. Research DR 2007/09, Dept. Estadística e Investigación Operativa. Universidad Politécnica de Cataluca, Barcelona, España.

Press, W.H., Teukolsky, S.A., Vetterling, W. T., and Flannery, B. P. (2007). *Numerical Recipes: The Art of Scientific Computing*, 3rd ed. Cambridge University Press, New York.

Pujari, A. K. (2001). The GPS algorithm. In *Data Mining Techniques*, pp. 256–260. Universities Press, India.

Quinlan, J. R. (1993). *C4.5: Programs for Machine Learning.* Morgan Kaufmann, Menlo Park, CA.

Quinlan, J. R. (1996). Improved use of continuous attributes in c4.5. *Journal of Artificial Intelligence Research*, 4: 77–90.

Richardson, M., and Domingos, P. (2002). *The intelligent surfer: Probabilistic combination link and content information in PageRank.* http://research.microsoft.com/en-us/um/people/mattri/papers/nips2002/qd-pagerank.pdf.

Seifert, J.W. (2004). *Data mining: An overview.* Congressional Research Service. The Library of Congress. CRS Report for Congress.

Shakhnarovich, G., Darrell, T., and Indyk, P. (2005). *Nearest-Neighbor Methods in Learning and Vision.* MIT Press, Cambridge.

Tukey, J. (1977). *Exploratory Data Analysis.* Addison-Wesley, Reading, MA.

Wu, X., and Kumar, V. (2009). *The Top Ten Algorithms in Data Mining.* Chapman & Hall, Boca Raton, FL.

Wu, X., Kumar, V., Quinlan J.R., et al (2008). Top 10 algorithms in data mining. *Knowledge and Information Systems*, 14(1): 1–37.

Zhang, T. (1997). Birch: A new data clustering algorithm and its applications. *Data Mining and Knowledge Discovery*, 1(2): 141–182.

Zhao, C. M., and Luan, J. L. (2006). Data mining: Going beyond traditional statistics. *New Directions for Institutional Research,* 131: 7–16.

Zwikael, O. (2007). Quality management: A key process in the service industries. *The Service Industries Journal,* 27(8): 1007–1020.

5

Software for Solving Real Service-Management Problems

Elena Pérez-Bernabeu, Horia Demian, Miguel Angel Sellés, and María Madela Abrudan

CONTENTS

5.1 Introduction

The relevance of the services belonging to the tertiary sector increased significantly in industrial countries in the last century. Today in many industries, services are the most important business sector, amounting to more than 70% of national economy, and this increasing impact of the service sector is not just based on the growth of self-contained services (Ascher and Whichard 1987). Changes in producing industries involve modification of product ranges, and product-relating and product-supporting services have become an integral part of the tertiary sector (Langeard et al. 1981).

For both self-contained services and product-relating and product-supporting services, quality is essential for the offered services to be successful.

However, as a study has shown, more than 40% of the services introduced into the market failed to survive (Cooper and Edgett 1999). There are two main factors behind this high failure rate:

- The spontaneous and unsystematic procedure for service development
- Lack of service-specific methods and tools

Given this background and its complexity, the main question is how to develop service efficiently to ensure high-quality service processes. Consequently, the development process for services needs to be systemized or standardized in the same way as for physical products. Decision-making tools (DMTs) are very important in this context.

5.2 Decision-Making Tools (DMTs)

Decision making can be defined as the act of choosing from among a set of feasible courses of action. This definition contains several elements, all of which are important. First, decision making involves action, and action is either a change in the present situation or a deliberate choice to retain the present situation. Second, courses of action must be feasible; that is, the decision maker must be able to carry them out. If there are no feasible courses of action, no decision is possible. Third, there must be more than one course of action available. If there is only one feasible course of action, there is no decision to be made. Fourth, decision making involves selecting one course of action from several, which usually implies some limitation to the resources available. If all the available courses of action can be pursued simultaneously, no decision is needed. A decision becomes necessary only when more than one course of action exists and not all of them can be pursued simultaneously. Finally, decision making is an act, which means that the choice among courses of action must not be passive or be allowed to happen by default. If the situation is allowed to drift until all the courses of action but one are foreclosed, or until circumstances force the choice of a particular course, then there is no need for a decision maker.

Nowadays, by far the most popular decision-making processes are rational decision-making models. These models use a rational, logical, and sensible approach to decision making, which are typical decision-making tools and techniques of rational decision-making models. The idea is to gather as much information as possible, and then to analyze it and rationalize it, and to come up with the best solution. The decision-making process steps are designed to do just this. Evidently, the more analytical and rational the model, the more steps there are! These decision-making tools and techniques are the main focal point of this chapter.

5.3 Decision-Making Tools Software (DMTS)

Given the number of decision-making tools and techniques available, it is not surprising that programs have been written to help with the decision-making process, which are divided into three categories: decision-making software, decision support software, and decision tree software. In this chapter, decision-making software is limited to programs that are advertised to help make decisions and does not include decision support software or decision tree software. There are several decision-making software programs available; some are for personal use, but most are for business purposes. Using computers enables the decision maker to analyze large amounts of data very quickly. This means they can also make more important mistakes much faster. Many of these programs also allow us to present the information and the results in many formats, which can be useful if we need to convince others about your decision.

What this means is that the computer only does as it is told. The user inputs data, and the quality of these data determines the quality of the results. Remember that we are still making the decision even with software decision making.

The final decision is only as good as the information that has been inputted in the system. Remember that when humans are involved, a decision is not completely objective. Human bias, preference, judgment, and opinion can never be completely removed.

These programs are rational models, and rational decision making is not how humans make decisions naturally. Rational models organize the decision-making process so that it has a structure and an ordered sequence of steps ranging from identifying the problem and opportunities to reaching the final decision and the actions to be taken. A factor that often determines if software is available to service sector companies is the price, and some of them are quite expensive.

QFDcapture uses a rational model and generates decision matrices where options are compared and verified. Data about what is important, how to accomplish the outcome, and trade-offs can be inputted. There are various ways available to present data and to save projects for future use.

Logical Decisions® for Windows uses a decision analysis to break complex situations into their component parts for easier manageability. These parts can then be recombined logically; that is, each alternative is evaluated according to its pros and cons. This software also uses decision matrices and its demo walkthrough does not go into as many details as others do.

Decision Explorer® was designed by academics at the University of Bath and by Banxia® Software, and is quite technical. It is cognitive mapping software. These cognitive maps were designed for structuring and analyzing qualitative data. Each aspect of the decision, or concept, is entered as a short phrase. The causal mapping technique allows the viewer to see the cause-and-effect relationships among the various concepts. This program can also handle well over 2000 concepts, which makes it a good candidate in very complex situations.

Any spreadsheet program can also be used as decision-making software. They allow for rapid data sorting and data analysis, and can be useful for correlating large amounts of data. They can also be easily utilized for generating decision matrix diagrams.

5.4 Service Sector Analysis

Given increasing needs for improved direction and management of service organizations, operations management researchers and practitioners have started implementing integrated information systems that have been developed in manufacturing sectors. Although you might think that DMTs are implemented to improve service delivery management, the project scope does not often encompass the functions related to the service itself (Cigref 2006). It is, therefore, necessary to define what is meant by DMT implementations in the service sector.

The main reasons for DMT implementations in service companies are (Botta-Genoulaz and Millet 2006):

- To lessen the administrative workload
- To replace dispersed legacy systems
- To replace unreliable finance and materials management systems
- To improve visibility across the entire system
- For investment security, an important consideration, particularly among public sector services limited by financial constraints
- For real-time data processing

A common problem faced by service organizations when adopting DMTS is the misfit issue; that is, the gaps between the functionality offered by the package and that required by the adopting organization. As a result, companies have to choose between adapting to the new functionality and customizing the package.

The following subsections list the main characteristics of enterprise resource planning (ERP) solutions for services.

5.4.1 DMTS Implementation in Hospitals

Health care service is a patient-oriented service that requires a continuous interaction with customers. It uses facilities and equipment, and consumes a large volume of nursing care. Therefore, it becomes increasingly important for health care executives to understand what kind of facility, equipment, and workforce decisions are critical to achieve the commonly acknowledged goal of providing quality health service at a reasonable cost.

DMTS implementation concerns finance, materials, and in-patient management systems. However, existing human resources and outpatient management systems have been retained. Public sector-specific requirements revolve around reporting requirements to regulatory authorities, standard formulas, and government-dedicated processes (reimbursement to hospitals for services to patients). On the other hand, there are more problems related to patient care systems, where practices vary from country to country. In addition, patient care modules are specialized industry modules and are not well integrated with traditional modules, such as finance and materials management. Hospital management is provided with resource usage data that were not previously available.

An integrated approach to information technology applications in hospitals will bring more benefits than the traditional isolated systems. The goal of hospital information systems is to use computers and communications equipment to collect, store, process, retrieve, and communicate patient care and administrative data for all hospital affiliated activities, and to satisfy all authorized users' functional requirements. It can be found on vendor Web sites into which many hospitals have introduced DMT solutions.

A hospital workflow problem includes many aspects, such as resources (physicians, nurses, equipment, beds, etc.), time, and costs. The goal is to achieve a good assignment of resources to tasks. Depending on the assignment solution provided, it has an impact on the rest of the workflow as resources are limited (Prieditis et al. 2004).

5.4.2 DMTS Implementation in Banks

Banks live or die depending on the quality of their decisions. This affirmation combines the requirements of the Sarbanes–Oxley Act of 2002, with demands in terms of security, profitability, and growth, and a bank's survival may depend on the soundness of employee decisions. The value that banking executives place on good decisions is obvious if the amount of money spent on decision support software is any indication.

Yet substantial monetary investment in technology does not guarantee results. In a recent survey performed with chief financial officers, only a quarter of them agreed that their decision support technology met their needs and expectations (Blaskovich and McAllister 2008). This all-too-common result has encouraged academic and industry researchers to make a considerable effort toward research into the success and failure of decision support technology. Although this research provides a good deal of important data, findings are not complete. In particular, previous research has been less diligent in exploring the critical relationship between decision support technologies and organizational objectives.

Traditionally, banks have implemented DMTS for a specific purpose, including management of investments and analyses of credit portfolios. Although DMTS is valid for specific purposes and outcomes, we contend that banks should expand their viewpoints by considering how decision support technologies improve decisions and generate benefits for the organization as a whole. While it may seem obvious that the primary reason to implement decision support technologies is to improve decisions, it may not be clear whether and how technology provides organization-wide benefits. This broader viewpoint enables organizations to use decision support technology more effectively and to improve evaluations of the expected and actual benefits obtained through technology.

The considerable challenges facing the banking industry fuel the need for good decisions and DMT technology. The actual dramatic change in this industry's core business means that the future of successful banking is more dependent on providing knowledge-based solutions and innovative products to secure customer relationships. Unprecedented regulatory demands require banks to collect, store, and distribute data with greater transparency and predictability. Specifically, bank personnel must be competent and have timely accessibility to all the necessary data in order to maintain compliance with the ever-growing list of banking regulations. Finally, the intense growth and merger activity that financial institutions have been undergoing in the past 10 to 15 years has dramatically altered the competitive landscape of this industry. Smaller banks are no longer able to match economies of scale and cannot access the capital that megabanks possess. Therefore, smaller banks are differentiating themselves by developing world-class management expertise and analytical capabilities, while also providing superior customer service. DMTS plays an instrumental role in the banking industry's ability to survive and thrive in this new environment.

5.4.3 DMTS Implementation in Other Services

The commercial restaurant business has been slow to adopt the most current DMTs. These have been typically viewed as an additional cost of doing business rather than an investment in future profitability. In general, increased costs are avoided in an industry characterized by low profit margins.

Restaurants initially focused on cutting costs (as did other hospitality firms) by automating their back-office functions, including payroll, accounting, and inventory systems. These are some of the difficulties faced by restaurant firms today in terms of them updating their DMTs because each company typically installed its own proprietary software, with little concern about the prospective integration of these systems and no thought as to how well systems would perform as the firm expanded. Implementation of DMTS in restaurant chains generates management reports on critical performance indicators and provides data for improved marketing in restaurant firms. Software tracks guest history, helps manage human resources and finances, and evaluates customer satisfaction. Yet, in fact, restaurants develop DMT applications for areas such as guest history, human resource management, and finance management. However, no firm seems to have a long-term perspective that considers the interrelationships among various system components. An opportunity for the future lies in function integration and the development of a centralized system that transmits and shares data from all applications.

Demand for DMTS capability will vary among industries due to the different levels of information intensity, marketplace volatility, business unit synergies, and the strategy formation process. Finance, for instance, is defined to have more DMT infrastructure given the greater integration between business units. In the finance industry, firms wish to gain a picture of the entire relationship with the customer. Concurrently, finance firms are undergoing rapid change due to the finance industry's ongoing internationalization and to the need to build infrastructure to support clients whose businesses are expanding globally.

Government organizations are increasingly adopting DMT systems for various benefits such as integrated real-time information, better administration, and result-based management. Given their social obligations, higher legislative and public accountability, and unique culture, government organizations face many specific challenges in the transition to enterprise systems.

5.4.4 Analysis of Commercial DMTS for the Service Sector

Knowledge is the main source of not only competitive advantage in the knowledge economy but also of the innovation of those organizations involved in the services sector (Zamfir 2010). It is really about transforming data into a useful data management process. We must bear in mind, however, that the speed with which information circulates, as well as accessibility, searches, and storage forms, are all critical factors to be considered when streamlining decision making.

By taking into account the likelihood of achieving results, several models have been developed to optimize decisions starting by classifying decisions: decisions under conditions of certainty, risk, and uncertainty. The elements that distinguish these decision types are very important for all managers and software developers.

Although the development of computer applications has raised no problems in terms of the use of established models, should there be economic consequences known with certainty (for example, ELECTRE), the real challenge for software developers is to optimize the decisions of category risk and uncertainty.

In the service sector, as in any other field of economics, all computer applications are based on the distinction made between definitions of risk and uncertainty:

- Risk is present when future events occur with a measurable probability. Risk implies a negative outcome.
- Uncertainty is present when the likelihood of future events is indefinite or incalculable (Knight 1921).

5.5 Monte Carlo Simulation

Monte Carlo methods are a class of computational algorithms that rely on repeated random sampling to compute their results. These methods are most suited to computer calculations and tend to be used when computing an exact result with a deterministic algorithm if infeasible. One important fact in Monte Carlo simulation is the sampling used in the simulations. This sampling can be generated according to a distribution model, and there is a lot of distribution that can be used as normal distribution, uniform distribution, triangular distribution, and so on. According to Savage et al. (2006), "simulations without acceptable input distributions are like light bulbs without electricity," who observe very well that "only a few people within an organization have the expertise to estimate probability distributions, and even fewer have the managerial authority to get their estimates accepted on an enterprise-wide basis." One of these authors' ideas was to manage probability distributions centrally and to replace the classical probability distributions with stochastic libraries as pregenerated random trials that approach stochastic inputs or which can be the results of simulation and optimization models.

ProbabilityManagement.org has presented DIST™, which provides Monte Carlo simulation with a data structure containing thousands of Monte Carlo trials. The main advantages of DIST is that it can be generated by experts to represent virtually any type of probability distribution to then be distributed to others in a standardized format.

5.6 SAS

SAS is a leading developer of software for optimizing decisions for services. Since it started business in 1976 as a provider of software solutions, SAS now has customers at more than 50,000 sites worldwide.

Regarding financial services, SAS has more than three decades of experience working with financial institutions all over the world. Nowadays, SAS solutions are used by more than 3100 financial institutions worldwide, including 97% of banks in FORTUNE Global 500 (SAS 2011).

The concerns of the SAS software developers dedicated to risk management in services are based on several basic concepts. Gary Cokins, Manager of Global Advisory Services at SAS, considers: "With potentially hundreds of risks that can be identified, dealing with them may seem daunting. Let's break it down into more manageable chunks and start by categorizing various risks. Risks could be grouped in any of a number of ways: external and internal; controllable and uncontrollable; or insurable and uninsurable." The four alternative types include: market and price risk, credit risk, operational risk, and legal risk (SAS 2011).

SAS considers that the real advantages of the solutions it develops are integrated, comprehensive data management; one of the most powerful predictive analytics available; user-friendly; self-service reporting; a transparent environment for managing the entire process from identifying the risk, to measuring, mitigating, and monitoring it on an ongoing basis.

In the services domain, SAS Risk Management for Banking—an integrated, end-to-end solution that includes data management, preconfigured risk analytics and reporting—delivers functionality for all major risk types, as well as data management and reporting. This enables business units within banks to independently and separately calculate measures of risk. The benefits of using this program are that it supports an integrated risk management strategy, enables innovation, improves competitive advantage, provides more control over and ownership of risk management data, and cuts the total cost of ownership (SAS 2011).

SAS for Hotels is a useful software that redesigns service offerings, improves retention, launches loyalty programs or prices to maximize revenue and profit, and has in-depth knowledge of consumer behavior, market conditions, and competitor influence.

For example, SAS developed Revenue Optimization for Hotels to effectively optimize revenue and profits. It is based on the idea that hotels need pricing solutions that provide world-class data management, forecasting, and optimization.

5.7 Risk Solver Platform

Risk Solver Platform® is developed by Frontline Systems and is fully integrated with Microsoft Excel®. It can be used for linear and nonlinear models, conventional optimization, decision tree models, risk analysis, and Monte Carlo simulations. ProbabilityManagement.org considers this package the first, and as of this date, the most powerful interactive Monte Carlo simulation package for Microsoft Excel. For simple simulations involving a few variables, it can perform 100,000 trials essentially instantly, as the user changes parameters in the model.

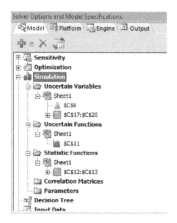

FIGURE 5.1
Model Panel in Risk Solver Platform software.

A Model Panel consisting of four tabs is displayed to the right of the screen as seen in Figure 5.1. The Model tab is used to view the description of the current model, which can be of an optimization, simulation, or a decision tree. The platform tab is used for defining parameters for the optimization model, parameters for the simulation model, or parameters for the decision tree model. The Engine tab is used for specifying engines that will be used in problem solving. A nonlinear engine or a risk solver engine can be selected, or it automatically selects which engine to use to solve the problem. Which engine should we use? The recommendation is to try them all and use that which best performs with the model (Risk Solver Platform Reference guide, version 10.0).

Users need to be familiar with Microsoft Excel to use this platform. The platform comes with a lot of useful examples for optimization, simulation, and simulation and optimization, like the Simple Business Plan Forecast Model with uncertainty. There are examples of illustrating how to model the growth of investment funds given uncertain return rates; model multiperiod inventory problems with uncertain demand; use Monte Carlo simulation to estimate the workforce levels required to meet demand; and use of Monte Carlo simulation in airline yield management. This example also shows the use of a parameterized simulation model to arrive at good decision policies under uncertainty.

It is very important to mention that a useful package named Risk Solver Engine can be used to develop and deploy application to users. Risk Solver Platform comes with a set of functions that can be used to define optimization models and Monte Carlo simulation models. According to their documentation, most of these functions are to be used for Monte Carlo simulation models or for the uncertain elements of stochastic optimization models. Here it is used to solve a problem when placing a reservation at a hotel under the following conditions: the hotel has 50 rooms, and the price for a room is

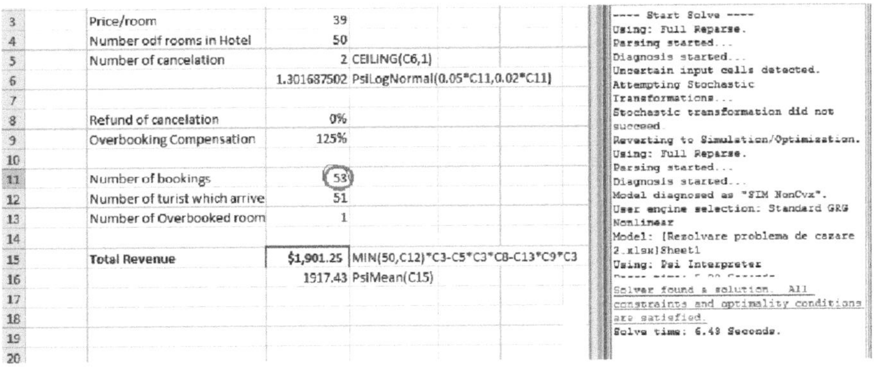

3	Price/room	39		
4	Number odf rooms in Hotel	50		
5	Number of cancelation	2	CEILING(C6,1)	
6		1.301687502	PsiLogNormal(0.05*C11,0.02*C11)	
7				
8	Refund of cancelation	0%		
9	Overbooking Compensation	125%		
10				
11	Number of bookings	53		
12	Number of turist which arrive	51		
13	Number of Overbooked room	1		
14				
15	**Total Revenue**	$1,901.25	MIN(50,C12)*C3-C5*C3*C8-C13*C9*C3	
16		1517.43	PsiMean(C15)	
17				
18				
19				
20				

```
---- Start Solve ----
Using: Full Reparse.
Parsing started...
Diagnosis started...
Uncertain input cells detected.
Attempting Stochastic
Transformations...
Stochastic transformation did not
succeed.
Reverting to Simulation/Optimization.
Using: Full Reparse.
Parsing started...
Diagnosis started...
Model diagnosed as "SIM NonCvx".
User engine selection: Standard GRG
Nonlinear
Model: [Rezolvare problema de cazare
2.xlsx]Sheet1
Using: Psi Interpreter
Solver found a solution. All
constraints and optimality conditions
are satisfied
Solve time: 6.49 Seconds.
```

FIGURE 5.2
Created model with parameters and final solution using Risk Solver Platform.

39 Euros. Should a client wish to cancel the reservation, the hotel does not refund the price of the first night of accommodation. Should a tourist show up and no room is available, a 25% compensation for the price of a room is paid by the hotel for the tourist to seek accommodation in another hotel. The model can be created in under 5 minutes. This model is preferred because the evolution of simulation for each particular case can be seen accurately and the algorithm can be well understood.

To solve this problem, Risk Solver simulated all 50 cases with 10,000 situations in 0.25 seconds. Another very useful aspect is that it displays the results that can be viewed and analyzed in a different window. Multiple simulations can be combined in the analysis, or the results for only one simulation can be viewed. The same problem can be solved if simulation and optimization model is defined.

A maximum objective is defined with a condition; the booking number has to be an integer (it is not possible to book 5.34 rooms). We fix C11 as a variable in the optimization section, as seen in the model in Figure 5.2. Other parameters of the engine are 1 simulation and 1000 trials per simulation. The solution to our problem is 53 rooms and was achieved in 6.5 seconds.

5.8 DecisionTools Suite Platform

DecisionTools Suite is another software platform developed by Palisade that can be used in simulations and optimizations. Some of the tools included in this suite are @RISKtool, for Monte Carlo simulation; PrecisionTree, which is used for decision analyses based on decision tree models; NeuralTools, employed for prediction and forecasting based on neural networks; and @RISKOptimizer, for optimization with simulation. It is important to

FIGURE 5.3
DecisionTools Suite Platform ribbon in Microsoft Excel.

FIGURE 5.4
Presented model with parameters using @RISK.

mention that these tools are created for use in Microsoft Excel. According to the official Web site (www.palisade.com), this tool is used in more than 400 academic institutions worldwide. Video tutorials are presented on the site and help understand how to work with these tools. Immediately after installation, a new ribbon appears in Microsoft Excel. This ribbon is presented in Figure 5.3.

DecisionTools Suite comes with a set of functions that can be used for defining optimization models and Monte Carlo simulation models as Risk Solver Platform does. Yet if we create models with some of these functions, we can use the SWAP Function options, which allow our models to be presented on computers that do not have @RISK installed.

If we want to use @RISKOptimizer, we must first create a model. Based on this model, we can define our optimization goal, which can have maximum, minimum, or specific values. We can also define different ranges for different cells that are to be used in simulation, and we are able to define constraints. Another useful option in defining a constraint is the possibility of adding a description to that constraint. The model is more easily understood when we want to adjust it after a while. We have attempted to solve the same problem of placing a reservation with a hotel using the @RISKOptimizer tool. The model appears in Figure 5.4.

The Risk Log Normal function from @Risk is employed, which is similar to PSI Log Normal from Risk Solver Platform. A variable is defined

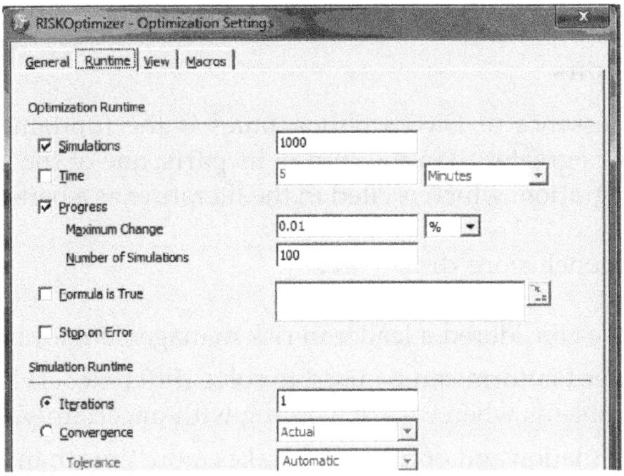

FIGURE 5.5
Optimization settings.

Results	
Valid Simulations	113
Total Simulations	113
Original Value	$1,482.00
+ soft constraint penalties	$0.00
= result	$1,482.00
Best Value Found	$1,950.00
+ soft constraint penalties	$0.00
= result	$1,950.00
Best Simulation Number	12
Time to Find Best Value	0:00:09
Reason Optimization Stopped	Progress condition
Time Optimization Started	7/24/2011 14:10
Time Optimization Finished	7/24/2011 14:11
Total Optimization Time	0:01:09
Adjustable Cell Values	'Sheet1'!C11
Original	40
Best	53

FIGURE 5.6
Final solution obtained with @RISKOptimizer.

in C11, which can take integer values from 40 to 60 (E11<=C11<=F11), and it also defines C15 as a maximum mean value, this being the equivalent of the PSI Mean function from Risk Solver Platform. Other set engine parameters are 1 simulation and 1000 trials per simulation, as seen in Figure 5.5.

After 9 seconds, the results are displayed in a new window with lots of explanations. The best solution we obtained is 53 rooms for reservations, as seen in Figure 5.6. @RISKOptimizer provided the same results as Risk Solver Platform.

5.9 Conclusions

Although the essence of DMTS philosophies is the fundamental premise that the whole is greater than the sum of its parts, one of the main topics is functional integration, which is cited in the literature as a benefit of the decision system.

The specific conclusions drawn are:

- SAS can be considered a leader in risk management for banking.
- Risk Solver Platform can be used to solve different services optimization problems when we are working with uncertainty.
- Using simulation and optimization takes more time than Risk Solver Platform to solve our problem, but it is easier to define the model and to interpret the results.
- @RISKOptimizer is somewhat slower for our simulations than Risk Solver Platform.
- In @RISKOptimizer, constraints can be created with a description for each one, which is most useful.

References

Ascher, B., and Whichard, O. G. 1987. Improving services trade data. In *Emergence of the Service Economy*, Orio Giarini, ed., 255–281. Oxford: Pergamon Press.

Blaskovich, J. L., and McAllister, B. P. 2008. Taking an organizational viewpoint of decision support software, *Bank Accounting and Finance*, February 1.

Botta-Genoulaz, V., and Millet, P. A. 2006. An investigation into the use of ERP systems in the service sector. *International Journal of Production Economics*, 99: 202–221.

Cigref. 2006. 1999—Retours d'expérience ERP. http://www.cigref.fr.

Cooper, R. G., and Edgett, S. J. 1999. *Product Development for the Service Sector: Lessons from Market Leaders*. Cambridge, UK: Perseus.

Knight, F. H. 1921. *Risk, Uncertainty, and Profit*. Boston, MA: Hart, Schaffner and Marx; Houghton Mifflin Co.

Langeard, E., Bateson, J. E. G., Lovelock, C. H., and Eiglier, P. 1981. *Service Marketing: New Insights from Consumers and Managers*. Cambridge, MA: Marketing Science Institute.

Prieditis, A., Dalal, M., and Arcilla A. 2004. Simulation-based real-time resource allocation for hospital workflows. International Health Sciences Simulation Conference, San Diego, CA.

SAS. 2011. http://www.sas.com.

Savage, S., Scholtes, S., and Zweidler, S. 2006. Probability management, Part 2. *OR/MS Today*, 33(2).

Zamfir, A. 2010. *Management of Services within the Knowledge-Based Society*. Bucharest: ASE, pp. 83–85.

Section 2

Decision Making in Health Services

6

Logistics and Retail Pharmacies: Providing Cost-Effective Home Delivery of Prescriptions

Yi-Chin Huang, Michael J. Fry, and Alex C. Lin

CONTENTS

6.1 Introduction

The retail pharmacy industry in the United States is dominated by a few large drugstore chains and pharmacies associated with grocery and discount stores. CVS, Walgreens, and Rite Aid are the three largest drugstore chains; combined they hold more than 35% of the total market share in the United States. Grocery stores and discount chains such as Kroger and Walmart also make up a large portion of retail pharmacy sales. This makes it very challenging for smaller drugstore chains and independent pharmacies to compete, especially on accessibility. Many smaller retail pharmacy chains emphasize high customer service levels and additional services not available at the larger chains in order to compete. One additional service offered by some smaller pharmacies is home delivery of prescriptions. Many customers receiving prescriptions are elderly or ill. Traveling to the pharmacy to pick up prescriptions can be an added nuisance and, in some cases, impossible for these patients.

In this chapter we examine the use of traditional operations-research tools and methodologies to enable small drugstore chains to provide home delivery prescription services at low cost. We use versions of a traveling salesperson problem (TSP) model and the vehicle routing problem (VRP) to solve a home-delivery problem for Clark's Pharmacy, a small pharmacy chain located in the Midwest United States. Clark's Pharmacy desired to provide a high level of service to their customers but at minimum cost in terms of vehicles, delivery drivers, fuel usage, and so on. This chapter demonstrates the powerful use of operations-research methods in service industries, even for companies of relatively modest size.

6.2 Background

Clark's Pharmacy is a local chain of pharmacy stores headquartered in Dayton, Ohio. The pharmacy was established in 1965. At the time of this project, Clark's Pharmacy owned nine drug stores, seven of which provide prescription home delivery service. These seven stores are referred to as Rx1, Rx2, Rx3, Rx4, Rx5, Rx6, and Rx7 in this chapter. Each store has its own patients and needs to deliver up to dozens of prescriptions each business day.

Clark's Pharmacy was interested in evaluating several strategies for providing home delivery of prescriptions. We partnered with Clark's to analyze its problem and to use operations research to help them evaluate these different strategies and choose the best method for providing this service. The management of Clark's Pharmacy had several questions. The current operation was to assign one delivery vehicle to each store; hence, this requires seven total vehicles. Clark's management wanted to know if one vehicle could serve two or more stores. If so, which stores should be served by a single vehicle? What cost savings would this provide? Management also wanted to evaluate the savings offered by a centralized solution where all deliveries would originate from a single depot. Currently, each store handled home deliveries for its own customers. Clark's wondered if significant savings could be gained by consolidating deliveries to fewer locations.

We agreed to help the management team of Clark's Pharmacy answer these questions using operations-research tools and models. First, we had to gather the necessary data. We required historical data on home delivery of prescriptions: date of delivery, location of delivery, store assigned for delivery, and so on. These data were collected by one of the authors over multiple days of personal observations. We also needed to gather the appropriate cost data. The main delivery costs come from drivers' pay, vehicle purchase costs, vehicle repairs, insurance, and fuel. Drivers' pay depends on how many drivers are employed. One driver is hired to drive one vehicle, so how many drivers are needed depends on the number of vehicles required. Fewer vehicles equate to decreased vehicle purchases, repair, and insurance costs. Fuel cost depends on gasoline price and distance traveled by the vehicles. Since we assume gasoline price to be static in our analysis, distance traveled determines fuel costs. In all, the delivery cost depends on number of vehicles needed and travel distance. All cost data were provided by Clark's Pharmacy.

We compare three possible strategies for Clark's Pharmacy to provide home delivery of prescriptions. Scenario I represents a completely decentralized solution. Here, each store maintains one vehicle to deliver prescriptions for its own customers. Each store's vehicle picks up the prescriptions to be delivered that day, delivers them to the customers, and then returns to the store. Scenario II is a hybrid solution where we group several pharmacies together to share a single vehicle. Scenario III is the centralized solution. In this scenario, a central depot is used for all prescriptions. All vehicles begin and end the day at this central depot.

For each scenario, we formulate a model to minimize transportation costs. We then compare the costs required under each scenario in terms of vehicles, fuel, repairs, labor (driver wages), and so on. We discuss our findings as well as several specific challenges faced in this project.

6.3 Modeling Background: Traveling Salesperson Problems (TSPs) and Vehicle Routing Problems (VRPs)

Scenario I (decentralized scenario) and Scenario II (hybrid) can be modeled using the classical traveling salesperson problem (TSP). A TSP can be described as a problem where a salesperson must visit a variety of cities for business purposes. The traveling cost may vary upon different routes, and distance is often used as a proxy for travel costs. The objective of the TSP is to find the lowest-cost tour for the salesperson. The optimization model is linear, but it is hard to solve for even a moderate number of cities because of its computational complexity. This difficulty is caused by the fact that there are many possible routes. Given n cities and assuming the salesperson starts the tour from city 1, there are a total of $(n - 1)!$ route options for the salesperson. Obviously, solving a TSP to optimality is difficult due to the combinatorial complexity. Most successful optimization methods use some form of cutting planes to reduce the feasible space that must be searched. Lawler et al. (1985) and Applegate et al. (2007) provide excellent overviews of the TSP from its origins to the advanced solution methods. Laporte (1992), Reinelt (1994), and many others discuss exact and heuristic solution methods for the classical TSP.

Various relaxations of TSPs exist that include all feasible solutions of a TSP; thus, solving the relaxed problem provides a lower bound for the optimal solution. From this relaxed problem, violated constraints (cuts) in the full problem are added and the solution is improved until the upper bound is found. Several relaxations such as N-path relaxation, assignment relaxation, one-tree relaxation, linear programming relaxation, and others have been developed (Hoffman and Padberg 1996).

There are a variety of open source and commercial solvers available to handle TSPs as well. Concorde is a TSP solver that is freely available through Georgia Tech University's Web site at www.tsp.gatech.edu/concorde.html. It can solve TSPs with over 15,000 locations and is generally considered one of the best TSP solution methods available. Many other freely available TSP solvers are available online including those that use optimization models such as SYMPHONY (www.branchandcut.org/) and branch-and-bound algorithms (Little, Murty, Sweeney, and Karel 1963; see http://optimierung. mathematik.uni-kl.de/old/ORSEP/contents.html). TSP solvers have also been developed for more general software such as Mathematica (Skiena 1990; see www.cs.uiowa.edu/~sriram/Combinatorica/NewCombinatorica.m) and Google maps (see www.gebweb.net/optimap/). Many open-source TSP solvers are also available that are based on heuristics. The Concorde solver can also implement several well-known heuristics for TSPs including the k-opt heuristic (Helsgaun 2000). Other heuristic approaches include evolutionary computation (Frick 1996) used in the pgapack code available at www.rz.uni-karlsruhe. de/~lh71/, simulated annealing used in the Mathematica-based program Operations_Research 3.0 (see http://web.archive.org/web/20080527080625/ http://www.softas.de/op_researchframe.htm) and others.

The biggest source of combinatorial complexity in TSPs is eliminating subtours. A subtour exists when there are multiple tours instead of one large tour through all locations to be visited. In this project, we use max-flow cuts for subtour elimination (see Toth and Vigo 2002, for an overview of methods; see Padberg and Hong 1980, for details on applying max-flow formulations to identify subtours). A max-flow model is solved to find a subtour violation, then a constraint to prevent this subtour violation is appended to the TSP; this process is repeated until no further subtours are found. To apply a standard TSP model to our problem, some revision is needed to reflect current delivery operations. This is because the vehicle must go to the drug store to pick up prescriptions before visiting patients. A simple revision is made to our model to consider this requirement.

Scenario III requires the scheduling of multiple vehicles from the same starting location (depot). Therefore, we must use a vehicle routing problem to optimize this problem. The VRP schedules customers for delivery by a fleet of vehicles. It is a classic problem closely related to the TSP, but it is even more difficult to solve. Many heuristic solution techniques exist to solve standard VRPs including those presented in Clarke and Wright (1962), Solomon (1987), and various meta-heuristic approaches (for example, see Gendreau et al.

1992; Gendreau et al. 1994; Gendreau et al. 2002; Laporte and Semet 2002; Braysy and Gendreau 2005a, 2005b). In this chapter we employ max-flow formulations to identify subtours and solve the VRP. The max-flow model will be solved iteratively for each vehicle used for deliveries.

Similar to the solvers mentioned earlier for TSPs, many commercial and open-source programs exist for the VRP. Many of the largest technology companies provide routing software that includes VRP solvers. IBM/ILOG has developed Transportation Analyst, and UPS Logistics has Roadnet. Other well-known commercially available solvers include the TruckStops software from MicroAnalytics, the Route Planning Suite from Descartes Systems Groups and others (see Partyka and Hall 2010, for a more complete listing of commercially available VRP solvers). Professor Larry Snyder provides free software for not-for-profit purposes that implement a randomized version of the Clarke and Wright heuristic at http://coral.ie.lehigh.edu/~larry/software/vrp-solver/.

In the scenario considered here, all facility locations are known before routing occurs since we are dealing with existing facilities. It is possible to solve such problems using multiple depot vehicle routing formulations such as those described by Cordreau et al. (1997), Laporte et al. (1988), and others. Such models typically utilize heuristic solution methods such as tabu search. While it is possible to formulate Scenario I and Scenario II using these types of models, it does not match Clark's current or desired operational capabilities captured in these scenarios. For Scenario I (decentralized solution), Clark's wishes to keep each customer assignment to his or her existing pharmacy. Thus, this scenario effectively decouples into separate problems. Similarly, in Scenario II, other operational concerns helped us identify the two locations that would be used to fill customer orders in the hybrid scenario. The geographic locations of these two locations made it easy to identify which customers would be assigned to each location, easing the need for a more complex formulation.

Prescription deliveries for Clark's Pharmacy are generally known 24 hours in advance. Therefore, in this setting, as long as each TSP or VRP is solved within these 24 hours, the delivery locations are deterministic in nature. However, other researchers have examined vehicle routing problems with random demands (for example, see Bertsimas et al. 1990, Bertsimas 1992, and Laporte et al. 1994). Laporte et al. (1992), Lambert et al. (1993), and Kenyon and Morton (2003) examine vehicle routing problems where the travel and service times are stochastic. Although it is true that travel and service times can also be stochastic in our scenario, we relax these concerns in our analysis. We do this for several reasons. First, it makes the problem much more analytically tractable. Second, and more important, little is gained by including the stochastic travel and service times in our model. There are no target completion times for each delivery as in Laporte et al. (1992) and Lambert et al. (1993), and the total length of delivery does not greatly impact the costs incurred by Clark's Pharmacy. Third, very little data are available to provide estimates of the stochastic travel and service times on this network. And finally, Clark's Pharmacy believes that its drivers have enough experience and familiarity with the network to minimize the impacts of randomness by dynamically adjusting their routes to traffic and time-of-day conditions.

A related class of research examines dynamic vehicle-routing problems (Psaraftis 1988) where delivery requests can evolve in real-time (see also Gendreau and Potvin 1998; Bertsimas and van Ryzin 1991). While Clark's occasionally has last-minute delivery requests that must be incorporated into their deliveries, such occurrences are rare. Thus, for simplicity these requests were not included in our models.

6.4 Data Preparation

This section describes the data used in this project. All data were collected during a 1-month time frame (May 2008) through either personal observation or driver-generated delivery logs that were verified for accuracy.

Delivery data—These data include drug store locations and their delivery destinations. For example, Table 6.1 shows the address of Rx6 (the first row in the table) and the delivery points (patients' locations) for one particular day during the data collection period. Besides the addresses, the table also includes latitude and longitude of every location. (All delivery addresses have been disguised here and lat/long values removed from the table to protect customer confidentiality.) The Web site Geo-code Transfer Web Service (www.batchgeocode.com/) was used to translate actual addresses into latitudes and longitudes. These data are then used to calculate distances between locations as described later. In this example, five prescriptions are delivered on the observed date for the store in Rx6.

TABLE 6.1

Rx6 and Delivery Points Addresses for One Day in May 2008

Address	City	State	Latitude	Longitude
1200 W Elm Street	Tipp City	OH	XX.XXX	YY.YYY
1741 Windstar Dr	Tipp City	OH	XX.XXX	YY.YYY
482 Avenue K Apt 13	Tipp City	OH	XX.XXX	YY.YYY
721 Euclid St	Tipp City	OH	XX.XXX	YY.YYY
43 Whitewater Cr	Tipp City	OH	XX.XXX	YY.YYY
408 Newtown Rd	Tipp City	OH	XX.XXX	YY.YYY

TABLE 6.2

Costs of Delivery in April 2008

	Rx1	Rx2	Rx3	Rx4	Rx5	Rx6	Rx7
Driver's pay	$2,322.01	$3,469.38	$3,349.41	$761.05	$1,314.32	$1,663.00	$1,665.26
Driver's taxes	$495.62	$582.76	$1,411.30	$66.79	$121.01	$133.23	$163.65
Vehicle repair	$2,375.20	$87.56	$152.37	$102.64	$143.70	$0.00	$0.00
Fuel	$606.33	$342.59	$1,582.64	$664.48	$711.16	$505.92	$78.54
Insurance	$246.61	$257.36	$257.36	$218.62	$209.23	$119.13	$119.13
Payments	$0.00	$910.72	$910.72	$0.00	$0.00	$0.00	$0.00
Total	$6,045.77	$5,650.37	$7,663.78	$1,813.58	$2,499.42	$2,421.29	$2,026.58

Cost data—These data provide the costs that Clark's Pharmacy incurred on delivery in April 2008. This was the best data that was available for this problem, so we had to assume that similar monthly costs would apply going forward. (Costs in Table 6.2 have been disguised at the request of the company.)

Clark's Pharmacy informed us that a single vehicle could handle at most 40 deliveries per day. An examination of the daily deliveries required each day at each pharmacy shows that a single vehicle can accommodate all deliveries for that store. Furthermore, we estimated that we might be able to process all deliveries with as few as three vehicles if we combined locations. We will return to this analysis in our comparison of scenarios.

6.5 Scenario I: Decentralized Scenario Models

In Scenario I each store has its own vehicle to provide home delivery of prescriptions. The optimal route for a vehicle can be found by solving a TSP model as shown next.

6.5.1 General TSP Model

Indices: i, j, nodes (customer, pharmacy locations)

Parameters: n = number of nodes; D_{ij} = distance between nodes i and j

Decision variables: x_{ij} = 1 if the vehicle should travel between nodes i and j, else = 0

Objective function: $\min \sum\limits_{i=1}^{n} \sum\limits_{j=1}^{n} D_{ij} x_{ij}, \ i \neq j$

Subject to:

$$\sum_{i=1}^{n} x_{ij} = 1, j = 1, 2, n \tag{6.1}$$

$$\sum_{j=1}^{n} x_{ij} = 1, i = 1, 2, n \tag{6.2}$$

$$\sum_{i \in S} \sum_{j \in S} x_{ij} \leq |S| - 1 \text{ for every subset } S \tag{6.3}$$

$$x_{ij} = 0, 1 \text{ for all } i, j, i \neq j$$

The objective function is to minimize delivery tour length. Constraint set (6.1) states that each node must be traveled to–from other nodes; constraint set (6.2) means that each node must be traveled from–to other nodes. The entire travel must be a single tour and cannot be split based on subtour elimination constraints in (6.3). Distance values, D_{ij}, are calculated from the latitude and longitude values of customers and pharmacies. To approximate actual road distance traveled, we use the approximation from Simchi-Levi et al. (2007), $D_{ij} = \rho * 69\sqrt{(lat(i) - lat(j))^2 + (lon(i) - lon(j))^2}$, where ρ is a coefficient used to inflate straight-line distances to approximate road distances. Simchi-Levi et al. (2007) suggest a value of $\rho=1.3$ for metropolitan locations. We find that a value of $\rho=1.35$ fits best for our data set,[*] so we use $\rho=1.35$.

The difficulty in solving this model is that there are many possible subtours. For a tour with n points, there are $(2^n - 1 - 1 - n)$ subtour elimination constraints required. For example, a problem with 10 retail store locations would have almost 1024 subtour elimination constraints. In this chapter, the TSP model is solved without subtour elimination constraints, and subtours are identified by solving the following max-flow model. The max-flow model is solved for each terminal node, $t = 2, ..., n$, to find violated subtours. This process is repeated until no further subtours are found.

6.5.2 Max-Flow Model

Indices: i, j, k = nodes; source node is 1, terminal node is t

---

[*] This was determined by comparing several of the tours generated from our solution and comparing the distance using the approximation shown with the road miles generated from Google Maps (http://maps.google.com). The value of $\rho = 1.35$ minimized the sum of the squared deviations between the approximate distance and the road miles for these routes.

Parameters: x_{ij} = capacity between node i and node j

Decision variables: z_{ij} = flow between nodes i and j

For each node $t = 2, ..., n$, maximize the flow from 1 to t

Objective function: max $\sum_{i}^{n} z_{1i}$

Subject to:

$$\sum_{i}^{n} z_{ij} - \sum_{k}^{n} z_{jk} = 0 \text{ for each node } j \qquad (6.4)$$

$$\sum_{j}^{n} z_{1j} = \sum_{i}^{n} z_{it} \qquad (6.5)$$

$$-x_{ij} \le z_{ij} \le x_{ij} \text{ for all } i, j, i \ne j \qquad (6.6)$$

Constraints (6.4) provide conservation of flow. Flow in should be equal to flow out for each node; flow out of the source node 1 should be equal to flow into the terminal node t in constraint (6.5). Since the TSP model is symmetric, z_{ij} could be <0, which is allowed through constraint set (6.6). x_{ij} are the variables in the relaxed TSP model (eliminating subtour constraints). The solutions x_{ij} are fixed and become parameters in max-flow model.

This model should be solved for each terminal node ($t = 2, ..., n$) to find violated subtours. In general, when a violated subtour is found, the objective value of the max-flow problem is 0. The dual prices of constraints (6.4) in the max-flow model tell us which nodes could have more capacity. We use AMPL to formulate our model (for example, see Pataki 2003) and CPLEX 11.2 to solve the models. Solution times for Scenario I models are all less than 5 seconds.

6.6 Scenario II: Hybrid Scenario Models

In Scenario II, multiple pharmacy locations may share a vehicle. Based on the historical data, it was clear that the most likely locations to share a vehicle were the Rx6 and Rx7 locations (based on volume of deliveries and geographic proximity). This situation is somewhat more complex than a standard TSP. Under Scenario II, we must guarantee that the vehicle visits the customer's designated pharmacy *before* visiting that customer to ensure that the vehicle has picked up the prescription for delivery. Thus, we must add constraints to our previous TSP model; this is shown next in the Revised TSP model.

6.6.1 Revised TSP Model

Indices: i, j, nodes

Parameters: n = number of nodes, 1 and 2 are pharmacies Rx6 and Rx7, set $n2$ represents customers of pharmacy Rx6; D_{ij} = distance between point i and j

Decision variables: x_{ij} = 1 if the vehicle should travel between points i and j, else = 0

Objective function: $\min \sum\limits_{i=1}^{n} \sum\limits_{j=1}^{n} D_{ij} x_{ij}, i \neq j$

Subject to:

$$\sum_{i=1}^{n} x_{ij} = 1, j = 1, 2, n \tag{6.7}$$

$$\sum_{j=1}^{n} x_{ij} = 1, i = 1, 2, n \tag{6.8}$$

$$\sum_{i}^{n2} x_{2i} = 1 \tag{6.9}$$

$$x_{i2} = 0 \text{ for each } i \text{ in } n2 \tag{6.10}$$

$$\sum_{i \in S} \sum_{j \in S} x_{ij} \leq |S| - 1, \text{ for every subset } S \tag{6.11}$$

$$x_{ij} = 0, 1 \text{ for all i, j, } i \neq j$$

In this revised TSP model, constraint (6.9) and constraint (6.10) are new. Constraint (6.9) forces the next stop after visiting Rx7 to be one of its patients; constraint (6.10) ensures that vehicles do not travel from customers of Rx7 (i.e., customers in set $n2$) to Rx7. The addition of constraints (6.9) and (6.10), together with the other constraints, ensure that the vehicle visits Rx7 before the customers of Rx7. Most Scenario II runtimes were similar to Scenario I (on the order of a few seconds), but for a few instances the runtime was closer to 1 minute.

6.7 Scenario III: Centralized Scenario Model

In Scenario III, a central depot is used for all vehicles. The location of the central depot was chosen as Rx1 through discussions with the management team from Clark Pharmacy. Rx1 is centrally located relative to the other pharmacy locations and has a large patient base. Because we now must assign multiple vehicles to routes, we must employ a VRP model to find the optimal routing of vehicles to provide home delivery of prescriptions in Scenario III.

6.7.1 Vehicle Routing Problem

Indices: i, j, node; v, vehicle

Parameters: n = number of nodes; $s = 1, \ldots, 7$ are pharmacies, set Ps represents customers of pharmacy s

k = number of vehicles; D_{ij} = distance between nodes i and j

Decision variables: $x_{ijv} = 1$ if the vehicle v travel between nodes i and j, else = 0; $y_{iv} = 1$ if node i is served by vehicle v, else = 0.

Objective function: $\min \sum\limits_{i=1}^{n} \sum\limits_{j=1}^{n} D_{ij} \sum\limits_{v=1}^{k} x_{ijv}, i \neq j$

Subject to:

$$\sum_{v=1}^{k} y_{iv} = 1, i = 2, 3, \ldots, n \tag{6.12}$$

$$\sum_{v=1}^{k} y_{1v} = k \tag{6.13}$$

$$\sum_{j=1}^{n} x_{ijv} = y_{iv}, i = 1, 2, \ldots n; v = 1, 2, \ldots l, k \tag{6.14}$$

$$\sum_{j=1}^{n} x_{jiv} = y_{iv}, i = 1, 2, \ldots, n; v = 1, 2, \ldots, k \tag{6.15}$$

$$\sum_{i=1}^{n} y_{iv} \leq 40, v = 1, 2, \ldots, k \tag{6.16}$$

$$\sum_{i \in Ps} \sum_{v=1}^{k} x_{siv} = 1, s = 2, ..., 7 \qquad (6.17)$$

$$\sum_{v=1}^{k} x_{isv} = 0, s = 2, ..., 7; i \in Ps \qquad (6.18)$$

$$\sum_{i \in S} \sum_{j \in S} x_{ijv} \leq |S| - 1, v = 1, ..., k; \text{ for every subset } S \qquad (6.19)$$

$$x_{ijv} = 0,1 \text{ for all } i, j, i \neq j \qquad (6.20)$$

$$y_{iv} = 0,1 \text{ for all } i, v \qquad (6.21)$$

In this VRP model, constraints (6.12), (6.14), and (6.15) ensure that each node is visited exactly once, and constraint (6.13) requires that k vehicles leave the depot (node 1). Vehicle capacity is specified in constraint set (6.16). Constraint set (6.17) indicates the next stop from a drugstore has to be one of its patients; no travel from a pharmacy's patients to that store is allowed in constraint set (6.18). Constraint set (6.19) provides subtour eliminations. Constraints (6.20) and (6.21) enforce binary values for the decision variables.

Similar to previous solution methods, the max-flow model will also be solved based on all the delivery points traveled by one vehicle. The solutions to the VRP model without subtour constraints, y_{iv}, is used to find all nodes served by each vehicle, v. For example, all nodes i for which $y_{i1} = 1$ are served by vehicle 1.

6.8 Solution Comparisons

From the data in Table 6.2, we can associate costs with the vehicle assigned to each drugstore in Scenario I. This is shown in Table 6.3 (again, actual costs have been disguised). The vehicle costs in Table 6.3 are estimated based on the costs and duration of ownership specified by Clark's Pharmacy. Fuel cost is excluded here because it is estimated separately based on distance in the following analysis. From these individual pharmacy values, we calculate a total average monthly fixed cost ($5,140.35). We use an average of the individual pharmacy costs because management believes that this would be most representative for comparing the different scenarios going forward.

TABLE 6.3

Estimated Vehicle Monthly Costs (Based on Data in April 2008)

Costs	Rx1	Rx2	Rx3	Rx4	Rx5	Rx6	Rx7	Average
Driver's pay	$3,061.13	$4,573.72	$4,415.56	$1,003.30	$1,732.68	$2,192.35	$2,195.33	
Driver's taxes	$653.38	$768.26	$1,860.53	$88.05	$159.53	$175.64	$215.74	
Vehicle repair	$3,131.25	$115.43	$200.87	$135.31	$189.44	$0.00	$0.00	
Insurance	$325.11	$339.28	$339.28	$288.21	$275.83	$157.05	$157.05	
Payments	$0.00	$1,200.61	$1,200.61	$0.00	$0.00	$0.00	$0.00	
Vehicle cost	$690.28	$690.28	$690.28	$690.28	$690.28	$690.28	$690.28	
Total fixed cost	$7,861.15	$7,687.58	$8,707.11	$2,205.15	$3,047.75	$3,215.32	$3,258.40	$5,140.35

TABLE 6.4

Delivery Costs Comparison in Two Scenarios for Rx6 and Rx7

Day	Scenario I Distance (mi)	Scenario II (Rx6 and Rx7) Distance (mi)	Scenario I Cost (Daily)	Scenario II (Rx6 and Rx7) Cost (Daily)	Cost Difference (Daily)
1	17.0	61.9	$373.71	$194.33	$179.38
2	73.2	94.9	$381.58	$198.95	$182.63
3	26.4	58.1	$375.03	$193.80	$181.23
4	22.6	65.3	$374.50	$194.82	$179.69
5	56.2	80.1	$379.21	$196.88	$182.33
6	82.6	114.4	$382.90	$201.68	$181.23
7	56.5	83.6	$379.25	$197.37	$181.89
8	48.7	87.3	$378.15	$197.90	$180.26
9	60.0	93.0	$379.74	$198.69	$181.05
10	51.8	74.5	$378.59	$196.09	$182.50
11	50.6	81.4	$378.42	$197.06	$181.36
12	51.2	89.5	$378.51	$198.20	$180.30
13	30.8	62.5	$375.65	$194.42	$181.23
14	57.2	70.7	$379.34	$195.56	$183.78
15	44.3	85.1	$377.54	$197.59	$179.95
16	46.5	76.0	$377.85	$196.31	$181.53
17	70.7	98.6	$381.23	$199.48	$181.75
18	139.2	131.0	$390.82	$204.01	$186.81

The average total fixed cost of $5,140.35 is equivalent to an average daily fixed cost of $185.67.

We assume that the price of gasoline is $2.79 per gallon (based on 2008 fuel prices) and that each vehicle averages 20 miles per gallon; the fuel cost per mile traveled is $0.14. Table 6.4 compares the costs (costs are again disguised, but relative differences between scenarios have been retained) under the optimal solutions for Scenarios I and II for the Rx6 and Rx7 pharmacy locations only. Costs shown are daily costs during May 2008.

TABLE 6.5

Daily Delivery Costs Comparison in Two Scenarios for Rx2 and Rx4

Day	Scenario I Distance (mi)	Scenario II (Rx and Rx4) Distance (mi)	Scenario I Cost	Scenario II (Rx and Rx4) Cost	Cost Difference
1	163.4	150.8	$394.11	$206.69	$187.43
2	150.8	112.8	$392.35	$201.41	$190.95
3	163.4	131.3	$394.11	$203.98	$190.13
4	207.7	144.8	$400.30	$205.87	$194.43
5	119.7	117.2	$388.05	$202.00	$186.04
6	131.0	94.9	$389.59	$198.89	$190.69
7	170.9	142.9	$395.15	$205.59	$189.56
8	130.7	124.7	$389.56	$203.04	$186.52
9	149.5	140.1	$392.20	$205.24	$186.96
10	127.2	107.8	$389.09	$200.72	$188.37
11	166.2	131.9	$394.52	$204.08	$190.44
12	187.9	164.0	$397.54	$208.57	$188.97
13	117.8	110.9	$387.80	$201.12	$186.67
14	169.3	151.1	$394.96	$206.75	$188.21
15	171.5	129.4	$395.27	$203.73	$191.54
16	100.5	83.3	$385.35	$197.29	$188.06
17	181.9	153.9	$396.72	$207.16	$189.56
18	167.1	153.9	$394.65	$207.16	$187.49
19	143.3	126.0	$391.32	$203.26	$188.06

Table 6.4 clearly shows that Scenario II may be preferable to Scenario I, at least when combining deliveries for the Rx6 and Rx7 pharmacies. Note this is true even though Scenario II requires greater miles traveled than Scenario I. In Scenario II, the single vehicle must service customers of both Rx6 and Rx7. Hence, it often has to travel a greater distance to do so. However, these additional travel costs are offset by the savings in vehicle costs. Scenario I requires two vehicles, whereas Scenario II requires only a single vehicle. Based on these results, Clark's management wanted to investigate other potential combinations of pharmacies. Examining volume of deliveries and geographic locations, we decided to compare the decentralized and hybrid scenario for combining Rx2 and Rx4 (Table 6.5), and Rx1, Rx3, and Rx5 (Table 6.6).

Table 6.5 and Table 6.6 confirm that Scenario II provides significantly lower costs for these additional combinations of pharmacies. Notice in Table 6.5 and Table 6.6 that Scenario II actually requires fewer miles traveled than Scenario I for many days. For these instances, our TSP solutions are able to generate routes that are more efficient for Scenario II than Scenario I. By removing the distance to and from the individual pharmacies, the combined routes are shorter. Thus, the savings provided by Scenario II are relatively larger.

TABLE 6.6

Daily Delivery Costs Comparison in Two Scenarios for Rx1, Rx3, and Rx5

Day	Scenario I Distance (mi)	Scenario II (Rx1, Rx3, Rx5) Distance (mi)	Scenario I Cost	Scenario II (Rx1, Rx3, Rx5) Cost	Cost Difference
1	245.7	228.1	$591.25	$217.49	$373.79
2	187.9	227.5	$583.20	$217.37	$365.84
3	225.9	254.2	$588.51	$221.11	$367.38
4	249.8	266.7	$591.81	$222.86	$368.95
5	279.0	240.6	$595.93	$219.25	$376.68
6	241.0	258.6	$590.62	$221.73	$368.89
7	182.5	219.9	$582.48	$216.33	$366.15
8	170.9	186.6	$580.82	$211.71	$369.11
9	172.5	191.3	$581.10	$212.34	$368.73
10	111.8	156.1	$572.62	$207.47	$365.15
11	140.7	181.3	$576.61	$210.96	$365.68
12	195.4	197.6	$584.24	$213.22	$371.02
13	170.0	182.2	$580.72	$211.08	$369.61
14	302.5	246.9	$599.20	$220.13	$379.06
15	270.8	225.9	$594.80	$217.18	$377.62
16	235.9	178.1	$589.90	$210.52	$379.38
17	206.1	212.7	$585.75	$215.36	$370.39
18	76.7	139.2	$567.69	$205.08	$362.60
19	271.1	241.6	$594.83	$219.35	$375.48
20	171.2	184.4	$580.88	$211.40	$369.48
21	224.6	218.7	$588.36	$216.17	$372.15

Scenario III (the centralized scenario) represents a benchmark solution that Clark's Pharmacy wanted to investigate for possible future operations. Scenario III would allow all home delivery prescriptions to be filled from a single central depot. Clark's expected this to minimize the number of vehicles required, but it would also require significant changes to operations to fill all home delivery prescriptions from a single location. To examine the potential benefits of the centralized solution, we chose 3 days from our data set that appear to be a "typical" daily delivery, maximum daily delivery, and minimum daily delivery in terms of numbers of customers. Due to the computational complexity of solving these relatively large VRPs to optimality in Scenario III, we set 12 hours as the program execution time limitation. At this time point, the relative MIP gap* was reported to be around 0.01% for each scenario.

---

* The "MIP gap" refers to the difference between the best integer objective function value, $\bar{z}$, found and the bound on the optimal solution, $\hat{z}$. The values reported here refer to the "relative MIP gap" reported by CPLEX, which is defined as $\dfrac{|\bar{z} - \hat{z}|}{\bar{z}}$.

TABLE 6.7

Comparison under Three Scenarios for Three Different Patient Volume Instances

	Patient Volume Instance	Number of Patients	Scenario I (7 Vehicles Required)	Scenario II (3 Vehicles Required)	Scenario III (3 Vehicles Required) (relmipgap = 1e–4)
Distance (miles)	Typical	61	426.0	440.8	387.4
	High	77	321.1	323.9	360.0
	Low	50	534.1	509.3	355.0
Daily cost ($)	Typical	61	$1,359.11	$618.48	$611.26
	High	77	$1,344.63	$602.34	$607.40
	Low	50	$1,374.45	$628.29	$606.30

Table 6.7 shows the comparison under the different scenarios. Scenario II shown in Table 6.7 represents all the combinations examined in Tables 6.4 to 6.6: Rx6 and Rx7 are combined; Rx2 and Rx4 are combined; and Rx1, Rx3, and Rx5 are combined. Clearly, Scenario I results in much higher costs than either Scenario II or Scenario III. (Given the small MIP gap reported in Scenario III we can be quite confident that the costs shown here are near the optimal costs in the centralized scenario.) Scenario III reduces the costs from Scenario I by approximately 55% in each instance. This is largely driven by the large savings in reducing the number of vehicles required from seven to three. However, Scenario II (hybrid scenario) is very competitive with scenario III. For our "typical instance," Scenario III results in costs that are only about 1.1% less than scenario III. For the "high" and "low" patient-volume instances, Scenario II costs are actually slightly lower here due to the fewer miles traveled (both Scenario II and Scenario III require three vehicles). We should emphasize here that all of our results may be specific to Clark Pharmacy's network. The relative preference for Scenarios I, II, or III could be different when considered for a different network.

6.9 Conclusions

This chapter presents an applied case study of designing a prescription home-delivery service for a small pharmacy chain. We utilize classical operations-research methods to compare several delivery strategies. We formulate the home delivery problem as a fairly small instance traveling salesman or vehicle routing problem. We then compare the costs of the different methods based on vehicle, transportation, and labor costs. Our analysis indicates that a decentralized solution where each pharmacy maintains its own delivery vehicle is considerably more expensive than sharing vehicles between stores. However, we also find that a hybrid scenario where we share a vehicle between two or three stores is very competitive, and sometimes preferable, to a completely centralized scenario using a single delivery depot.

Our analysis indicates that the hybrid scenario may be preferable due to the easier transition from the current decentralized scenario and ability to keep a "neighborhood feel" to the pharmacy's operations. The hybrid solution may have additional, less quantifiable benefits for Clark's Pharmacy through more consistent customer interactions with the same delivery drivers and more flexibility to respond to emergency orders, because the delivery drivers will tend to be stationed closer to the majority of their customers. We should also remark here that the locations used in our hybrid scenario were chosen through interaction with Clark Pharmacy management based on the stores that they deemed viable for such a scenario. However, a straightforward extension to the models used here would be to use some sort of multilocation gravity model (for example, see Tanset et al. 1983) or clustering analysis (see Mulvey and Crowder 1979, for a classic approach to this problem, and Klose and Drexl 2005, for a more recent review of work in this area) to determine which two (or more) locations would be best served to be included in a hybrid model.

Small to midsize chain pharmacies such as Clark's cannot compete with the larger drugstore chains and supermarkets solely on price. They must offer superior customer service and innovative services. Home deliveries represent one such service that is typically not offered by larger drugstore chains and supermarkets. Operations-research tools have provided value-added analysis in this project for Clark's Pharmacy and it has allowed them to be more confident in their decisions regarding how to offer home delivery services in the future. The problem instances examined here are quite modest in size for TSPs and VRPs. Yet, even here we encounter computational challenges when trying to find optimal solutions. It is likely that larger problems would require heuristic solution methods to be of value to practitioners to use on a daily basis for prescription home deliveries. Another challenge identified in this project that is common in working with small-to-medium-size companies is that much of the data required for our analysis had not been actively collected and recorded by the company prior to our study. Thus, considerable initial effort had to be spent in simple data collection so that our OR models could be applied. Although this required additional time and effort, it also gave us excellent opportunities to become aware of many of the intricacies in modeling this problem for Clark's Pharmacy.

References

Applegate, D. L., R. Bixby, V. Chvatal, and W. Cook. 2007. *The Travelling Salesman Problem: A Computational Study*. Princeton, NJ: Princeton University Press.

Bertsimas, D. J. 1992. A vehicle routing problem with stochastic demand. *Operations Research* 40:574–85.

Bertsimas, D. J., P. Jaillet, and A. R. Odoni. 1990. *A priori* optimization. *Operations Research* 38:1019–1033.

Bertsimas, D. J., and G. van Ryzin. 1991. A stochastic and dynamic vehicle routing problem in the Euclidean plane. *Operations Research* 39:601–15.

Braysy, O., and M. Gendreau. 2005a. Vehicle routing problem with time windows, part I: Route construction and local search algorithms. *Transportation Science*, 39:104–18.

Braysy, O., and M. Gendreau. 2005b. Vehicle routing problem with time windows, part II: Route construction and local search algorithms. *Transportation Science* 39:119–39.

Clarke, G., and J. W. Wright. 1964. Scheduling of vehicles from a central depot to a number of delivery points. *Operations Research* 12:568–81.

Cordeau, J.-F., M. Gendreau, and G. Laporte. 1997. A tabu search heuristic for periodic and multi-depot vehicle routing problems. *Networks* 30:105–119.

Frick, A. 1996. TSPGA—An evolution program for the symmetric traveling salesman problem, in *EUFIT98—6th European Congress on Intelligent Techniques and Soft Computing*, ed. H. J. Zimmerman. Aachen: Mainz-Verlag.

Gendreau, M., A. Hertz, and G. Laporte. 1992. New insertion and postoptimization procedures for the traveling salesman problem. *Operations Research* 40:1086–94.

Gendreau, M., A. Hertz, and G. Laporte. 1994. A tabu search heuristic for the vehicle routing problem. *Management Science* 40, 1276–90.

Gendreau, M., G. Laporte, and J.-Y. Potvin. 2002. Metaheuristics for the capacitated VRP. In *The Vehicle Routing Problem*, ed. P. Toth and D. Vigo. Philadelphia, PA: Society for Industrial and Applied Mathematics.

Gendreau, M., and J.-Y. Potvin. 1998. Dynamic vehicle routing and dispatching. In *Fleet Management and Logistics*, ed. T. G. Crainic and G. Laporte. Boston: Kluwer, 115–26.

Helsgaun, K. 2000. An effective implementation of the Lin-Kernighan traveling salesman heuristic. *European Journal of Operational Research* 126:106–30.

Hoffman, K. L., and M. Padberg. 1996. Traveling salesman problem. In *Encyclopedia of Operations Research and Management Science*, ed. S. I. Gass and C. M. Harris. Norwell, MA: Kluwer Academic.

Kenyon, A. S., and D. P. Morton. 2003. Stochastic vehicle routing with random travel times. *Transportation Science* 37:69–82.

Klose, A., and A. Drexl. 2005. Facility location models for distribution system design. *European Journal of Operational Research* 1:4–29.

Lambert, V., G. Laporte, and F. Louveaux. 1993. Designing collection routes through bank branches. *Computers and Operations Research* 20:783–91.

Laporte, G. 1992. The traveling salesman problem: An overview of exact and approximate algorithms. *European Journal of Operational Research* 59:231–47.

Laporte, G., F. Louveaux, and H. Mercure. 1992. The vehicle routing problem with stochastic travel times. *Transportation Science* 26:161–70.

Laporte, G., Y. Nobert, and S. Taillefer. 1988. Solving a family of multi-depot vehicle routing and location-routing problems. *Transportation Science* 22:161–172.

Laporte, G., F. Louveaux, and H. Mercure. 1994. *A priori* optimization of the probabilistic traveling salesman problem. *Operations Research* 42:543–49.

Laporte, G. and F. Semet. 2002. Classical heuristics for the capacitated VRP. In *The Vehicle Routing Problem*, ed. P. Toth and D. Vigo. Philadelphia, PA: Society for Industrial and Applied Mathematics.

Lawler, E. L., J. Lenstra, A. Rinnooy-Kan, and D. Shmoys. 1985. *The Traveling Salesman Problem*. New York, NY: John Wiley & Sons.

Little, J. D. C., K. Murty, D. Sweeney, and C. Karel. 1963. An algorithm for the traveling salesman problem. *Operations Research* 21:972–89.

Mulvey, J. M., and H. P. Crowder. 1979. Cluster analysis: An application of Lagrangian relaxation. *Management Science* 25:329–40.

Padberg, M. W., and S. Hong. 1980. On the symmetric travelling salesman problem: A computational study. *Mathematical Programming Studies* 12:78–107.

Partyka, J., and R. Hall. 2010. On the road to connectivity. *OR/MS Today* 37(1).

Pataki, G. 2003. Teaching integer programming formulations using the traveling salesman problem. *SIAM Review* 45:116–23.

Psaraftis, H. N. 1988. Dynamic vehicle routing problems. In *Vehicle Routing: Methods and Studies*, ed. B. L. Golden and A. A. Assad. Amsterdam: North-Holland, 223–48.

Reinelt, G. 1994. *The Traveling Salesman: Computational Solutions for TSP Applications*. Berlin: Springer-Verlag.

Skiena, S. S. 1990. *Implementing Discrete Mathematics: Combinatorics and Graph Theory in Mathematica*. Redwood City, CA: Addison-Wesley.

Solomon, M. M. 1987. Algorithms for the vehicle routing and scheduling problems with time window constraints. *Operations Research* 35:254–65.

Simchi-Levi, D., P. Kaminsky, and E. Simchi-Levi. 2007. *Designing and Managing the Supply Chain*, 3rd ed. Boston: McGraw-Hill/Irwin.

Tanset, B. C., R. L. Francis, and T. J. Lowe. 1983. Location on networks: A survey. Part I: The *p*-center and *p*-median problems. *Management Science* 29:482–97.

Toth, P., and D. Vigo. 2002. An overview of vehicle routing problems. In *The Vehicle Routing Problem*, ed. P. Toth and D. Vigo. Philadelphia, PA: Society for Industrial and Applied Mathematics.

7

Improving Pharmacy Operations Using Computer Simulation

Alberto Isla, Alex C. Lin, and W. David Kelton

CONTENTS

7.1 Introduction

All firms produce and store large quantities of data, and the analysis of these data has become invaluable in supporting the decision-management process. In recent years computers with extremely high processing power have become affordable, thus facilitating the data mining of business data. Huge databases are commonly scrutinized for correlations between apparently unrelated variables, which may explain what appears to be random behavior.

There are some situations, though, in which data mining cannot answer the questions at hand, especially when firms wonder what would happen if some important change were implemented in their customer-service settings, in their production facilities, or in their supply-chain system. Often it is not feasible to try out those changes in a controlled environment prior to company-wide implementation. In these cases, computer simulation can be a useful approach.

In this chapter, we will illustrate a case in which computer simulation was a useful decision-making tool for business. Our firm was a major community (retail) pharmacy chain in the United States. The main objective of this chapter is to show how computer simulation can be used to improve various operational aspects in a retail pharmacy setting. Pharmacy managers may know that operations are inefficient and patient wait times are excessive, but they often find it difficult to pinpoint the root causes of those inefficiencies because pharmacy activities are complex. Intuitively they may have a sense of possible solutions to those inefficiencies, but they are afraid of the impact the changes could have on key performance parameters. Unfortunately, few of those potential improvement strategies can be tested in real-world pharmacy settings due to patient-safety considerations and operational issues. A computer-simulation approach provides a means to analyze and compare the results from several improvement strategies in an effective way.

Although our analysis involves a community pharmacy, most of the issues described in this chapter are common to many health care settings (like hospitals, emergency rooms, outpatient centers, etc.), in which patients arrive in a random way and a few professionals (doctors, nurses, etc.) serve them by performing a complex set of activities (triage, consultation, x-ray, etc.) with scarce resources (beds, operating rooms, etc.). In this chapter we will first summarize the activities that take place in a pharmacy on a daily basis to illustrate the complexity of the process. Next, we will outline the steps required prior to collecting data and building the model. We will continue by stating some of the biggest challenges and difficulties that are faced when building a simulation model. Next, we will highlight the main performance statistics that pharmacy managers want to learn from the simulation model. Finally, we explain why the previously mentioned statistics can best be obtained through the use of computer simulation and report some projects that are currently being carried out in community pharmacies using computer simulation.

Limited examples of the use of computer simulation to improve pharmacy operations can be found in the literature. A computer-simulation approach was applied to determine the optimum staffing patterns in an outpatient pharmacy (Vemuri 1984). A computer-simulation study was conducted for a major U.S. drugstore chain to determine the impact of some operational and layout changes (Lin et al. 1996). Computer simulation was applied to assist in changing a hospital's drug distribution system from a decentralized pharmacy satellite model to a centralized robotic drug-dispensing configuration (Buchanan 2003).

There are many different types of computer simulation. Normally in health care settings, discrete event simulation (DES) is used. DES software manages queues of events and triggers them as designed. Arena is the simulation package used for this analysis and the simulation-related vocabulary should be understandable by users of any other DES package.

7.2 Brief Description of the Operation Procedures in a Community Pharmacy

A community pharmacy in the United States is a retail store with the main purpose of dispensing prescribed drugs and offering specialized counseling to the people in its neighborhood. Prescriptions need to be filled in a careful and accurate way. All of the procedures in the pharmacy are subject to strict legislation. Most often, community pharmacies are located in big stores, which in addition to the dispensary where medications are stored and dispensed, offer a retail storefront typically selling diverse products including over-the-counter drugs. This chapter will focus just on the dispensary of drugs and prescriptions, which we will call the "pharmacy," and will ignore the other retail store services.

A prescription drug cannot be delivered to a customer if a pharmacist has not previously inspected it. That is why at least one pharmacist (usually called a "Registered Pharmacist," RPh) is required to be on-duty at all times at the pharmacy. The staff may also include trained pharmacy technicians. The pharmacist is responsible for the quality of service provided by the pharmacy, and hence she must supervise the technicians to ensure prescription safety. Some states have limited by law the number of technicians who can be supervised by a pharmacist, allowing ratios of 2:1, 3:1, or 4:1. Pharmacies increasingly also rely on different types of automation to assist with prescription filling.

Each day the pharmacy needs to fill and dispense many orders for drugs. Each order consists of one or more prescriptions. The average number of prescriptions per order in community pharmacies is between two and three. Orders can arrive at the pharmacy in different ways. The patient can bring them in person to either the drop-off window or drive-through window (if available). Also,

TABLE 7.1

List of Activities at the Pharmacy

Type of Activity	List of Activities	RPh-Tasks	Comments
Direct Activities	1. Receiving 2. Data Entry 3. Filling 4. Inspection 5. Packing & Storing 6. Dispensing	Inspection is an RPh task.	For each prescription, the direct activities always happen in this order.
Indirect Activities	Phone calls Inventory management Management tasks Customer relations Special services Miscellaneous tasks	Phone calls with physicians, patient counseling and staff management are RPh tasks.	
Unproductive Activities	Coffee or chatting breaks Idle time		

the doctor's office can deliver an order directly to the pharmacy via electronic means (this is called an ePrescription), fax, or phone call. Many community pharmacies have also a courtesy-refill program that produces orders to refill prescriptions automatically, for patients who have previously signed up for that program, a few days before they are scheduled to run out.

Table 7.1 includes a list of the main tasks at a pharmacy. Figure 7.1 is a flow diagram of the main activities at the pharmacy. During the order-receiving process, the technician asks the patient when she plans to pick up the order. This allows the pharmacy to give higher priority to the orders that are expected to be picked up sooner. Most pharmacies place the forms and vials corresponding to each order in a basket, and they use baskets of different colors to identify the higher- and lower-priority orders. Once the order has been received, its basket moves to the data-entry stage, where all the information about the order is entered into the pharmacy records, and the order is processed with the paying entity (most often the insurance company). At the end of the data-entry section the vial label is printed and the order basket is moved to the filling counter. In the filling stage the drug bottle is retrieved from the storage shelves and scanned, the tablets or capsules are counted (or the drug is poured when it is liquid), the main and auxiliary labels are attached onto the vial, and the vial is capped. Next the order moves on to the inspection section, where the pharmacist reviews all the information, including the patient's drug-utilization history, and checks the order against the scanned prescription, the drug bottle, and the tablets or capsules inside the vial. In the next section, packing and storing, the vials and bottles corresponding to the different prescriptions in the order are placed in a bag together with an information pamphlet, and the bag is placed in the will-call area. When the patient comes to the pharmacy to pick up her order either at the pick-up window or at the drive-through window, the dispensing activity

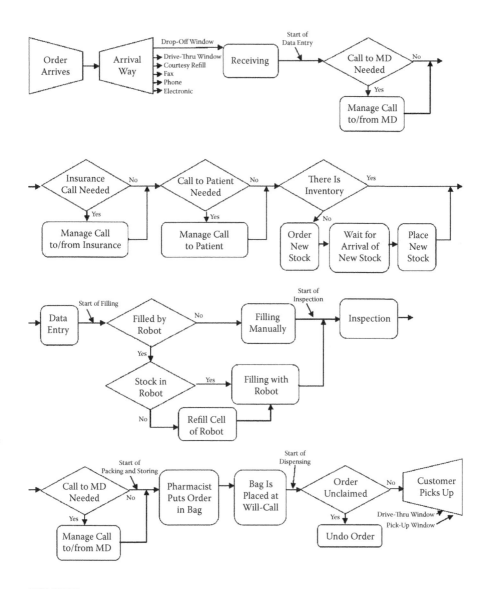

FIGURE 7.1
Flow diagram of tasks related to prescription filling.

takes place: the bag is taken from the will-call area, the order is delivered to the patient, the receipt or the vial's bar code is scanned, and the patient pays.

All of the previous activities, receiving, data entry, filling, inspection, packing and storing, and dispensing, are referred to as *direct activities* because they happen for all orders. But also many *indirect activities* take place at the pharmacy. Indirect activities are necessary but not directly related to filling orders. For example, many phone calls are made between the pharmacy and insurance companies, doctors' offices, and customers. Some other indirect tasks include inventory management (reviewing inventory, ordering new inventory, placing new stock), management tasks (reconciling unclaimed orders, scheduling, training, handling job applications, etc.), customer relations (conversations with customers, answering questions about over-the-counter drugs), special services (diabetic shots, flu shots, etc.), and other miscellaneous tasks (attending to a printer, routine cleaning, restocking inventory of vials, etc.). In addition to direct and indirect activities, the staff spends some time in *unproductive activities*, such as coffee breaks, chatting with peers, or just being idle.

7.3 Preparing for the Data-Collection Process

The data-collection process is of paramount importance for the relevance of the analysis. Since no retrospective data are available, the only way to learn how the staff spends their time is observing them working at the pharmacy for long periods. Methods based on asking the pharmacy staff how much time they think they devote to each activity do not produce reliable conclusions. As a consequence, collecting data for these kinds of projects is neither inexpensive nor fast. Prior to starting data collection, several important decisions need to be made.

7.3.1 Work Sampling versus Time Study

How will on-site observations be made at the pharmacies? Several procedures are available to perform this kind of work measurement, but the two most widely used are *work sampling* and *time study*. Work sampling was originally developed by L. H. C. Tippett (1935) and consists of recording snapshots at fixed or random time intervals, including the activity task that a worker is doing at that precise time. Time study, on the other hand, consists of directly measuring the duration of each work process or task. The direct outcomes of a time study are the duration times required for the different repetitions of a task, whereas the direct outputs from a work-sampling study are the proportion of time that each worker spends on each activity.

At first glance it might seem that time study would be the best choice for this project because it measures the time required for each task, thus allowing an easy calculation of both the average and the shape of the probability distribution of the time required for each activity. This probability distribution is precisely the input information that needs to be entered in the simulation model in order to characterize the time required for each task. Therefore, the outputs from time study are a perfect match to the inputs to the simulation model.

Pharmacists and technicians work in such a way that it makes it very difficult to use a time-study methodology successfully. Some of the tasks performed by the pharmacy staff are very short (for example, data entry for a refill prescription can last only a few seconds), and quite often there is overlap between the same activity done for two or more different prescriptions (for example, a technician can print out the labels for several different prescriptions in a row). In addition, there are frequent interruptions (like customers showing up or incoming phone calls) that split some activities into at least two pieces. All of these make it much easier for an observer to record what activity a worker is doing when a beeper goes off, than trying to time the start and finish points of each activity. Also, time study provides no information about the amount of time devoted throughout the day to each task.

Ideally the project should include observations made with both techniques. But this is not feasible in most cases due to budgetary constraints. Using both work-measurement approaches creates technical difficulties since interapproach biases will show up, due to the fact that the observations with each technique will be performed either the same day by different observers or by the same observer but on different days. This interapproach bias often cancels out most of the improvement achieved by using both methods.

Therefore, we need to choose one, and in doing so two important project parameters come into play: money and time. In order to produce statistically valid results, work sampling requires approximately one-fifth of the observation time required for a time study. This is in part due to the fact that when using work sampling a single observer can simultaneously record the activities of several pharmacy workers because all of them spend the workday in a circumscribed area (Finkler et al. 1993). By using work sampling with observations at fixed intervals of one minute, a high level of accuracy is reached in just a few days of observation. This is the main reason why work sampling has become the most frequent work-measurement technique reported in the literature and also the approach that we have used in our projects.

7.3.2 Different Sources of Data

In addition to on-site observations, other sources of data will be needed, including the pharmacy records corresponding to the same days in which the observations were performed (once personal data about patients have been removed due to legal restrictions). Pharmacy records store a lot of information about each prescription (arrival time, way of arrival, times at which the prescription went through the major steps, filling time, expected pick-up time, actual pick-up time, etc.). Having available these standard operating procedures of the pharmacy also allows us to better understand the workflow in the pharmacy and the level of priority assigned to each task. This knowledge will be extremely helpful in building a simulation model that accurately mirrors the actual flow of activities at the pharmacy.

7.3.3 Making a List of the Output Statistics

A detailed plan for the entire project should be made prior to starting data collection. First, we should create a list of all the output statistics that need to be produced by the simulation model, like average customer wait time at the drop-off window, since we need to design the simulation model so that it can produce the desired outputs. It is also common for some other outputs to be used to validate the model. Researchers need to consider beforehand how they are going to obtain them as well. Collecting data is an expensive and long process, so we cannot afford to overlook measuring some important parameter.

7.3.4 Deciding the Set of Categories to Be Observed

Next, all the categories to be observed in the work-sampling procedure need to be identified. For example, if one of the output statistics we need is the proportion of time that pharmacists spend inspecting prescriptions, one of the work-sampling categories needs to be "inspecting prescriptions." Each category needs to be fully described. For example, the category "inspecting prescriptions" will include the following actions: walk to inspection station; obtain vial and drug bottle from basket; scan label and drug bottle; review information from computer; check against prescription, drug bottle, and tablets/capsules inside vial; recount narcotic drugs; scan biometric; and file script. In order to get accurate conclusions, the list of categories must be mutually exclusive, so that there is not an activity that can be included in two different categories.

Making a list of the desired output statistics prior to deciding the set of observation categories helps prevent mistakes in the experimental design. This can be better explained with an example. If we want to learn the arrival pattern of customers to drop off orders, it will suffice to define a category called "receiving" that will be selected every time a technician is observed receiving an order from a customer either at the drop-off window or at the drive-through window. But if we are instead trying to assess the impact of closing the drive-through window, two different categories, "receiving at drop-off window" and "receiving at drive-through window" will be needed. In other words, the model now needs to be sensitive to the type of window, and this requires more detailed data collection.

Deciding the appropriate number of categories is important. Having just a few broad categories, each one including a long list of activities, will not allow the researcher to generate detailed output statistics from the simulation model. On the other hand, having many narrow categories will increase the number of observation errors. To obtain statistically meaningful conclusions from the work-sampling approach, more observation days will be required if more categories are being included in the study. Thus, measuring at a too-detailed level, which gives data that will not be used in further analysis, will unnecessarily increase the cost and duration of the project.

7.3.5 Deciding the Length and Sites of Study

The number of required observation days can be obtained using statistical formulas. We will not go into the technical details here because there is extensive literature about this, but we would like to point out several facts that researchers need to take into consideration.

First, the number of samples of data that will need to be collected ("sample size") will depend on the desired precision. The higher the precision demanded, the larger the sample size required.

Second, the average length of the intervals at which the work-sampling measurements will be taken should be as short as possible. We recommend intervals between 1 and 2 minutes, depending on the number of pharmacists and technicians that each observer needs to follow.

Third, if some of the predefined task categories rarely occur, a very large sample size will be necessary to capture the frequency of that category with an adequate level of precision. Thus, researchers should avoid observing for a very infrequent task category unless it is essential for the purpose of the project.

Finally, the pharmacies and days in which the researchers will make the observations need to be representative for the purpose of the study. There are important differences among pharmacies with different workloads, with and without automation, with and without a drive-through window, and so forth. It is recommended that observations be split between several similar pharmacies so that the results are more representative of the entire population of pharmacies having the same characteristics.

7.4 Challenges When Building the Simulation Model

The next step is to build the computer simulation model. Table 7.2 includes a list of commonly used simulation terms and their definitions. Even if the researchers have complex understanding of the operations at the pharmacy and all the data have been correctly collected, building a reliable simulation model will present a number of challenges. Let's review some of the main ones and how they can be overcome.

7.4.1 Data from Different Sources Need to Be Combined

The direct result from work sampling is the percent of time that each worker devotes to each activity. For example, work sampling will tell us that the pharmacist spends 4 hours (240 minutes) per day inspecting prescriptions. But the input information that needs to be entered into the simulation model is the probability distribution of the time spent inspecting a prescription. To specify this, we need to look at the pharmacy records as well. If the records

TABLE 7.2

Glossary of Simulation Terms

Computer Simulation Model	Computer program that tries to simulate an abstract model of a particular system or environment.
Scenario	Simulation model with a specific set of input parameters. Different scenarios of the same simulation model can sometimes be created by changing only the input parameters (task durations, arrival rates of entities, etc.) but not the logic involved, but in other cases it is necessary to change the logic to create the desired alternative scenarios.
Replication	Individual runs of one scenario of the simulation model. It is recommended that a number of independent replications for the same scenario be run in order to get more precise output statistics.
Input parameters	Set of variables that can be modified in order to define each different scenario of the simulation model. Typical input parameters are entity-arrival patterns, task durations, probabilities of some events, etc.
Control statistics	Performance parameters used by firms and organizations.
Output statistics	Values of the performance parameters that are obtained for each scenario after the simulation model has been run.
Probability distribution	Most of the input parameters in the simulation model will be generated as draws from input probability distributions (Kelton et al. 2010). Some of the most frequently used input probability distributions in simulation are exponential, triangular, and uniform.
Resources	Simulated representation of the material and human resources in a real setting. In the simulation model, resources need to be seized in order to perform some process or task.
Entities	Entities are the dynamic objects in the simulation. They are created, move around, change status, affect and are affected by other entities and the state of the system, and affect the output performance measurements. Most entities represent "real" things in a simulation (Kelton et al. 2010).
Attributes	To individualize entities, you attach attributes to them. An attribute is a common characteristic of all entities but with a specific value that can differ from one entity to another (Kelton et al. 2010).
(Task) Modules	A module is each one of the little pieces with which a simulation model is built. There are different types of modules to simulate different actions, like decide, create, dispose, wait, route, hold, seize, release, etc. Some of the modules are used to represent the processes or tasks happening in the real setting. Those are referred to as "task modules."
Queue	When an entity cannot move on, perhaps because it needs to seize a unit of a resource that is tied up by another entity, it needs a place to wait, which is the purpose of a queue (Kelton et al. 2010). Simulation programs offer many different alternatives to handle entities waiting in queues.
Model animation	Graphic representation of the elements of the simulation model (entities, attributes, resources, queues, etc.) so that what happens during a simulation run can be visually observed.
Model verification	Checking that the flowchart and assumptions that formed the abstract representation of the real setting have correctly been translated into the computer model and making sure that it performs as intended.
Model validation	A model will be validated when the researchers (and most important, the managers who will base their decisions on the model) gain a high level of confidence in the model and firmly believe that the outcomes that will be obtained from the model will be correct.

show that 240 prescriptions were inspected during that day, then we can calculate the average time spent inspecting a prescription, which is 240 minutes/240 prescriptions = 1 minute per prescription. But of course it will not take exactly 1 minute to inspect each of the 240 prescriptions. The pharmacist will inspect some prescriptions in a shorter time than others, depending on factors such as previous knowledge of the patient, the existence of a warning from the drug utilization review system, or the potential toxicity of the drug. Since we do not know the histogram of the time spent for all the individual inspections because we did not perform a time study, some assumptions need to be made here in order to determine the shape of the probability distribution to be specified in the simulation model. In general it is a fair approximation to use a triangular distribution slightly skewed to the right, with a mean coinciding with the average duration (1 minute) and with a standard deviation higher or lower depending on assumptions made based on the observations.

Pharmacy records can also be useful for estimating the number of prescriptions per order, the proportion of prescriptions that are filled by the automation robot (if available), the proportion of prescriptions that cannot be filled because of an out-of-inventory situation or that are only partially filled, and so forth. Finally, discussions with pharmacy managers will be necessary to learn some relevant information, like the firm's policy regarding staff resting time throughout each shift.

7.4.2 Learning Customers' Arrival Patterns

Another important challenge is specifying the arrival patterns of orders and customers to the pharmacy. Knowing the arrival time and the expected pickup time of each order is extremely helpful for determining the best staffing schedules. It was already mentioned that orders can arrive at the pharmacy in different ways. The distribution per hour of the day of customers arriving to drop off and pick up orders at the pharmacy windows can be obtained from direct observation at the pharmacy site. Also, pharmacy records show the exact time when every order was picked up, thus allowing a different way to obtain the pick-up pattern. But the arrival patterns of the orders arriving via fax, e-mail, or phone can be much tougher to identify. Pharmacy records usually store the time at which each prescription was entered in the system, but that is not necessarily the time when the prescription arrived. For example, a fax order could be sitting on the fax machine for hours before it is entered into the system. Pharmacy records most commonly show the arrival mode of each order, so we need to make some assumptions to come up with an arrival distribution.

The arrival pattern of customers coming to the pharmacy to pick up their orders is different from the arrival pattern of customers dropping off prescriptions. Usually, in community pharmacies most orders are dropped off in the morning and in the afternoon, and are picked up in the afternoon and evening hours. Currently identifying the right pick-up pattern is more important than identifying the right drop-off pattern because dispensing is much more time-intensive than receiving. It takes on average between two and three times longer to serve a customer at the pick-up window than a customer at the drop-off window, since the technician needs to find the bag with the order in the will-call area and sometimes the customer also wants to receive counseling from the pharmacist. Besides, only about half of the prescriptions are brought in, but almost all of them are picked up by the customers themselves. In short, a lot of staff time will be required to serve the customers during the hours of the day at which the pick-up rate peaks.

We can view the customer queue at the pharmacy windows in terms of standard queuing-theoretic terminology. The arrival process of the customers to the pharmacy is Poisson, that is, the interarrival times between successive customers follow an exponential distribution. The service-time distribution at the window is typically bell shaped although it has been simplified to a triangular distribution in the simulation model to prevent the long negative tails of normal distributions, which make no sense for a positive-duration value like service times. The number of servers is one in our model, since only one customer can be served at a time in each window. The number of places in the system, that is, the size of the queue, has no practical limit. The calling population (pool of external potential customers who might arrive) also has no natural limit either. And the queue's discipline is FCFS (first come, first served). By now the reader might think that we could treat this as a typical queuing problem and apply well-known formulas to learn customer waiting times. There is an important issue that prevents us from taking advantage of queuing theory and forces us to use simulation: the server (in queuing terms) or the resource (in simulation terms), that is the pharmacy employee, will be available in a shorter or longer time to serve the customer at the window depending on many complex factors (how many technicians are on duty, how many customers are waiting at the other windows, how many phone calls are received, or how many urgent prescriptions for waiting customers need to be filled, etc.). Here is where computer simulation comes into play. The model will assign the different tasks to the pharmacy staff following predefined levels of priority, so that every time a customer arrives at a window, the model will assign a resource to serve the customer as soon or as late as it happens in a real pharmacy.

7.4.3 Some Tasks Can Only Be Done by a Pharmacist

Once the required inputs are decided upon, the modeling stage can start. However, it also comes with some difficulties. Since the goal of this chapter is not to make the reader an expert in computer simulation, we will outline the problems and solutions without going into depth.

Some tasks can only be done by a pharmacist, like inspecting the filled prescriptions, advising physicians on drug interactions, or counseling patients. All other tasks at the pharmacy can be done by both pharmacists and technicians, for example, receiving orders from the patient or having phone calls with the insurance companies. In other words, pharmacists can perform all the tasks in the pharmacy but technicians can do only some of them. We will call "RPh tasks" those that can be performed only by a pharmacist, and "tech tasks" are the ones that can be performed by any staff. A pharmacist gives higher priority to the RPh tasks. When there are no such tasks pending at the pharmacy, a pharmacist usually helps technicians with the tech tasks.

Modeling this properly represents an interesting challenge. Pharmacists and technicians are treated as resources by the simulation model. Resources can be seized by entities (in our model the entities are the prescriptions) in order to perform some task. The best way to model this is to have all tech tasks seize a resource from an ordered set of resources (rather than seize a specific single resource). This resource set consists of first the technicians and then the pharmacists. The resource-selection rule used to choose between the resources in this set would be that a tech task first tries to seize a technician and will move on in the set to try to seize a pharmacist only if all technicians are busy.

7.4.4 Some Tasks Have Higher Priority

Another modeling problem intimately related to the previous one is the assignment of priorities when there are several entities (prescriptions) trying to seize a single resource (technician or pharmacist). We need to prevent a prescription entity from seizing a pharmacist to perform a tech task when there are RPh tasks pending. The way to do this is to make RPh tasks seize pharmacists with a higher priority level than when a resource set is seized to perform a tech task. It is necessary to assign priorities for deciding the order in which different tasks will seize a human resource. For example, the model can have at one point in time four entities waiting at the queues of four different tasks wanting to seize a pharmacist. The first one is a prescription waiting to be inspected. The second one is a phone call to a doctor's office waiting to be done. The third one is a phone call to the voice-mail system to check if there are new orders from doctors' offices. The last one is a customer waiting to receive counseling at the pick-up window. If all these entities are trying to seize the pharmacist with the same level of priority, they will be ordered depending on the time at which each entity arrived at its queue, that is, it will be an FCFS queuing system. This is not a good solution, because the customer waiting for counseling might have been the last one to arrive but needs to be served first. It would not be a good idea to have the customer waiting at the pick-up window while the pharmacist makes a call to the voice mail to see if there are new messages. Therefore the task "customer waiting for consultation with pharmacist" needs a higher priority than the other tasks and so on.

7.4.5 Some Orders Have Higher Priority: Simulating the Red Basket

At this point, we have already assigned a level of priority to each task in the model. The tasks with the highest priority will seize available resources first. All the RPh tasks have higher levels of priority than do the tech tasks. Unfortunately, this only partially solves the problem at hand. Let's see what happens in the following example. A patient drops off an order and says that she will pick it up 8 hours later. Immediately after her, another patient drops off an order and says she will be waiting at the pharmacy to pick it up when it is ready. Pharmacies typically commit to have an order ready within 15 minutes if the patient decides to wait at the pharmacy. When the second customer leaves the drop-off window, the technician puts her order in a special basket (usually red) and that order is given highest priority throughout the entire process, so that it is entered into the system, filled, and inspected before any other order already in process for a nonwaiting patient. But in the simulation model, the entity representing the waiting customer order will enter the process after the entity representing the order from the previous customer and will have to wait until the first order is processed unless we simulate the red basket.

One good and flexible way to do this is by using attributes. The entities in a simulation model can carry attributes, which are values specified for each one of the entities in the model. For example, some attributes for the entities in our model (which represent the prescriptions) could be "insurance company" or "customer ID." In order to fix the red-basket problem, we can create an attribute called *due_time*, which will record for each entity the expected pick-up time by the customer. The simulation model allows us to sort all the entities wanting to seize a common resource by their values stored in that attribute. In this way, orders with a closest *due_time* will seize the resource first, as happens with the orders traveling in a red basket. In fact, this also is what happens in well-organized pharmacies with the orders for nonwaiting customers, in which the technicians write down the expected pick-up time given by the customer and place the basket of each new order in the corresponding slot in the queue, so that the most urgent are filled first.

Finally, it will be necessary to create two shared queues in the simulation model, the first one among all the tech tasks and the second one among all the RPh tasks, so that the system of priorities we have just designed applies homogeneously all over the model. The explanation of what would happen without a shared queue is a little technical, so we will spare the reader. The use of two shared queues instead of many individual queues will impose some limitations if the researcher wants to build a realistic graphic animation of the model or if she wants to get some statistics about the individual queues. But still it is the best way to mimic the red-basket orders.

7.4.6 Choosing the Right Entities

At this point the reader will have started to appreciate the attention that must be paid to every little detail when building a simulation model. We will finish this section by stating a different type of modeling issue regarding the election of the right entities in the simulation model. The entities are the main actors in the model. They are created, assigned attributes to carry, seize resources, and travel through all the different task modules in the model until they are disposed. Figuring out what the entities represent is probably the first thing you need to do in modeling a system (Kelton et al. 2010). When building a simulation model of a community pharmacy, we have essentially three options regarding specification of the entities: treating the orders as the entities, choosing the individual prescriptions as the entities, or building a model with both types of entities (orders and prescriptions). Each one of these alternatives has advantages and drawbacks.

Choosing the orders as the entities in the pharmacy simulation makes sense since the orders are the real items that arrive, are produced, and leave the pharmacy. A customer dropping off one or more prescriptions represents an arriving order to the pharmacy. In most community pharmacies in the United States, the technician takes a basket when she receives the order, places the information about the whole order in the basket, and the order "travels" throughout the entire process in the basket, regardless of the number of prescriptions in the order. For example, when the order needs to be filled, the technician fills the vials for all the prescriptions that are included in the order. Or when the order needs to be inspected, the pharmacist takes the basket and inspects all the prescriptions in it. Finally, the customer picks up the entire order. An important drawback of having orders as the model entities has to do with long processing times. For example, it will typically take approximately 9 minutes to fill a five-prescription order. If some higher-priority task shows up while the technician is filling the order, for example, a customer arrives at one of the windows, the technician will interrupt the order-filling process to serve the customer. Due to technical reasons beyond the scope of this chapter, modeling these interruptions makes the model much more complex. However, running the simulation model without allowing interruptions would be a serious misrepresentation of the actual working procedures in the pharmacy and would prevent the researcher from collecting accurately important statistics such as the wait times of customers at the windows because their values would be nonsense (in the model the customer should have to wait 9 minutes until the technician finishes filling the five prescriptions of the order).

Using individual prescriptions as our main entities in the model will avoid the previous problem because the time required for the tasks will be much shorter. This approach has a few other advantages. Most common statistics are expressed in "per-prescription" units rather than in "per-order" units, like the time per prescription. Also researchers would be able to treat differently the prescriptions in a single order. For example, an order may consist of a new prescription and a refill. The data-entry process for a refill prescription is usually shorter and refill prescriptions usually have fewer problems with insurance companies or with the prescribing doctors because most of the potential problems were already fixed and approved when the patient brought that prescription to the pharmacy for the first time. Simulating these features of the individual prescriptions is much easier when the prescriptions and not the orders are the entities in the simulation model. Also, some orders cannot be completed because of lack of inventory or some other problem with one of their prescriptions. When this happens, the prescription with a problem is moved out of the original order and the remaining prescriptions continue being processed (this is called a partially filled order). This is also much more easily simulated when prescriptions are used as entities instead of orders. Using the prescriptions as the entities also has some disadvantages. Two of the main quality-of-service parameters are the wait time and the service time at the pharmacy windows. Working with prescriptions as entities makes the collection of statistics measured in "per-customer" units much more complicated.

The best solution is a mixed one, in which both orders and prescriptions are used as entities depending on the section of the model. This solution forces the researcher to make some complicated modifications in the model: each order entity will need to be broken into the prescription entities after the order has been received, and all the prescription entities will need to be packed together again into the order entity after they have been inspected, so that the finished order entity can be picked up by the customer. By doing this, the researcher will be able to take advantage of the pros of both approaches. Since some processing times are related to prescriptions and other ones are related to orders, such a mixed solution will always use the most natural processing times. Some readers will wisely argue that this mixed model could misrepresent reality, because in an actual community pharmacy the prescriptions belonging to the same order travel together in a basket and are processed together, while in the model each one of the prescriptions belonging to the same order will be represented by independent entities that will be processed separately. But a closer look at our model shows that all the entities in the same order are created together and have the same value in their *due_time* attribute. Since we are using this attribute to assign the priorities in all the queues in the model, all prescriptions in the same order will be processed one after the other. Since they are processed one at a time, the technician or the pharmacist will do something different between processing two prescriptions from the same order only if some urgent task shows up, such as a customer wanting to drop off an order. And this is precisely what happens in a real community pharmacy. Therefore, the way our model is built is a highly accurate mirror of the actual pharmacy setting.

7.5 Description of the Most Useful Statistics That Can Be Obtained from the Simulation Model

At this point we have built the simulation model, and we are ready to run it and produce the desired output statistics. We outline next some of the statistics that are found most useful by pharmacy managers.

7.5.1 Customer Wait Times

Customer wait times are definitely among the most relevant pieces of information that can be obtained from a simulation model. Wait times are one of the key customer-satisfaction drivers. Nobody likes to arrive at the pharmacy and find a long queue of customers. There are different customer wait times in a community pharmacy: at the drop-off window, at the pick-up window, and at the drive-through window.

It is important to be aware that the individual wait times of the customers at the queues at the pharmacy windows are not independent events. On the contrary, a high level of positive autocorrelation will exist. For example, if a customer arrives at a peak hour and needs to wait a long time, it is very likely that the wait times of the customers who arrived immediately before and after her will also be long. Due to this, it is necessary to make each scenario of the model simulate many days and run many independent replications in order to get an accurate (unbiased) and precise average wait time at each of the windows. Figure 7.2 shows a typical output from a computer simulation, with the minimum, average, and maximum values for the wait time being analyzed, and also the half width of the 95% confidence interval for the expected value. But this is not adequate to describe the wait times accurately at the pharmacy windows. For example, pharmacy managers might accept a scenario in which the average wait time is 1 minute. But if they knew that 80% of the customers did not have to wait at all and 20% of the customers needed to wait 5 minutes or more, then the scenario would very likely become unacceptable, because the managers know that many of the customers having to wait 5 minutes would probably go to a different pharmacy next time. Therefore, in order to provide comprehensive and

Waiting Time	Average	Half Width	Minimum Value	Maximum Value
Q01 Customer waiting at the drop-off window	0.5325	0.087144964	0.00	8.3400
Q02 Customer waiting for drop-off at the drive-through window	0.6072	0.116785404	0.00	8.5940

FIGURE 7.2
Typical statistical output from a simulation model.

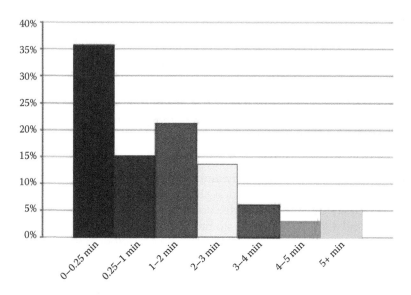

FIGURE 7.3

Histogram for the wait times at the pickup window. The first column was computed for the customers who waited less than 15 seconds. The second column represents the customers who waited between 15 seconds and 1 minute, and so on.

meaningful information to pharmacy managers, the simulation model will have to produce a histogram for the customer wait times at each of the pharmacy windows. Figure 7.3 shows one of these histograms. The x-axis represents the customer wait time and the y-axis is the percentage of customers in each wait-time category. It is recommended that the histogram be created in such a way that the first column represents the proportion of customers who had to wait only a few seconds until some staff started to assist them. This information is especially valuable for pharmacy managers.

7.5.2 Actual Time Duration per Section

The actual time required per section is the time it takes a prescription to move from the start point to the end point of a defined section of interest. It should not be confused with the total processing time. For example, if it takes only 1.5 minutes to do the data entry of a prescription, but the prescription needed to wait 50 minutes from the moment it entered the queue at the data-entry section until it was actually processed, then the processing time was 1.5 minutes and the actual time duration was 51.5 minutes. We need to define the start and end points of each section and prepare the simulation model to record time stamps for each prescription as it passes through those points. By analyzing those time stamps, the model will compute the actual time durations for all prescriptions at each section, which will help identify the slowest sections at the pharmacy, which could represent a bottleneck.

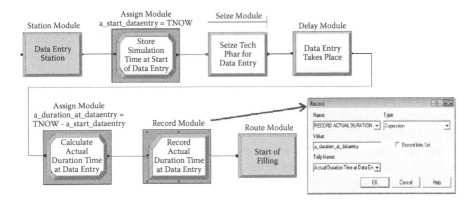

FIGURE 7.4
Design of actual time duration statistic.

The way to build the model so that the actual time durations are computed is shown in Figure 7.4. An Assign module is used at the start of each section (in this case, data entry) to store the current simulation time (TNOW in Arena) when the prescription entity enters the data-entry stage in an attribute called, say, *a_start_dataentry*. At the end of the section another Assign module calculates the actual duration time of the prescription in the data-entry stage and stores it in an attribute called *a_duration_at_dataentry*. It is important to note that all these values are stored in attributes so that they travel individually with each entity. If (global) variables were used instead, their values would be updated every time a new prescription enters or leaves the section, which is not what we want. Finally, a Record module produces an output statistic in the form of a tally variable from the value of the *a_duration_at_dataentry* attribute for each entity.

7.5.3 Time in Process and Time at Production

The time in process and the time at production are two aggregates of the actual duration times for several sections. The time in process measures the total length that the prescription stays at the pharmacy, from the moment at which the customer arrives at the queue of the drop-off window, until the customer finishes picking up the order. The time at production measures the time duration from the start of data entry (considered to happen when the customer leaves the drop-off window) until the moment the prescription is ready to be bagged and placed on the will-call shelves. The time at production is a valuable parameter to pharmacy managers, because it gives information about the actual time it takes to fill a prescription, including all the queues at the different sections of the entire process. Pharmacies target a time at production of approximately 15 minutes for orders brought by customers who decide to wait at the pharmacy until their orders are ready.

7.5.4 Proportion of Prescriptions Finished on Time

The proportion of prescriptions finished on time is also a key customer-satisfaction measure. A customer who goes to the pharmacy to pick up her order and learns that it is not ready yet will most likely be disappointed. We describe next a way to make the simulation model generate this statistic. If the customer says that she wants to come back 4 hours later, the prescription entities brought by that customer will all contain the value of 4 hours in an attribute called *target_prod_time* and the time at which the customer leaves the drop-off window in another attribute (*start_prod*). When a customer decides to wait at the pharmacy the value stored in the attribute *target_prod_time* will be 15 minutes. At that moment the production starts, the prescriptions are placed at the queue of data entry, and will later go through filling and inspection. When each prescription is finally ready to be bagged in the will-call area the production finishes and the total time at production is computed by measuring the time passed since the time or value stored in the attribute *start_prod* and compared to the value stored in the attribute *target_prod_time*, which in our example was 4 hours. That comparison lets us know if the prescription was finished on time.

7.5.5 Work in Process

Work in process (WIP) statistics are useful for measuring the workload in every step of the entire process in the pharmacy and to identify bottlenecks. To get these statistics for each section, we need to create tracking variables (counters) in the simulation model whose value will increase by one unit every time a prescription enters the section and will decrease by one unit every time a prescription leaves the section. Next, we need to create the statistics that will report the time-persistent average, minimum, and maximum value of these tracking variables. In addition to these numerical statistics, we should create plots of the tracking variables because they provide a visual snapshot of the evolution of the WIP statistics during the day. Figure 7.5 shows a plot (produced by Arena) of the tracking variable created for the data-entry section. The x-axis in the plot represents the first 2 days of the simulated run (12 hours = 720 minutes per day). The y-axis represents the number of prescriptions that were simultaneously at the data-entry section at each moment. The plot shows how the number of prescriptions at data entry is zero at the start of the first day, since the simulated pharmacy starts in an empty state when the simulation run begins, and how it evolves as time passes. We might want to ignore the first few days of each run, allowing the simulated system to reach a steady state (assuming our interest is in steady-state performance). This is because the first few days are a transition from the initial empty state to something more typical of long-run steady-state conditions. These initial few days are called a warm-up period. The model will start collecting output statistics only when the warm-up period ends. These

Work in Process
Prescriptions at Data Entry

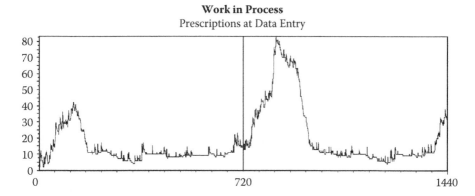

FIGURE 7.5
Plot of work in process for data entry.

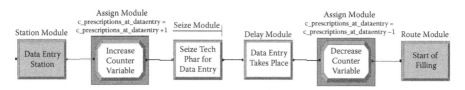

FIGURE 7.6
Design of WIP statistics. Updating of values to the WIP counter variables.

plots show the workload pattern during the different hours of the day and are valuable if pharmacy management wants to improve staff scheduling.

Figure 7.6 shows how to update the values of the WIP tracking variables for the same section (Data Entry). As soon as a prescription enters the section, an Assign module increases the value of the counter *c_prescriptions_at_ dataentry* by one unit. When the prescription goes through the entire section (Figure 7.6 shows only the most relevant Arena modules of the section) and leaves it, another Assign module will decrease the value of the counter by 1. This way, the tracking variable will always store the number of prescriptions present at the Data Entry at the current simulation time. The best way to extract useful information from this variable is defining an output in the Statistic data module as in Figure 7.7. The statistic will treat the counter as a time-persistent variable, so that the output statistic will report its minimum, maximum, and continuous-time average values throughout the entire simulation run.

A couple of aggregated WIP statistics are also of interest to the pharmacy managers. The *prescriptions in process* are all those that are at the pharmacy right now, including the ones carried by the customers waiting in line for the drop-off window, and the ones stored in bags on the will-call shelves just waiting until the customer comes to pick them up. The *prescriptions at production* are those that are at some point of the production process right now. The

Name	Type	Expression	Report Label	Output File
Number of Prescriptions at Data Entry	Time -Persistent	c_drugs_at_dataentry	Number of Prescriptions at Data Entry	

FIGURE 7.7
Design of WIP output statistic.

production process starts at data entry and finishes at the end of inspection. The prescriptions at production are a more meaningful statistic because they can be controlled by changing the staff scheduling. On the other hand, the prescriptions in process depend mainly on the number of orders and on the arrival time and pick-up time of customers, so better staff scheduling will have very little impact on this parameter.

7.5.6 Statistics by Prescription and by Order

All the previous output statistics, except the wait times, have been described in terms of individual prescriptions, but they can also be computed for each order. For example, in addition to the proportion of prescriptions finished on time, the simulation model can also calculate the proportion of orders finished on time. One order will be considered to be finished late if at least one of the prescriptions belonging to the order was finished late. The time at production of an order is the time at production of the prescription of that order that took the longest to be filled.

7.5.7 Statistics for Error-Free Prescriptions

Some prescriptions encounter some kind of error or problem that prevents them from being processed at the regular pace. We call them *error-found prescriptions*. The insurance company might reject the cost of the drug, or there might be an issue with the physician because the drug-utilization review produces a warning. Sometimes these prescriptions get stuck at the pharmacy waiting for an approval from the insurance company or from the doctor's office. Since the solution to these problems depends on the response time of third parties, the pharmacy staff can do little to prevent those prescriptions from being finished late. This delay does not really represent a flaw in the pharmacy process that the managers should try to correct, and hence it makes sense to make the simulation model calculate the output statistics without considering these error-found prescriptions. Otherwise, statistics such as the average time at production or the proportion of prescriptions ready on time will be negatively affected by the error-found prescriptions, which serve as extreme outliers, and the data will become nonmeaningful for pharmacy management. Therefore, it is recommended that the model calculate statistics only for error-free prescriptions and error-free orders. An order will be considered error-free when all its prescriptions are error-free.

7.5.8 Statistics for Special Types of Prescriptions or Orders

Some other statistics regarding specific types of prescriptions are of interest. For example, pharmacy managers might be interested in the statistics for the orders consisting of one prescription, or three prescriptions, or fewer than five prescriptions, and so on; or those for the orders that arrived between 10 a.m. and noon; or perhaps the statistics for the orders consisting of only refill prescriptions. All these statistics and many others can be obtained in the simulation model, so that their values can be seen for each different scenario. But there is one specific subset of the orders whose statistics are always appealing to the pharmacy managers: the orders brought by customers who wait at the pharmacy (we call them *waiting orders*). Many customers like to wait at the pharmacy, so they appreciate it when their orders are ready in a short time. Offering fast service for waiting customers is a great way to attract more customers to the pharmacy. In fact, some chain drugstores even compensate the customer if her order is not ready within 15 minutes with some kind of discount or gift card, and promote this commitment as part of its value proposition. It is hence important that the simulation model measures statistics related to the waiting orders, like the proportion of waiting orders finished on time (within 15 minutes), or the average time at production for waiting orders with different numbers of prescriptions, or the work-in-process statistics of waiting orders at production.

7.6 Advantages of Using Computer Simulation

We have already described the key control statistics in which pharmacy managers might have interest. But we have not expanded upon why we said in the Introduction of this chapter that computer simulation is a great tool for supporting managerial decision making.

Pharmacy managers have easier ways to estimate all the control statistics that we have mentioned before without using a complex computer-simulation model. The wait time at the pharmacy windows can be observed by direct observation at some sites for a number of days. Also, pharmacy records include time stamps at different moments of the process, including the start of data entry and the sale of the order to the picking-up customers. Therefore, the time at production and many of the actual time durations per section and the WIP statistics can be estimated directly from analyzing the pharmacy records. The pharmacy records also tag the prescriptions that are waiting for a problem resolution with the physician or with the insurance company, and create queues of prescriptions waiting for an outgoing or for an incoming phone call. Therefore, the statistics regarding only the error-free prescriptions can also be computed from the analysis of pharmacy data.

In short, by analyzing all the data included in the pharmacy records and for the on-site observations, we can learn what has happened in the pharmacy. For example, the researcher (and hence the pharmacy managers) can see that during a particular day a pharmacist worked 12 hours, and 3 technicians worked a total of 24 hours, 200 prescriptions arrived at the pharmacy and 180 were finished with an average time at production of 4.5 hours per prescription; and 86% of the orders were finished on time. As we said, all the important control statistics are observable without simulation. But the pharmacy managers may want to reduce the 14% of the orders that are finished late. How can they do it? Most likely by scheduling more staff. But increasing human resources is expensive, so this is something that needs to be done carefully. The pharmacy managers will wonder if the best thing to do is add 4 or 8 more hours of technician time, if she should start her shift in the morning or perhaps around noon, or if it is maybe better to have two pharmacists overlapping in the middle hours of the day instead of adding technician time. They will wonder if perhaps improvement could be achieved by just modifying the shifts of the currently scheduled technicians. In other words, the managers come up with multiple alternatives representing potential improvements. In order to pick the best one, they need to estimate the impact of each alternative on the proportion of orders finished late.

Without simulation, the managers will not be able to forecast which one of the possible alternatives will be better. This is because all the performance statistics, like the proportion of orders finished on time, are the result of a very complex interaction among many queues and scarce resources in the pharmacy. When we say "queues" we are referring not only to the queues of customers waiting at each window, but also to the queues formed by prescriptions at each one of the different processes that take place at the pharmacy. Because all the queues are interconnected and all the delays in the queues are transferred throughout the entire process, a small change in the resources can have a big impact on global performance. That is why sometimes an important improvement can be obtained by simply making a technician start her shift a couple of hours sooner or later. Computer simulation replicates all those queues, treats them with the appropriate system of priorities, mimics the random arrival patterns of customers to the pharmacy throughout the day, and mirrors the scarce resources to produce all the desired output statistics. Each one of the alternatives will turn into a different scenario of the model. By running these scenarios, we will learn how the different performance statistics are affected with each alternative.

Pharmacy managers might be tempted to try out different staff schedules at the real pharmacy and see which one is better. But this method has several important limitations. First, due to the random behavior of the customers, each schedule will need to be tried out for many days to obtain statistically meaningful measurements of the performance parameters. Second, if the schedule happens to be a bad one, the performance at the pharmacy, and therefore the quality of service, might deteriorate and the pharmacy might risk losing customers and, even worse, the number of errors in prescriptions could increase as a consequence of making the staff work under pressure. Also, the pharmacy managers may try to forecast what would happen at one site by analyzing data from another site in which the staff time is higher. Again, this is seldom a good idea, because there could be significant differences between the two sites being compared, regarding the arrival patterns of customers, the proportion of orders that are generated by the courtesy-refill system, the proportion of customers who wait at the pharmacy, the number of resources (such as automation, cash registers, or drive-through windows), or the layout of the pharmacy, and so on—in order to make meaningful comparisons between two such sites. But chain-drugstore managers know that it is really tough to find two sites with such level of coincidence. As a consequence, simulation remains the best way to measure the impact on the performance parameters if any of the potential alternatives is to be implemented.

In summary, the available data give pharmacy managers performance statistics that allow them to identify the ones that need to be improved. When the managers analyze all the potential corrective actions that could be made in the pharmacy, a lot of what-if questions arise. The managers wonder if the corrective actions will result in better performance. The answers to those questions require the use of computer simulation. Current performance statistics will be used to validate the simulation model. In order to do this, the input parameters (number of prescriptions, number of hours of staff, etc.) will be set as they actually happened on the observation days. The model is then run, and the output statistics from the model are compared to the actual performance statistics observed from pharmacy records. If the difference between them is small enough, the model is deemed validated, that is, the researcher can be comfortable that the model is giving an accurate representation of pharmacy activities. It then can be used to run different scenarios, to determine the impact on performance parameters, and find the answers to the what-if questions.

7.7 Practical Examples in Which Computer Simulation Can Help Pharmacy Managers Make Better Operational Decisions

We have identified three types of improvement projects that pharmacy managers are trying to implement with the help of computer simulation: (1) finding the appropriate configuration of material and human resources; (2) making modifications to the standard operating procedures at the pharmacy in order to improve performance; and (3) analyzing the convenience of implementing new systems and setups like centralized administrative or filling services, automated on-site filling machines, or a new policy for managing levels of inventory for drugs at the pharmacy. Next we discuss some of these topics to illustrate why they are important for pharmacy managers.

Appropriate staff scheduling is the key to success. The community pharmacy business in the United States has become highly competitive. In such an environment, good quality of service has become paramount for keeping the customer base. Pharmacy managers need to find a way to keep errors and customers' wait times low but at the same time keep labor costs as low as possible. In other words, they need to schedule the smallest staff possible at which the quality of service can still be guaranteed. This is not an easy task because, as we have previously described, pharmacy facilities have become highly complex systems. Computer simulation is a great tool to predict the minimum amount of labor time required to meet performance targets. Community pharmacies are mature businesses, so that all the big chains competing in the marketplace have very similar cost structures. However, use of computer simulation will give an advantage to a company over its competitors.

Modification in the utilization of the pharmacy windows is also a type of relevant analysis that can be performed with use of computer simulation. Many pharmacy managers wonder if they should allow only one or maybe two customers to be served at the drop-off window at the same time, or if the pharmacy should have a drive-through window, or if installing an additional cashier would significantly benefit pharmacy performance. All these changes will have a direct impact on the customers' wait times and an indirect impact on the way in which the staff spends its time, thus affecting the times at production. Computer simulation will allow the managers to predict the impact of these changes so they will be in a much better position to make an informed decision.

The workflow of activities performed in community pharmacies to fill an order is the result of an evolution that has taken place over several decades. Every little detail has been adjusted in order to reduce inefficiencies as much as possible and minimize the time at production for each prescription. Pharmacy managers continue to search for potential changes to their standard operating procedures that could yield additional savings. For example, they wonder if they should make changes to the layout of the pharmacy, or to the way in which the different drug bottles are stored in the shelves, or to the way in which the work is organized and distributed among the different technicians, and so on. Computer simulation can evaluate such changes so a good decision can be made.

Automation is implemented in pharmacies to reduce staff costs, improve efficiency, and reduce the number of dispensing errors. The most typical automation systems used in community pharmacies are robots that take the tablets/capsules from a large container, count them, put them into the vial, print the label, and attach it to the vial. In other words, they take care of a major part of the filling process. Computer simulation can help the pharmacy manager analyze the impact of implementing an automation system in the pharmacy and estimate its profitability. The simulation model can predict how much the performance parameters will improve after automation is installed and set up if the staff levels remain unchanged. It can also show how much the staff can be reduced after installing the automation while maintaining the main quality-of-service parameters. Almost all automation systems will improve the pharmacy efficiency thus producing staff savings. But implementation of automation will be recommended only if its total cost (including acquisition and maintenance) is lower than the total costs it saves (including labor and cost of additional errors). All of these costs can be entered into the simulation model, so that we can see if the installation of the automation system is worthwhile. It might also be of interest to determine workload that is necessary in a pharmacy to justify the automation system.

Many chain drugstores utilize central filling facilities to serve to the pharmacies in the area. The service is mainly of two types: central processing of the prescriptions (especially for those that require negotiation with the insurance company or with the physician), and central filling and inspection of prescriptions. In both cases, the advantage of using central filling services is the gain in efficiency due to economies of scale. For example, technicians at the central filling facilities will be experts in dealing with insurance companies and utilize each phone call to address several issues occurring in different pharmacies. Central filling facilities can afford to utilize large automation systems that fill prescriptions in a very efficient way. Using these central filling facilities saves work for the community pharmacies, so that they will be able to get by with less staff. On the other hand, new costs and inefficiencies will result from use of the central facility. Computer simulation can help managers determine when a central filling facility will make financial sense, to what extent it will speed up the process at the local pharmacies, and how it should be configured. Some chain drugstores use a different system: the technicians at a pharmacy can see the queues of prescriptions waiting to be processed at other pharmacies, and if they have time available, they can take control of them and process them themselves. That way, they are using the periods of the day in which they have low workload to help their colleagues at a different pharmacy. Again, the impact and convenience of this collaborative procedure can be analyzed using computer simulation.

It was mentioned in the Introduction that other types of health care settings have performance issues that are similar to those at community pharmacies. Therefore, their managers could take advantage of simulation modeling to improve their staff scheduling, their facility layout, their standard operating procedures, and so forth. In fact, there is extensive literature on use of computer simulation in health care settings to improve levels of staffing (Spry and Lawley 2005), to improve performance in a local clinic (Weng and Houshmand 1999), to improve staffing schedules of physicians in an emergency department (Rossetti et al. 1999), to improve the availability of beds in an intensive care unit (Cahill and Render 1999), to improve the process in a cancer treatment center (Sepúlveda et al. 1999), to improve appointment scheduling (Wijewickrama and Takakuwa 2005) to reduce patient waiting time (Wijewickrama and Takakuwa 2006) in an outpatient department of internal medicine, and to improve the inventory control of perishable drugs in hospitals (Vila-Parrish et al. 2008).

7.8 Conclusions

Community pharmacies have very complex operational aspects. Operations-research professionals have previously applied analytical techniques to improve processes at pharmacies. However, these techniques ignore in great part the multiple interactions among subsystems, which is a common feature in pharmacies. The community pharmacy business has become highly competitive, with a number of big chains competing aggressively. Customers can easily change pharmacies, so providing excellent service and innovative services is essential to maintain the customer base and attract new customers. Two driving forces for change that are currently more often pursued by pharmacy managers, cost reduction and risk reduction, can be difficult to reconcile. Any reduction in the level-of-service provision is potentially dangerous; patients can change to another pharmacy if drugs are not available when needed, and even worse, patients are put at risk if errors are made in drug delivery. Staff-reducing strategies can result in a great deal of disruption. It is therefore much safer and less disruptive to investigate the impact of potential changes prior to their introduction by using simulation modeling. In addition, it can help pharmacy managers make better operational decisions, implement better innovative services, and obtain a competitive advantage.

Computer simulation is an excellent and flexible tool to model different types of settings and environments. The more complex the setting being simulated, the less likely it is that a decision can be validly based on traditional data analysis, and the more advantageous it is to build a computer simulation. They provide insight into the working of a system and can be used to predict the outcome of a change. This is particularly useful when the system is very complex or when experimentation is not possible or is associated with high costs due to the disruption of the practice setting in which the system is implemented.

References

Buchanan, E. Clyde. January/March 2003. Computer Simulation as a Basis for Pharmacy Reengineering. *Nursing Administration Quarterly,* 27(1), 33–40.

Cahill, William, and Marta Render. 1999. Dynamic Simulation Modeling of ICU Bed Availability. *Proceedings of the 1999 Winter Simulation Conference,* P. A. Farrington, H. B. Nembhard, D. T. Sturrock, and G. W. Evans, eds., pp. 1573–1576.

Finkler, Steven A., James R. Knickman, Gerry Hendrickson, Mack Lipkin, Jr., and Warren G. Thompson. 1993. A Comparison of Work-Sampling and Time-and-Motion Techniques for Studies in Health Services Research. *Health Services Research*, 28(5), 577–597.

Kelton, W. David, Randall P. Sadowski, and Nancy B. Swets. 2010. *Simulation with Arena*, 5th ed. McGraw-Hill.

Lin, A. C., R. Jang, D. Sedani, et al. 1996. Re-Engineering of Pharmacy Work System and Layout to Facilitate Patient Counseling. *American Journal of Health-System Pharmacy*, 53(13), 1558–1564.

Rossetti, M. D., Gregory F. Trzcinski, and Scott A. Syverud. 1999. Emergency Department Simulation and Determination of Optimal Attending Physician Staffing Schedules. *Proceedings of the 1999 Winter Simulation Conference*, P. A. Farrington, H. B. Nembhard, D. T. Sturrock, and G. W. Evans, eds., pp. 1532–1540.

Sepúlveda, José A., William J. Thompson, Felipe F. Baesler, María I. Alvarez, and Lonnie E. Cahoon, III. 1999. The Use of Simulation for Process Improvement in a Cancer Treatment Center. *Proceedings of the 1999 Winter Simulation Conference*, P. A. Farrington, H. B. Nembhard, D. T. Sturrock, and G. W. Evans, eds., pp. 1541–1548.

Spry, Charles W., and Mark A. Lawley. 2005. Evaluating Hospital Pharmacy Staffing and Work Scheduling Using Simulation. *Proceedings of the 2005 Winter Simulation Conference*. M. E. Kuhl, N. M. Steiger, F. B. Armstrong, and J. A. Joines, eds., pp. 2256–2263.

Tippett, L. H. C. 1935. Statistical Methods in Textile Research; Part 3: A Snap-Reading Method of Making Time Studies of Machines and Operatives in Factory Surveys. *The Journal for the Textile Institute Transactions*, 26, 51–55.

Vemuri, S. 1984. Simulated Analysis of Patient Waiting Time in an Outpatient Pharmacy. *American Journal of Hospital Pharmacy*, 41(6), 1127–1130.

Vila-Parrish, Ana R., Julie Simmons Ivy, and Russell E. King. 2008. A Simulation-Based Approach for Inventory Modeling of Perishable Pharmaceuticals. *Proceedings of the 2008 Winter Simulation Conference*. S. J. Mason, R. R. Hill, L. Mönch, O. Rose, T. Jefferson, J. W. Fowler eds., pp. 1532–1538.

Weng, Mark L., and Ali A. Houshmand. 1999. Healthcare Simulation: A Case Study at a Local Clinic. *Proceedings of the 1999 Winter Simulation Conference*. P. A. Farrington, H. B. Nembhard, D. T. Sturrock, and G. W. Evans, eds., pp. 1577–1584.

Wijewickrama, Athula, and Soemon Takakuwa. 2005. Simulation Analysis of Appointment Scheduling in an Outpatient Department of Internal Medicine. *Proceedings of the 2005 Winter Simulation Conference*. M. E. Kuhl, N. M. Steiger, F. B. Armstrong, and J. A. Joines, eds., pp. 2264–2273.

Wijewickrama, A. K. Athula, and Soemon Takakuwa. 2006. Simulation Analysis of an Outpatient Department of Internal Medicine in a University Hospital. *Proceedings of the 2006 Winter Simulation Conference*. L. F. Perrone, F. P. Wieland, J. Liu, B. G. Lawson, D. M. Nicol, and R. M. Fujimoto, eds., pp. 425–432.

8

Patient Prioritization in Emergency Departments: Decision Making under Uncertainty

Omar M. Ashour and Gül E. Okudan Kremer

CONTENTS

8.1 Introduction

At the turn of the millennium, the World Health Organization ranked the U.S. health system's performance at 37 out of 191 countries (WHO 2000). Six years later, the United States was number one in terms of health care spending per capita, but it ranked 43rd for adult female mortality and 42nd for adult male mortality (Doe 2009). In general, the performance of health care service can be assessed by overall responsiveness, measured in terms of patient waiting time and the quality of service. The longer the waiting time for medical intervention, the poorer the service is—a factor that can mean life or death in an emergency department (ED). EDs are considered to be vital components of the nation's health care safety net (Richardson and Hwang 2001a, 2001b; Weinick and Burstin 2001), and are responsible for 45% to 65% of all hospital admissions (Mahapatra et al. 2003). Most EDs in major areas are overcrowded; there were 119.2 million ED visits in 2006 (Pitts et al. 2008). Accordingly, the potential for performance problems in EDs is critical.

For patients, every minute spent waiting can make a big difference. Suppose, for example, that two patients—one 25 years old and the other 78 years old—arrive at an ED simultaneously and the triage nurse gives them the same priority level based on their status. Since a younger patient may have a higher chance of survival, the nurse might first send the older patient for further treatment and leave the younger one to wait. Or suppose that the nurse follows the first come, first serve (FCFS) practice, and suppose that the younger patient comes in 5 minutes before the older patient; the nurse would send the younger patient for further treatment before the older patient, who in fact might need treatment more urgently than the younger one. Even in this simple case comparison, it is easy to see how a decision pertaining to a patient might also impact another one; hence, we can recognize the interdependence of ED decisions.

Triage is a prehospital practice defined as "the preliminary clinical assessment process that sorts patients prior to full ED diagnosis and treatment ... [where] patients with the highest acuity are treated first" (Wuerz et al. 2000). Although it is a commonly used process that has an important effect on patient care, the elements of triage are not always well known (Cooper 2004). Many U.S. hospitals use the five-level Emergency Severity Index (ESI) to sort patients into distinct groups according to their anticipated clinical resources and operational needs. The ESI designates the most acutely ill patients as levels 1 or 2 (the highest levels) and uses the quantity of resources a patient needs to determine levels 3, 4, and 5 (Tanabe et al. 2004a). Level 1 and 2 patients usually require an immediate evaluation and treatment, whereas patients at levels 3 to 5 can be sent to the waiting area after going through the registration process (Gilboy et al. 2005).

In an ED setting, the triage nurse assigns an ESI level and then decides which patient will be treated first. Due to the complex nature of the triage process, neither the skills nor the contextual factors that are needed to make accurate ED triage decisions are known to this date (Göransson et al. 2008). Studies have shown that nurse decision making is mostly based on experience (Patel et al. 2008), knowledge, and intuition (Benner and Tanner 1987; Cone and Murray 2002; Gurney 2004; Andersson et al. 2006). However, Göransson et al. (2008) presented findings that reveal that nurse decision making during ED triage can be varied; some studies showed that there is a difference in decision making between expert and beginner-level nurses, whereas others found that both less experienced and more experienced triage nurse decision making was largely the same. Göransson et al. (2006) also investigated the quality of triage nurse decisions. Using the Canadian Triage and Acuity Scale (CTAS) in Swedish EDs, they studied the relationship between the personal characteristics (e.g., triage experience of the registered nurses [RNs]) and the accuracy in their acuity rating of patient scenarios. They concluded there is no apparent relationship between the personal characteristics of RNs and their ability to triage, and they proposed that relevant decision making might be affected by intrapersonal characteristics (e.g., RNs' decision making strategies). Göransson et al. (2006, 2008) recommended more research be conducted on RN decision making during the triage process to identify the essential characteristics of the triage nurse during the process.

In general, patient prioritization is a decision-making problem. Decisions in EDs often involve significant uncertainty with regard to the type of patient illness or injury (Claudio and Okudan 2010). Similar symptoms for different illness types and uncertainty due to subjective variables, as well as individual differences such as ESI and pain level variables, tend to complicate triage decision making. The uncertainty in the decision-making process can also come from inadequate understanding, incomplete information, and undifferentiated alternatives (Krishnamurthy 2005). Inconsistency in decisions made among nurses themselves can also complicate the matter. For example, Fields et al. (2009) investigated the discrepancies in decisions made across nurses in three clinical settings using Spearman's rank correlation comparison method. The results showed differences in patient rankings among nurses at different hospitals, and even among nurses working at the same hospital.

Given these conditions, the ESI algorithm currently in use at hospitals is lacking in many ways. First, it does not consider the prioritization of patients who are assigned to wait (i.e., levels 3, 4, and 5); rather, it assumes an FCFS routine. Neither does it take into account any descriptive variables (i.e., age and gender) nor the pain level. Beyond these issues, our interviews revealed that while we could ascertain that the relative importance of the vital signs changes depending on the primary patient complaint, the current ESI algorithm also fails to account for this shift. These problems are addressed in this chapter through a structured process that takes into account uncertainty consideration. We first use the multiattribute utility theory (MAUT) and then address the use of the fuzzy analytic hierarchy process (FAHP) and MAUT in tandem as decision-making tools in a complementary fashion.

This chapter is organized as follows. In Section 8.2, we introduce the triage process along with a review of the relevant literature pointing to its limitations. In Sections 8.3 and 8.4, we apply rigorous decision-making tools (MAUT and FAHP) to the process of prioritizing patients in EDs. Section 8.3 presents the MAUT-based approach to extend the patient sorting and ranking to include pain level, age, and gender as additional rank differentiators. Using the FAHP and MAUT-based triage approach, Section 8.4 considers the shifts in relative importance of the vital signs with respect to different chief complaints, along with other factors (age, gender, and pain level). Both approaches are applied to an actual data set from Susquehanna Health's Williamsport Hospital in Williamsport, Pennsylvania. Finally, Section 8.5 discusses and summarizes the chapter findings.

8.2 The Science of Triage and Relevant Literature

Many patients who come to an ED are nonurgent patients (Buesching et al. 1985; Patel et al. 2008). Therefore, to use resources appropriately, patients undergo a triage interview upon their arrival. Triage has been in place, formally or informally, since the first ED opened (Mahapatra et al. 2003). In the United States, triage was used as far back as the 1960s (Andersson et al. 2006). The purpose of the triage interview is to assign the patient to one of several queues, each having an associated maximum time until the patients in them see a physician (Guterman et al. 1993). The interview aims to sort the incoming patients by severity of illness or injury (Andersson et al. 2006; Beveridge 1998).

Prioritization of patients in EDs based on their health situation (status) and the likelihood of survival establishes the triage process as a dynamic decision-making problem (Claudio and Okudan 2010). During it, the quality and timeliness of information collected in the first interview with the patient must be maximized (Gifford et al. 1980). Thus, the qualifications and personal qualities of the triage nurses are vital to an effective triage (Andersson et al. 2006).

One of the most commonly used triage systems in the United States—the five-level ESI—sorts ED patients into five clinically distinct groups, as shown in Figure 8.1. These levels are different with respect to resource and operational needs (Zimmermann 2001; Gilboy et al. 2005). The most acutely ill patients are placed at ESI levels 1 or 2, the highest acuity levels. With less demanding conditions, ESI levels 3, 4, and 5 are assigned based on the number of needed resources (Tanabe et al. 2004a). For example, a patient at level 2 might be taken immediately to the treatment area, while patients at level 4 might be assigned to wait (Wuerz 2001; Wuerz et al. 2001; Eitel et al. 2003; Gilboy et al. 2005; Tanabe et al. 2004a; Tanabe et al. 2004b). During the triage

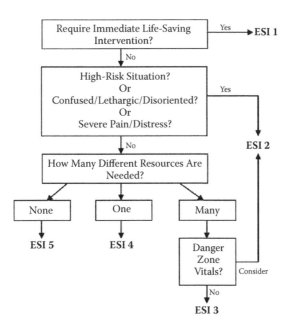

FIGURE 8.1
ESI triage algorithm, version 4. (Modified from Gilboy, N., P. Tanabe, D. A. Travers, A. M. Rosenau, and D. R. Eitel, 2005, *Emergency Severity Index, Version 4: Implementation Handbook*, AHRQ Publication No. 05-0046-2, Rockville, MD: Agency for Healthcare Research and Quality.)

process, the nurse records the patient's vital signs and requests information about current illness, past medical history, and other relevant information such as allergies and immunization status. Then the nurse decides whether the patient requires immediate evaluation and treatment or can wait (Claudio and Okudan 2010).

Numerous attempts have been made to improve ED care services: minimizing waiting time intervals (Spaite et al. 2002; Takakuwa and Shiozaki 2004) and improving patient satisfaction (Spaite et al. 2002), developing a reliable decision-support system in order to improve patient waiting times and service quality problems (Mahapatra et al. 2003), developing expert systems to help triage nurses assign patient categories (Pedro et al. 2004; Padmanabhan et al. 2006), and others. The challenge for triage nurses is to prioritize and rank nonurgent patients (Andersson et al. 2006) in order to recognize who is most in need of care. Some U.S. hospitals use a three-level triage that sorts the patients based on the question: "How long can this patient wait to be seen?" (Mahapatra et al. 2003). The five-level ESI is based not only on "Who should be seen first?" but also on "What will this patient need?" (Tanabe et al. 2004a). However, despite the fact that this system sorts patients and prioritizes them based on the severity of their illness or injury, the patient waiting period or the order of treating them, especially for the patients with the same acuity level, is rarely investigated (Claudio and Okudan 2010). Tanabe et al.

(2005) stated that physicians and nurses face a serious limitation when using ESI—that is, they cannot differentiate how acutely ill level 2 patients are when confronted with several level 2 patients in the waiting room. Moreover, in clinical studies, two distinct groups of ESI level 2 patients have been identified: those who can safely wait for physician evaluation for at least 10 minutes without clinical deterioration, and those who cannot wait (Tanabe et al. 2004a). Given that less than 3% of the ED patients in clinical studies are typically classified as ESI level 1 (Wuerz et al. 2000; Wuerz et al. 2001; Tanabe et al. 2004a), it is important to improve the capability of the triage system to make differentiations among those waiting patients with the same ESI classification since they comprise the largest proportion of patients.

Discussion thus far points to the complexity of patient prioritization, particularly within the same ESI level. Prioritization for the "time to be seen" is very critical for a patient, and is directly related to the patient's current and future health risk, especially if triage evaluation is delayed due to ED crowding (Cooper 2004). Recent research investigated the use of the utility theory to prioritize patients having the same acuity level in an ED (Claudio and Okudan 2010). In this study, the authors applied utility theory in patient prioritization to a hypothetical example. They explained the choice of utility theory to compensate for the inherent uncertainty in ED settings and that the utility theory can account for uncertainty.

The following section presents a solution to the patient prioritization problem in ED settings using MAUT by including patient age, gender, and pain level, along with the assigned ESI. For example, while vital signs (temperature, pulse, respiration rate, and blood pressure) were considered for patient ranking in the Claudio and Okudan (2010) study, data on patient age, gender, and pain level were neglected. Further, these variables are not considered explicitly in widely used ESI algorithms either. Accordingly, our FAHP–MAUT approach extends the work in the previous section to include vital signs and their relative importance to change with respect to a patient's chief complaint. As noted by Claudio and Okudan (2010), the main purpose of using decision making tools in the ED setting is not to mimic the choice of human decision makers (i.e., physician, nurse, and so forth) but to help them make fair decisions in a short amount of time; and so is our aim here.

8.3 Utility Function-Based Patient Prioritization Using Emergency Severity Index and Descriptive Variables

This section presents the application of MAUT as a way to prioritize patients in ED settings. First, the problem is stated and then the solution approach is introduced.

8.3.1 Triage and the Subsequent Patient Ranking Problem

It is important to clarify the problem before we describe the approaches developed to solve it. As discussed earlier, the triage nurse in an ED takes on patients with different illnesses and injuries, and based on the assessment of several factors (i.e., vital signs, complaints, and pain level, etc.) assigns an ESI level. Then, the nurse must decide which patient will be treated first. The widely used five-level ESI algorithm takes into account most vital signs in an assessment of the acuity level (e.g., respiration rate, oxygen saturation, and blood pressure, etc.). Approximately 3% of the patients are assigned to the highest acuity level and thus receive immediate service; the rest of the patients must wait. Among the waiting patients, those with lower ESI levels but showing a higher acuity level will precede the others in receiving care. However, among patients waiting with the same ESI level, there are no clear differentiators to establish a clear prioritization. Clinical observations attest to the difficulty of this situation (for example, Tanabe et al. 2005). Due to the dynamic and uncertain nature of the overall triage process in addition to the differentiation difficulty, methods are needed that can help the triage nurse to be efficient (without increasing potential bias) in prioritizing among patients with the same acuity classification.

8.3.2 Proposed Approach and Rationale

The goal of the proposed approach is to help triage nurses make decisions more efficiently and more easily, taking into account their intuitive judgment and preferences but still minimizing potential bias. Due to the inherent uncertainty in this decision problem, the utility theory has been chosen to help nurses with patient prioritization. The proposed approach aggregates patient age, gender, and pain level along with the assigned ESI to create a clear ranking among waiting patients. Our attributes are not mutually and preferentially independent; thus, the multiplicative form of the aggregate utility function is used instead of using the simpler-to-implement weighted average method (Krishnamurthy 2005).

The problem of patient prioritization is a decision-making problem with inherent uncertainty. As noted earlier, uncertainty can originate from inadequate understanding, incomplete information, and undifferentiated alternatives (Krishnamurthy 2005). MAUT takes into consideration the uncertainty and the trade-offs at various attribute levels (Keeney and Raiffa 1993). It is used to make choices from among a number of alternatives on the basis of two or more attributes. The choice of the decision maker (DM) is based on maximizing the utility function, which depends on specified attributes or variables (Keeney and Raiffa 1993).

DMs can have different attitudes toward risk; they can be risk averse, risk prone, or risk neutral. A single utility function (SUF) is built for each attribute, taking into account the risk attitude of the DM for a specific attribute, while capturing the full range of uncertainty; this makes it different from other methods such as minimax, maximin, and minimax regret (Thurston 2005).

Von Neumann-Morgenstern axioms provide the basis for the utility theory preference structures. The principal idea is to determine the indifference point between a standard lottery and the certainty equivalent (CE) of an event by changing certainty or probability over the lottery factor. The CE of a lottery is the certain payoff, less than the best payoff, at which the DM is willing to give up the probabilistic best and the worst payoffs (Keeney and Raiffa 1993; Krishnamurthy 2005). Within the patient prioritization problem, the DM is the triage nurse. The nurse would be risk averse with respect to one attribute if he or she is willing to accept a lower consequence value (CE) than the expected consequence and, conversely, would be considered risk prone if he or she expects a higher CE value than the expected consequence (Claudio and Okudan 2010).

The utility function has been described with different analytical forms; the exponential function form provides a constant risk premium in all conditions, and it works well in the whole attribute range (Krikwood 1997), and hence is used in the proposed approach. The additive and the multiplicative forms are available for aggregating SUFs. The multiplicative form is used when the mutually and preferentially independence condition is not satisfied (Krishnamurthy 2005). This section applies the utility theory approach in order to build a multiplicative utility function as a tool to aid the triage nurse's prioritization process.

Determining the ESI is an essential part of the triage process and hence is included in our formulation. Rationale for the selection of the remaining three attributes is provided next.

8.3.2.1 Patient Age

The age range for patients in this study is 18 to 90 years. Triage nurses typically give higher priority to elderly patients in comparison to younger patients. Confirming the appropriateness of this, Singal et al. (1992) pointed out that elderly patients have a high level of acuity and use more resources than younger adults. In another study, Ettinger et al. (1987) compared the elderly patients with the nonelderly and concluded that the elderly are more likely than the nonelderly to have an emergent diagnosis, to arrive by ambulance, to be admitted to the hospital, and to have a medical illness. They also have a higher proportion of cardiac and pulmonary disease. Thus, it is assumed that younger patients can wait longer than elderly patients. Suppose a triage nurse has received two patients; one is 78 and the other is 25, and both are males with the same pain level and ESI level. The nurse would give the older patient a higher priority than the younger patient. Thus, an older patient has a higher utility value than a younger patient. The SUF for this attribute is nonlinear over its range. Current triage practice does not take into account age while prioritizing patients.

8.3.2.2 Gender

It is assumed that females have a higher utility value than males. The utility function is assumed to be linear over the range of gender. Thus, we have two values: for males, the utility function will return a value of 0, and it will return a 1 for females. Prior clinical work has shown that even though triage algorithms do not explicitly take into account patient's gender, variability in triage decision making may be partially explained by patient gender (O'Cathain et al. 2004), attesting to the fact that triage nurses do take it into account.

8.3.2.3 Pain Level

Pain level is a subjective attribute; patients assess their feelings using a 0 to 10 scale. The SUF is nonlinear over the range of pain level; a higher pain level receives a higher utility value. This attribute is considered to be a source of uncertainty as it relies on a patient's judgment. For example, two patients may present with the same injuries or illness but one might describe his pain level as a 5 and the other might select level 9. Pain level is a primary factor in triage assessment, and hence it is included here.

In the next section, we demonstrate the application of the utility-theory-based patient prioritization as proposed.

8.3.3 Application

The data set was collected from the ED of Susquehanna Health's Williamsport Hospital. The hospital's triage policy is as follows. First, a nurse assesses a patient's vital signs and asks relevant questions about the patient's illness or injury to determine the ESL level. The nurse may also determine the need for x-rays prior to seeing a care provider in order to eliminate a delay. If a patient's illness or injury is life threatening, the patient is taken to an exam room and treated immediately. Otherwise, the patient is seated in the waiting area until registered (Susquehanna Health 2010).

Table 8.1 shows a portion of the data set. In it, each patient record presents the following: the assigned ESI level, age, gender, and pain level.

The exponential distribution is used to generate the utility functions. Equations (8.1) through (8.4) that are necessary to formulate the SUFs are provided next. In the formulas, A and B are factors that scale the utility values ($U_i(x_i)$) between 0 and 1 (Krikwood 1997; Thevenot et al. 2007).

$$U_i\left(x_i\right) = A - B.\exp(-x_i / RT) \tag{8.1}$$

$$A = \left(\exp\left(-\min\left(x_i\right)/RT\right)\right)/\left[\left(\exp\left(-\min\left(x_i\right)/RT\right)\right)-\left(\exp\left(-\max\left(x_i\right)/RT\right)\right)\right] \tag{8.2}$$

TABLE 8.1

Patient Data

Patient	ESI level	Age	Gender*	Pain Level
1	2	53	M	10
2	3	18	F	9
3	3	48	M	4
4	3	76	F	3
5	3	20	M	7
6	3	57	M	7
7	3	75	M	8
8	3	49	F	9
9	2	54	F	20
10	2	45	M	5
11	2	55	F	9
12	3	47	F	10
13	3	22	M	0
14	3	58	F	0
15	3	37	F	8
16	2	48	F	0
17	2	48	F	10
18	2	62	F	0
19	3	19	F	2
20	3	57	M	2
21	2	73	F	0

* M, male; F, female.

$$B = 1/\left[\left(\exp\left(-Min(x_i)/RT\right)\right)-\left(\exp\left(-Max(x_i)/RT\right)\right)\right] \qquad (8.3)$$

$$RT_i = -CE_i / \ln\left(\left(-0.5U_i\left(Max(x_i)\right)-0.5U_i\left(Min(x_i)\right)+A\right)/B\right) \qquad (8.4)$$

where RT is risk tolerance; $Min(x_i)$ is the minimum value for attribute i across all alternatives; $Max(x_i)$ is the maximum value for attribute i across all alternatives; and CE is the certainty equivalent.

As part of our solution, the triage nurse is identified as the DM (and for any clinical input, several nurses were contacted). Not all our included attributes are mutually and preferentially independent; that is, the ESI level is not independent from the others, and hence a multiplicative aggregation form is adopted. The sign of risk tolerance (RT) depends on the risk behavior of the DM regarding a specific attribute. RT is negative for risk-prone behavior and positive for risk-averse behavior. The value of CE is calculated using a lottery with a 50% probability to derive the best or the worst alternatives. The DM was asked a

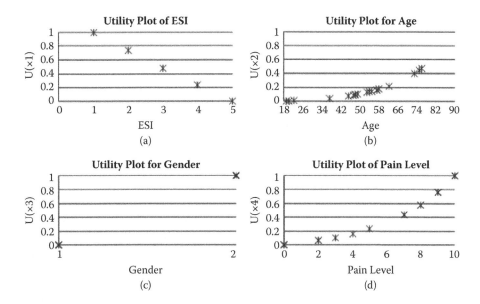

FIGURE 8.2
Utility plots for (a) ESI level, (b) age, (c) gender, and (d) pain level.

series of lottery questions to discern the utility functions. These functions are related to a specific DM, and they may be updated for different nurses, as has been done in other studies (for example, Claudio and Okudan 2010).

The following sections illustrate the procedure of constructing the utility functions for all attributes.

8.3.3.1 Emergency Severity Index (ESI)

The risk attitude is identified to be risk neutral for the ESI attribute. In other words, a patient with a lower criticality status receives a high ESI level and thus a lower utility value. Because risk attitude is neutral, the utility function would be linear, as seen in Figure 8.2a. The utility function for an ESI level is given in the following equation:

$$U_1(x_1) = -0.25x_1 + 1.25 \tag{8.5}$$

8.3.3.2 Patient Age

The risk attitude is identified to be risk prone for the age attribute. This means that the DM would give higher utility values to the patients between the ages of 77 to 90 than to those aged 18 to 77 years. Consequently, the risk tolerance is negative.

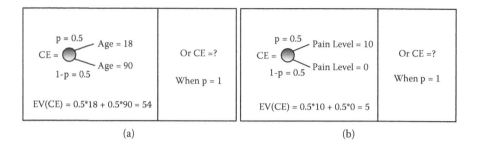

FIGURE 8.3
Lotteries built to discern the certainty equivalent of (a) age and (b) pain level.

Figure 8.3a shows the lottery question presented to the triage nurse. The nurse determined that the CE would equal 77, which is greater than the expected value of *CE* (54) and less than the maximum consequence value (90). Accordingly, after solving Equations (8.2) through (8.4), the value of *RT* is found to be –19.437, yielding the utility function for age as in Equation (8.6). Figure 8.2b illustrates the age utility function graph for the data in Table 8.1.

$$U_2(x_2) = -0.0252 + 0.0100 \ {}^*exp(x_2/19.437) \tag{8.6}$$

8.3.3.3 Gender

The risk attitude is identified to be risk neutral toward gender. A female patient has a higher utility value than a male patient. To build the utility function, a utility value of 0 is assigned for male and 1 for female. The utility function would be linear (see Figure 8.2c). The utility function equation for gender is given in the following equation:

$$U_3(x_3) = x_3 - 1 \tag{8.7}$$

8.3.3.4 Pain Level

Similar to the age case conditions, the DM has a risk-prone attitude toward the pain level and would give higher utility values if the pain level increases from a certain level to its maximum value than if it increases from zero to that certain value. Therefore, the RT has a negative value.

Figure 8.3b shows the lottery question presented to the nurse. The nurse identified the CE for the pain level to be between 7 and 8, which is greater than the expected value (5) and less than the maximum consequence value (10). After solving Equations (8.2) through (8.4), the RT is determined to be –4.102. The utility function for pain level is given in Equation (8.8). Figure 8.2d presents the utility function graph for the pain level.

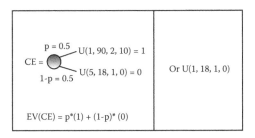

FIGURE 8.4
ESI level scaling factor lottery.

$$U_4(x_4) = -0.0957 + 0.0957 * exp(x_4 / 4.102) \qquad (8.8)$$

8.3.3.5 Attribute Trade-Offs

In this section, the attribute-scaling parameters are calculated as described by Keeney and Raiffa (1993). The first step is to rank the attributes in order of importance. The DM provided the following ranking in decreasing importance: ESI level (scaling parameter k_1), pain level (k_4), age (k_2), and gender (k_3). In order to determine k_i values, we start with the most important attribute (k_1), the ESI. Using the lottery questions, the indifference point of two choices is found as shown in Figure 8.4.

The DM should find the value of probability, p, such that the DM is indifferent between the two choices. The attributes in Figure 8.4 are arranged in the same order as in Table 8.1. The DM indicated that the value $p = 0.80$ establishes a condition of indifference between the two choices. Therefore, the value of $k_1 = 0.80$:

$$p*(1) + (1 - p)*(0) = k_1*(1) + k_2*(0) + k_3*(0) + k_4*(0) => k_1 = p = 0.80 \qquad (8.9)$$

Next, decision problems are constructed to find the other scaling factors. The first decision problem compares ESI level and age:

$$U \text{ (ESI level, 18)} \sim U \text{ (5, Age)} \qquad (8.10)$$

Equation (8.10) shows two choices. The first choice includes an unknown value for the ESI level and the lowest value of age. The second choice includes the maximum value for the ESI level and an unknown value of age. After a comparison, the DM needs to make these two choices indifferent by choosing the unknown values of ESI level and age. In this situation, the DM would be indifferent if the ESI level is 3 and the age is 85 years. Equation (8.6) and Equation (8.8) are used to calculate the utility values of ESI level and age, respectively, in order to build the relationship between k_1 and k_2 as shown next:

$$U \text{ (3, 18)} \sim U \text{ (5, 85)} \qquad (8.11)$$

$$k_1^*U_1(3) + k_2^*U_2(18) = k_1^*U_1(5) + k_2^*U_2(85) \qquad (8.12)$$

$$k_1^* \, 0.50 + k_2^*(0) = k_1^*(0) + k_2^* \, 0.767457 \qquad (8.13)$$

$$k_2 = 0.5212 \qquad (8.14)$$

Similarly, a decision problem is built between ESI level and pain level:

$$U \text{ (ESI level, 0)} \sim U \text{ (5, pain level)} \qquad (8.15)$$

The DM is indifferent if the ESI level is 3 and the pain level is 8. The utility values are calculated using Equations (8.5) and (8.8). The value of k_4 is calculated as follows:

$$U \, (3, 0) \sim U \, (5, 8) \qquad (8.16)$$

$$k_1^*U_1(3) + k_4^*U_4(0) = k_1^*U_1(5) + k_4^*U_4(8) \qquad (8.17)$$

$$k_1^* \, 0.50 + k_4^*(0) = k_1^*(0) + k_4^* \, 0.5772 \qquad (8.18)$$

$$k_4 = 0.6930 \qquad (8.19)$$

Finally, another decision problem is built between ESI level and gender:

$$U \text{ (ESI level, 1)} \sim U \text{ (5, Gender)} \qquad (8.20)$$

The DM is indifferent when the ESI level is 2 and gender is 2. The utility values are calculated using Equation (8.7) and Equation (8.8). The value of k_3 is calculated as follows:

$$U \, (3, 1) \sim U \, (5, 2) \qquad (8.21)$$

$$k_1^*U_1(3) + k_3^*U_3(1) = k_1^*U_1(5) + k_3^*U_3(2) \qquad (8.22)$$

$$k_1^* \, 0.50 + k_3^*(0) = k_1^*(0) + k_3^*(1) \qquad (8.23)$$

$$k_3 = 0.4000 \qquad (8.24)$$

As per the calculations, $k_1 = 0.8000$, $k_2 = 0.5212$, $k_3 = 0.4000$, and $k_4 = 0.6930$.

8.3.3.6 Aggregated Utility

The multiplicative form (Keeney and Raiffa 1993) is used to form the overall utility function:

$$U(x) = (1/K) \left[\prod_{i=1}^{n} \left(Kk_i U_i(x_i) + 1 \right) - 1 \right] \tag{8.25}$$

where $U(x)$ is the multiattribute utility of x; x_i is the performance level of attribute i; $U_i(x_i)$ is the single attribute utility for attribute i; $i = 1, 2, \ldots, n$ attributes; k_i is the attribute-scaling parameter for attribute i; and K is the normalizing constant.

The normalizing constant, K, is derived by the following relationship using the scaling parameters:

$$1 + K = \prod_{i=1}^{n} (1 + Kk_i) \tag{8.26}$$

Solving Equation (8.26) for K yields $K = -0.9793$. Substitution in Equation (8.25) yields the overall utility function for each patient (provided next).

$$U(x_1, x_2, x_3, x_4) = \left(\frac{1}{-0.9793} \right) * \left(\left\{ \left[-0.9793 * 0.8000 * (-0.2500x_1 + 1.2500) + 1 \right] * \right. \right.$$

$$\left[-0.9793 * 0.5212 * \left(-0.0252 + 0.0100 * \exp\left(\frac{x_2}{19.437} \right) \right) + 1 \right] * \left[-0.9793 * 0.4000 * \right. \tag{8.27}$$

$$\left. (x_3 - 1) + 1 \right] * \left[-0.9793 * 06930 * \left(-0.0957 + 0.0957 * \exp\left(\frac{x_4}{4.102} \right) \right) + 1 \right] \right\} - 1 \right)$$

where x_1 is the ESI level; x_2 the patient age; x_3 the patient gender; and x_4 the pain level.

Table 8.2 provides the summary of the attribute values for each patient, the utility value for each attribute, the calculated overall utility, and the ranking of the patients. The Spearman's rank correlation method is applied to measure the ranking association between variables (Upton and Cook 2002). The ranking association is tested between the overall utility ranking and the ESI utility, the age utility, the gender utility, and the pain level utility rankings individually. Additionally, SPSS was used to derive the ranking and the Spearman's correlation, which are presented in Table 8.3.

In general, a smaller value of correlation represents a smaller ranking association. According to the SPSS output in Table 8.3, there is a statistically high association between the overall utility ranking and the individual attribute utility rankings.

TABLE 8.2

Patient Ranking Based on Utility Functions

Patient	ESI level	Age	Gender*	Pain Level	U1(x1)	U2(x2)	U3(x3)	U4(x4)	U(x)	Ranking
1	2	53	M	10	0.7500	0.1276	0.0000	1.0000	0.8946	5
2	3	18	F	9	0.5000	0.0000	1.0000	0.7629	0.8389	7
3	3	48	M	4	0.5000	0.0929	0.0000	0.1581	0.4929	19
4	3	76	F	3	0.5000	0.4738	1.0000	0.1032	0.7547	12
5	3	20	M	7	0.5000	0.0027	0.0000	0.4316	0.5826	18
6	3	57	M	7	0.5000	0.1625	0.0000	0.4316	0.6184	17
7	3	75	M	8	0.5000	0.4488	0.0000	0.5772	0.7298	13
8	3	49	F	9	0.5000	0.0992	1.0000	0.7629	0.8482	6
9	2	54	F	10	0.7500	0.1357	1.0000	1.0000	0.9445	1
10	2	45	M	5	0.7500	0.0760	0.0000	0.2281	0.6790	14
11	2	55	F	9	0.7500	0.1442	1.0000	0.7629	0.9067	3
12	3	47	F	10	0.5000	0.0870	1.0000	1.0000	0.9051	4
13	3	22	M	0	0.5000	0.0058	0.0000	0.0000	0.4018	21
14	3	58	F	0	0.5000	0.1724	1.0000	0.0000	0.6766	15
15	3	37	F	8	0.5000	0.0419	1.0000	0.5772	0.7962	9
16	2	48	F	0	0.7500	0.0929	1.0000	0.0000	0.7771	11
17	2	48	F	10	0.7500	0.0929	1.0000	1.0000	0.9427	2
18	2	62	F	0	0.7500	0.2176	1.0000	0.0000	0.7934	10
19	3	19	F	2	0.5000	0.0013	1.0000	0.0601	0.6590	16
20	3	57	M	2	0.5000	0.1625	0.0000	0.0601	0.4748	20
21	2	73	F	0	0.7500	0.4024	1.0000	0.0000	0.8176	8

* M, male; F, female.

TABLE 8.3

Spearman's Correlation

			Rank of ESI	Rank of Age	Rank of Gender	Rank of Pain Level
Spearman's rho	Rank of Overall	Correlation coefficient	.448*	.179*	.439*	.643*
		Sig. (two-tailed)	.000	.000	.000	.000

* Correlation is significant at the 0.01 level (two-tailed).

8.4 Fuzzy Analytic Hierarchy Process and Utility Theory-Based Patient Sorting in Emergency Departments

Emergency records are those records of patients kept from the moment of arrival in the ED until the time of discharge to another area or home. In Section 8.3, we used the utility theory to prioritize ED patients. A clinical data set was used to build the overall utility function. Patient ages ranged from 18 to 90 years. Patients were ranked based on ESI and three descriptive variables: age, gender, and pain level. Equation (8.27) represents the overall utility function. Even though we have contributed to the triage decision-making problem, this solution has shortcomings. For example, while physiological variables influence patient symptoms (and hence the nurse decisions made during triage) they were not considered explicitly; they were implicitly covered within the ESI level. Patel et al. (2008) studied the decision-making process of nurses in the general ED and concluded that nurses' decisions are based on a generated hypothesis related to both the information given by the patient and to the awareness of single symptoms believed to be a characteristic of the diagnosis. Further, based on our interviews at clinical settings, it is essential to note that while we have ascertained that the relative importance of vital signs may change across different complaints, neither the complaints nor the relative importance shifts have previously been considered. In this section, we aim to extend the utility function presented in Section 8.3 to incorporate the relation between the patient complaints and the vital signs as well as the descriptive variables.

The relation between the vital signs and the complaints is operationalized as the changing relative importance of vital signs. What we mean by the "relation" is that a nurse might differently interpret the vital signs of two patients if one has chest pain and the other has a headache (i.e., the relative importance of vital signs might change depending on the patient's complaint). Consequently, the vital signs might contribute different weights to the overall utility value. The overall utility function aggregates the utilities of descriptive variables (age, gender, and pain level) and the "pretreated" values of the physiological variables (blood pressure, pulse, respiration rate, temperature, and oxygen saturation level). The values of the vital signs are treated using the fuzzy analytic hierarchy process, and the FAHP output is used in the overall utility function. The objective of the utility function is to quantify and minimize the associated clinical risk, and hence to arrive at an appropriate patient priority ranking.

8.4.1 Fuzzy Analytic Hierarchy Process

In ED settings, it is frequently difficult to ascertain accurate patient information because of the dynamic nature of the patient's status. For example, the vital signs change over time and assessment of certain variables, such as the pain level, is subjective (Fields et al. 2009). The use of fuzzy set theory allows decision makers to incorporate unquantifiable information, incomplete information, nonobtainable information, and partially ignorant facts into a decision model (Kulak et al. 2005), and hence is appropriate for such settings. The data relevant to the criteria (incomplete data) can be expressed as fuzzy data. This can be in linguistic terms, fuzzy sets, or fuzzy numbers. If the fuzzy data are in linguistic terms, they are transformed into fuzzy numbers. Then, these numbers (or fuzzy sets) are assigned crisp scores.

The analytic hierarchy process (AHP) is a systematic procedure designed to structure a complex problem in the form of a hierarchy and to evaluate a large number of quantitative and qualitative factors under multiple potentially conflicting criteria. Developed by Saaty (1980, 1994), the AHP has a number of shortcomings. Lee et al. (2001) summarized these as follows: (1) AHP is mainly used in crisp decision applications; (2) although the use of the discrete scale from 1 to 9 has the advantage of simplicity, AHP does not take into account the uncertainty associated with mapping one's judgment to a number; and (3) subjective judgments, selections, and preferences of decision makers exert a strong influence in the method.

Fuzzy logic is introduced to AHP to overcome these shortcomings. Therefore, when the input information or the relations between criteria are imprecise or uncertain, the adoption of fuzzy logic is recommended. The FAHP algorithm developed by Lee et al. (2001) involves the following steps: (1) construct a hierarchical structure for the problem to be solved; (2) establish the fuzzy judgment matrix, A, and weight vector, W; (3) calculate weight numbers using fuzzy scores of alternatives; and (4) rank the fuzzy scores to determine the optimum alternative.

8.4.2 Proposed Algorithm

The proposed decision algorithm starts by identifying the patient status as one would based on the current ESI algorithm (Gilboy et al. 2005). If the patient requires immediate intervention, he is considered to be in the "critical state." After this, the procedure progresses as follows: (1) Is the patient in need of immediate intervention? If the response is affirmative, he is a critical state patient. If no, he goes to the next step. (2) The triage nurse asks the patient about his complaint, pain level, age, and gender, and takes his or her vital signs. (3) The complaint and the vital signs data are treated using the FAHP as explained earlier to yield what we refer to as "pretreated" data. (4) The data from steps 2 and 3 are processed using the overall utility function to yield the utility value for each patient. (5) Patients with high utility values go to the treatment area first, and those with lower values are assigned to remain in the waiting room; they are subsequently treated in descending order of priority based on the overall utility values. This algorithm (shown in Figure 8.5) was applied to a clinical data set from Susquehanna Health's Williamsport Hospital, as shown in Table 8.4.

8.4.3 Patient Complaint

Every patient who comes to the ED has a chief complaint. These complaints are classified according to Claudio et al. (2009) into 17 categories as follows: (1) neurological complaints; (2) chest pain complaints; (3) abdomen/male; (4) abdomen/female; (5) seizure; (6) headache; (7) psychiatric complaints/

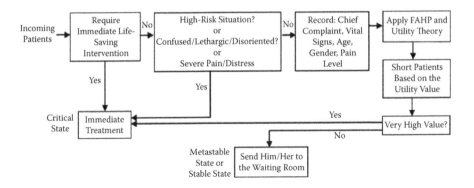

FIGURE 8.5
Proposed algorithm.

suicide attempt; (8) head/face trauma; (9) general medicine complaints; (10) respiratory complaints; (11) alleged assault; (12) multiple trauma; (13) motor vehicle crash; (14) extremity complaint/trauma; (15) back pain/injury; (16) skin rash/abscess; (17) eye, ear, nose, throat, and dental complaints.

As shown in step 2 of Figure 8.5, the triage nurse records the patient complaint along with the physiological and descriptive variables. After that the nurse assigns the ESI level for the patient. Beyond what is commonly applied during triage as prescribed by Gilboy et al. (2005), there is no systematic way to assign the ESI levels. Due to the presence of the uncertainty in taking such decisions, we have adopted the FAHP approach (Lee et al. 2001). Our selection of this approach stems from interviews conducted with expert triage nurses. Based on these interviews, we identified the relative importance of changes in vital signs given the patient complaint. In other words, the triage nurse assigns weights to the vital signs unequally based on the patient complaint, and then identifies the ESI level based on that weighting.

To ascertain how the relative importance weights of a patient's vital signs change over time, we conducted further interviews with expert triage nurses who rated each vital sign for its importance with respect to the patient complaint using the fuzzy number scale presented in Table 8.5. The hierarchy of this problem is illustrated in Figure 8.6. In the FAHP, vital signs are given different weights. In addition, each patient's vital signs are rated using the expert (triage nurse) identified, clinically meaningful value intervals and a corresponding fuzzy number. Table 8.6 shows an example of the fuzzy ratio scales for temperature. Then, the final score for each patient is calculated using the FAHP as described by Lee et al. (2001). These scores are then converted via a utility function into utility values in order to calculate the overall utility value (or ranking) for each patient. Another set of tables was used to rate the vital signs with respect to each chief complaint. Table 8.7 shows an example of the vital signs ratings with respect to the neurological complaint. We refer the reader to Ashour and Okudan (2010) for the full set of tables.

TABLE 8.4

Patient Data

	Complaint Description**	ESI Number	Age	Gender*	Pain Level	Systolic Blood Pressure	Diastolic Blood Pressure	Pulse	Respiration Rate	Temperature	SaO_2
1	CP	2	53	M	10	131	85	85	20	36.7	99
2	AF	3	18	F	9	130	77	96	16	36.6	99
3	T	3	48	M	4	98	68	96	18	37.8	97
4	GM	3	76	F	3	143	81	104	20	36.8	94
5	AM	3	20	M	7	117	79	69	20	36.3	100
6	AM	3	75	M	8	132	53	90	16	36.5	97
7	H	3	49	F	9	157	92	83	16	36.3	97
8	AF	2	54	F	10	159	99	70	32	37.1	98
9	CP	2	45	M	5	167	99	101	20	36.8	98
10	H	2	55	F	9	167	97	57	20	36.9	100
11	AF	3	47	F	10	125	76	74	20	36.5	99
12	PS	3	22	M	0	136	77	86	16	36.4	100
13	GM	3	58	F	0	117	73	65	22	35.6	96
14	NC	2	48	F	0	138	86	84	20	37	97
15	AF	2	48	F	10	122	77	64	20	36	98
16	CP	2	62	F	0	124	78	83	20	36.8	99
17	AF	3	19	F	2	113	68	74	18	36.7	98
18	GM	3	57	M	2	132	87	75	20	36.3	95
19	GM	2	73	F	0	147	80	84	14	37	97

* M, male; F, female.

** CP, chest pain complaints; AM, abdomen/male; AF, abdomen/female; Tm, extremity complaint/trauma; GM, general medicine complaints; H, headache; PS, psychiatric complaints/suicide attempt; NC, neurological complaints.

TABLE 8.5

Fuzzy Numbers for Linguistic Variables

Fuzzy Number	Attribute (Temperature, Pulse, …, etc.)
$\tilde{1}$	Low (L)
$\tilde{3}$	Relatively low (RL)
$\tilde{5}$	Medium (M)
$\tilde{7}$	Relatively high (RH)
$\tilde{9}$	High (H)

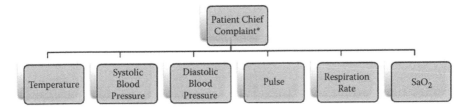

FIGURE 8.6
Problem hierarchy (*17 complaints).

TABLE 8.6

Fuzzy Ratio Scales for Temperature

Fuzzy Number	Temperature (°C)
Low (L)	(36.6–37)
Relatively low (RL)	(35–36.5)
Medium (M)	(37.1–37.3) or (33.9–34.9)
Relatively high (RH)	(37.4–37.8) or (32.2–33.8)
High (H)	(37.9 or greater) or (32.1 or lower)

TABLE 8.7

Neurological Complaints

Vital Sign	Rating
Temperature (°F)	L
Systolic blood pressure (mmHg)	H
Diastolic blood pressure (mmHg)	H
Pulse (beats/min)	H
Respiration rate (breaths/min)	H
Oxygen saturation level (%)	H

TABLE 8.8

Fuzzy Ratings of Patients with Respect to Each Criterion and the Weight Vectors

#	Systolic Blood Pressure	Diastolic Blood Pressure	Pulse	Respiration Rate	Temperature	SaO$_2$	Weight Vectors					
P1	3	3	1	1	1	1	9	9	9	9	1	9
P2	3	1	3	1	1	1	5	5	5	5	5	5
P3	1	1	3	1	7	1	5	5	5	5	1	5
P4	5	3	9	1	1	3	5	5	5	5	5	5
P5	1	1	1	1	3	1	5	5	5	5	5	5
P6	3	5	3	1	3	1	5	5	5	5	5	5
P7	5	5	1	1	3	1	5	5	5	3	1	3
P8	5	5	1	9	5	1	5	5	5	5	5	5
P9	7	5	9	1	1	1	9	9	9	9	1	9
P10	7	5	3	1	1	1	5	5	5	3	1	3
P11	3	1	1	1	3	1	5	5	5	5	5	5
P12	3	1	1	1	3	1	5	5	5	5	1	5
P13	1	1	1	5	3	1	5	5	5	5	5	5
P14	3	3	1	1	1	1	9	9	9	9	1	9
P15	3	1	1	1	3	1	5	5	5	5	5	5
P16	3	1	1	1	1	1	9	9	9	9	1	9
P17	1	1	1	1	1	1	5	5	5	5	5	5
P18	3	3	1	1	3	1	5	5	5	5	5	5
P19	5	1	1	3	1	1	5	5	5	5	5	5

8.4.4 Application

8.4.4.1 Fuzzy Analytic Hierarchy Process Implementation

In this problem, we have 19 alternatives (patients) and 6 criteria (vital signs). The fuzzy judgment matrix for all patients is provided in Table 8.8. A MATLAB® code was used to calculate the FAHP. After carrying out the fuzzy multiplication and addition, the fuzzy scores, the mean (pretreated value), and the standard deviation for all patients (listed in Table 8.9) are obtained. The utility function is built for the mean value in the next section. This way, the patient complaint, which is better represented due to the more appropriate consideration of vital signs, will be considered in the overall utility value.

8.4.4.2 Multiattribute Utility Theory Implementation

The exponential function is used to model the utility functions for the variables of age, pain level, gender, and the pretreated value (the mean) of the vital signs. The equations necessary to formulate the SUFs (Equations 8.1 through 8.4) are provided in Section 8.3.3. The attributes are not mutually preferentially independent; that is, a vital sign's pretreated value is not

TABLE 8.9

Mean and Standard Deviation of the Fuzzy Scores

#	Fuzzy Scores			Mean	Standard Deviation
P1	36	64	180	93.33	972
P2	18	50	154	74.00	843
P3	20	42	146	69.33	755
P4	42	100	224	122.00	1441
P5	18	40	140	66.00	705
P6	24	80	196	100.00	1283
P7	24	64	164	84.00	867
P8	54	120	252	142.00	1694
P9	120	148	288	185.33	1350
P10	30	82	186	99.33	1052
P11	18	50	154	74.00	843
P12	16	38	134	62.67	656
P13	24	60	168	84.00	936
P14	36	64	180	93.33	972
P15	18	50	154	74.00	843
P16	36	50	162	82.67	795
P17	18	30	126	58.00	584
P18	18	60	168	82.00	998
P19	24	60	168	84.00	936

independent from the others, and hence a multiplicative aggregation form is adopted.

The overall utility function combines all the physiological and descriptive variables. In Section 8.3, the overall utility function is calculated based only on the descriptive variables: age, gender, and pain level and the assigned ESI level. Here, the utility function is corrected by excluding the SUF of the ESI level and including all the SUFs of the other attributes plus the SUF of the pretreated value of the vital signs.

8.4.4.2.1 Single Attribute Utility Function of the Pretreated Value of Vital Signs

The DM has a risk-prone attitude toward the mean (pretreated value). The DM would give higher utility values if the mean increased from the CE to its maximum value in comparison to the increases from the smaller consequence values of the CE. Therefore, the RT has a negative value. Accordingly, the CE should be greater than the expected value of the utility function but less than the maximum consequence value. The expected utility value is calculated by setting all the rates with respect to each vital sign and the weight vector to $\tilde{5}$, which is the medium value. We infer that the CE for the mean is equal to 270. Solving Equations (8.1) through (8.4) iteratively, we find the $RT =$ −147.898. The utility function for the mean is given in Equation (8.28).

$$U_1(x_1) = -0.1149 + 0.0991 * \exp(x_1 / 147.898) \tag{8.28}$$

8.4.4.2.2 Single Utility Functions of Age, Gender, and Pain Level

Equations (8.6) through (8.8) in Section 8.3 represent the SUFs for patient age, gender, and pain level.

8.4.4.2.3 Attribute Trade-Offs

As illustrated in Section 8.3.3.5, the procedure from Keeney and Raiffa (1993) is used to calculate the attribute-scaling parameters. The DM ranked these attributes in order of importance: pain level (scaling parameter k_4), mean (k_1), age (k_2), and gender (k_3). In order to determine k_i values, we start with the most important attribute (k_4), the pain level. By using lottery questions, the indifference point of comparing two choices is also found, as shown in Figure 8.7. In the figure on the left, a lottery is presented where the best consequence is achieved with a probability of p; and on the right, the certain case for getting the best value for k_4 and the worst for rest of the attributes is presented. The goal is to find the value of probability, p, such that the DM is indifferent between the two choices. The attributes in Figure 8.7 are arranged in the same order (mean, age, gender, pain level). The DM indicates that the value of $p = 0.80$ makes him or her indifferent between the two choices. Therefore, the value of k_4 is 0.80. Due to space limitations, the detailed calculations are not shown here. Ashour and Okudan (2010) present the detailed results. The values of k_i are found to be: $k_1 = 0.756$; $k_2 = 0.587$; $k_3 = 0.182$ and $k_4 = 0.800$. Solving for K yields $K = -0.9804$. Accordingly, the overall utility function is

$$U(x_1, x_2, x_3, x_4) = \left(\frac{1}{-0.9804}\right) *$$

$$\left\{ \begin{matrix} \left[\left[-0.9804 * 0.756 * \left(-0.1149 + 0.0991 * \exp(\frac{x_1}{147.898})\right) + 1\right] * \\ \left[-0.9804 * 0.587 * \left(-0.0252 + 0.0100 * \exp\left(\frac{x_2}{19.437}\right)\right) + 1\right] * \\ \left[-0.9804 * 0.182 * (x_3 - 1) + 1\right] * \left[\begin{matrix} -0.9804 * 0.8000 * \\ \left(-0.0957 + 0.0957 * \exp\left(\frac{x_4}{4.102}\right)\right) + 1 \end{matrix} \right] \end{matrix} \right\} - 1 \tag{8.29}$$

where x_1 is the mean value (the FAHP output), x_2 is patient age, x_3 is patient gender, and x_4 is pain level.

Table 8.10 provides the summary of the values for each attribute and for each patient, the utility value for each attribute, the calculated overall utility,

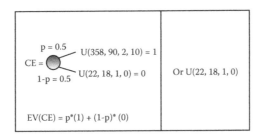

FIGURE 8.7
The pretreated value or the FAHP "mean" scaling factor lottery.

and the ranking of the patients. Table 8.10 also shows the ranking of the patients based on the results of Section 8.3, along with the new results. For both studies, the first and the last patient in the rank are the same (patient #8 and patient #12). To test the correlation between these rankings, Spearman's rank correlation method is applied. The ranking association is tested between the overall utility rankings for both results. SPSS was used to get the ranking of the patients and the Spearman's correlation shown in Table 8.11.

There is a statistically high association between the overall utility rankings, although the rankings are different due to the incorporation of the patient's chief complaint in the algorithm. The motivation behind correcting the previous function is that the chief complaint has an impact on the decision-making process, which might be gauged differently by different triage nurses, particularly among those with different levels of training or experience.

8.5 Discussion and Conclusions

The challenge of prioritizing patients in the EDs is a decision-making problem. Each decision is made based on a set of attributes. Uncertainty is an inherent quality within this kind of problem; certain patient condition attributes are subjective and are thus based on a triage nurse's knowledge, experience, and intuitions. To take into account this uncertainty, MAUT and FAHP methodologies have been applied to solve the patient prioritization problem during triage.

The goal of this study is not to make decisions for the DM; rather, it is to aid the triage nurse in prioritizing patients quickly and hence increase productivity. Decision-making tools can help a DM improve consistency, reliability, and repeatability; this means the same decision can be suggested for the same scenario. The proposed approaches can be used to rank patients based on ESI level, age, gender, and pain level, as well as based on vital signs and their relative importance with respect to the chief complaint. These approaches can reduce the stress and the strain on triage nurses and improve the service quality for patients.

TABLE 8.10

Patient Ranking Based on Utility Theory Function

#	ESI (x_1')	Mean (x_1)	Age (x_2)	Gender (x_3)	Pain Level (x_4)	$U(x_1')$	$U(x_1)$	$U(x_2)$	$U(x_3)$	$U(x_4)$	$U(X)$	Overall Rank	$U(X')$	Overall Rank*
P1	2	93.33	53	1	10	0.75	0.071	0.128	0.00	1.000	0.827	4	0.895	5
P2	3	74.00	18	2	9	0.5	0.048	0.000	1.00	0.763	0.696	7	0.839	7
P3	3	69.33	48	1	4	0.5	0.043	0.093	0.00	0.158	0.201	17	0.493	17
P4	3	122.00	76	2	3	0.5	0.111	0.474	1.00	0.103	0.506	9	0.755	11
P5	3	66.00	20	1	7	0.5	0.040	0.003	0.00	0.432	0.366	11	0.583	16
P6	3	100.00	75	1	8	0.5	0.080	0.449	0.00	0.577	0.630	8	0.730	12
P7	3	84.00	49	2	9	0.5	0.060	0.099	1.00	0.763	0.717	6	0.848	6
P8	2	142.00	54	2	20	0.75	0.144	0.136	1.00	1.000	0.871	1	0.945	1
P9	2	185.33	45	1	5	0.75	0.232	0.076	0.00	0.228	0.356	12	0.679	13
P10	2	99.33	55	2	9	0.75	0.079	0.144	1.00	0.763	0.729	5	0.907	3
P11	3	74.00	47	2	10	0.5	0.048	0.087	1.00	1.000	0.854	3	0.905	4
P12	3	62.67	22	1	0	0.5	0.036	0.006	0.00	0.000	0.031	19	0.402	19
P13	3	84.00	58	2	0	0.5	0.060	0.172	1.00	0.000	0.299	14	0.677	14
P14	2	93.33	48	2	0	0.75	0.071	0.093	1.00	0.000	0.269	15	0.777	10
P15	2	74.00	48	2	10	0.75	0.048	0.093	1.00	1.000	0.855	2	0.943	2
P16	2	82.67	62	2	0	0.75	0.058	0.218	1.00	0.000	0.319	13	0.793	9
P17	3	58.00	19	2	2	0.5	0.032	0.001	1.00	0.060	0.241	16	0.659	15
P18	3	82.00	57	1	2	0.5	0.058	0.163	0.00	0.060	0.176	18	0.475	18
P19	2	84.00	73	2	0	0.75	0.060	0.402	1.00	0.000	0.405	10	0.818	8

* The calculations are based on the results of Section 8.3.

TABLE 8.11

Spearman's Correlation

			Rank of MAUT
Spearman's rho	Rank of FAHP and MAUT	Correlation coefficient	.839**
		Sig. (two-tailed)	.000

** Correlation is significant at the 0.01 level (two-tailed).

With reference to the outputs shown in Table 8.2 and Table 8.10, some patients at ESI level 2 are ranked after patients at ESI level 3. For example, Table 8.10 lists patient #11 with a rank 3, while patient #1 has a rank 4, even though patient #1 has an ESI level 2 (meaning patient #1 has higher priority based on the ESI algorithm). This is because the utility values for the other variables are lower than those of the other patients. Studies have shown that the ESI level could be undertriaged or overtriaged (Wuerz et al. 2000; Wuerz 2001; Cooper 2004; Tanabe et al. 2004b). An underestimation (or overestimation) of emergency level occurs when the triage nurse allocates a triage level of higher (or lower) urgency than is required in a given case. An overtriage decision may not affect the overtriaged patient himself, but it might affect other patients by increasing their waiting time. On the contrary, an undertriaged patient might be affected because the waiting time for medical intervention will be increased. According to Considine et al. (2004) and Beveridge et al. (1999), overtriage and undertriage situations may occur in the Australian Triage Scale and in the Canadian Triage and Acuity Scale, respectively. Hence, the reasoning behind prioritizing certain patients at ESI level 2 after patients at ESI level 3 could be that the DM might overtriage these patients and give them higher priority than needed.

According to the ESI algorithm in Figure 8.1, emergency room prioritization does not take into account the patient age and gender in making the triage decision. In our proposed prioritization procedures, these two important descriptive factors—in addition to the ESI and the pain level—are integrated via MAUT to improve the health care system delivery in terms of productivity and quality. Then, the overall utility function is extended using FAHP to incorporate the physiological variables and integrate the FAHP output in the MAUT function. Thus, the accuracy and the repeatability of the utility function can be enhanced.

Acknowledgments

The authors thank Professor David Claudio and Miss Erica Fields for their help in data collection.

References

Andersson, A. K., M. Omberg, and M. Svedlund. 2006. Triage in the emergency department: A qualitative study of the factors which nurses consider when making decisions. *Nursing in Critical Care* 11(3): 136–145.

Ashour, O. M., and G. E. Okudan. 2010. Fuzzy AHP and utility theory based patient sorting in emergency departments. *International Journal of Collaborative Enterprise* 1(3): 332–358.

Benner, P., and C. Tanner. 1987. Clinical judgment: How expert nurses use intuition. *The American Journal of Nursing* 87: 23–31.

Beveridge, R. 1998. The Canadian triage and acuity scale: A new and critical element in health care reform. *Journal of Emergency Medicine* 16(3): 507–511.

Beveridge, R., J. Ducharme, L. Janes, S. Beaulieu, and S. Walter. 1999. Reliability of the Canadian emergency department triage and acuity scale: Interrater agreement. *Annals of Emergency Medicine* 34: 155–159.

Buesching, D. P., A. Jablonowski, E. Vesta, W. Dilts, C. Runge, J. Lund, and R. Porter. 1985. Inappropriate emergency department visits. *Annals of Emergency Medicine* 14(7): 672–676.

Claudio, D., and G. E. Okudan. 2010. Utility function based patient prioritization in the emergency department. *European Journal of Industrial Engineering* 4(1): 59–77.

Claudio, D., L. Ricondo, A. Freivalds, and G. E. Okudan. 2009. Physiological and descriptive variables as predictors for Emergency Severity Index. In *Proceedings of the IIE Annual Conference and Expo 2009*, May 30 – June 3 (IERC 2009), Miami, FL.

Cone, K. J., and R. Murray. 2002. Characteristics, insight, decision making, and preparation of ED triage nurses. *Journal of Emergency Nursing* 28(5): 401–406.

Considine, J., S. A. LeVasseur, and E. Villanueva. 2004. The Australasian Triage Scale: Examining emergency department nurses' performance using computer and paper scenarios. *Annals of Emergency Medicine* 44: 516–523.

Cooper, R. J. 2004. Emergency department triage: Why we need a research agenda. *Annals of Emergency Medicine* 44: 524–526.

Doe, J. 2009. Statistical Information System (WHOSIS). Geneva: World Health Organization.

Eitel, D., D. Travers, A. Rosenau, N. Gilboy, R. Wuerz. 2003. The Emergency Severity Index triage algorithm version 2 is reliable and valid. *Academic Emergency Medicine* 10: 1070–1080.

Ettinger, W. H., J. A. Casari, P. J. Coon, D. C. Muller, and K. Piazza-Appel. 1987. Patterns of use of the emergency department by elderly patients. *Journal of Gerontology* 42(6): 638–642.

Fields, E., D. Claudio, G. Okudan, C. Smith, and A. Freivalds. May 30- Jun 3, 2009. Triage decision making: Discrepancies in assigning the Emergency Severity Index. In *Proceedings of the IIE Annual Conference and Expo 2009*, (IERC 2009). Miami, FL.

Gifford, M. J., J. B. Franaszek, and G. Gibson. 1980. Emergency physicians' and patients' assessments: Urgency of need for medical care. *Annals of Emergency Medicine* 9: 502–507.

Gilboy, N., P. Tanabe, D. A. Travers, A. M. Rosenau, and D. R. Eitel. 2005. *Emergency Severity Index, Version 4: Implementation Handbook*. AHRQ Publication No. 05-0046-2. Rockville, MD: Agency for Healthcare Research and Quality.

Göransson, K. E., M. Ehnfors, M. E. Fonteyn, and A. Ehrenberg. 2008. Thinking strategies used by registered nurses during emergency department triage. *Journal of Advanced Nursing* 61(2): 163–172.

Göransson, K. E., A. Ehrenberg, B. Marklund, and M. Ehnfors. 2006. Emergency department triage: Is there a link between nurses' personal characteristics and accuracy in triage decisions? *Accident and Emergency Nursing* 14: 83–88.

Gurney, D. 2004. Exercises in critical thinking at triage: Prioritizing patients with similar acuities. *Journal of Emergency Nursing* 87(1): 514–516.

Guterman, J. J., N. J. Mankovich, J. Hiller. 1993. Assessing the effectiveness of a computer-based decision support system for emergency department triage. In *Proceedings of the Conference on Engineering in Medicine and Biology*, 594–595.

Keeney, R. L., and H. Raiffa. 1993. *Decisions with Multiple Objectives: Preferences and Value Tradeoffs.* Cambridge, UK: Cambridge University Press.

Kirkwood, C. W. 1997. *Strategic Decision Making: Multiobjective Decision Analysis with Spreadsheets.* Belmont, CA: Wadsworth.

Krishnamurthy, S. 2005. Normative decision making in engineering design. In *Decision Making in Engineering Design*, ed. K. E. Lewis, W. Chen, and L. C. Schmidt, 21–34. New York: ASME.

Kulak, O., M. B. Durmusoglu, and S. Tufekci. 2005. A complete cellular manufacturing system design methodology based on axiomatic design principles. *Computers and Industrial Engineering* 48(4): 765–778.

Lee, W. B., H. Lau, Z. Liu, and S. Tam. 2001. A fuzzy analytic hierarchy process approach in modular product design. *Expert Systems* 18(1): 32–42.

Mahapatra, S., C. P. Koelling, L. Patvivatsiri, B. Fraticelli, D. Eitel, and L. Grove. 2003. Pairing emergency severity index 5-level triage data with computer aided system design to improve emergency department access and throughput. In *Proceedings of the 2003 Winter Simulation Conference*, 1917–1925.

O'Cathain, A., J. Nicholl, F. Sampson, S. Walters, A. McDonnell, and J. Munro. 2004. Do different types of nurses give different triage decisions in NHS Direct? A mixed methods study. *Journal of Health Services Research and Policy* 9(4): 226–233.

Padmanabhan, N., F. Burstein, L. Churilov, J. Wassertheil, N. Hornblower, and N. Parker. 2006. A mobile emergency triage decision support system evaluation. In *Proceedings of the 39th Hawaii International Conference on System Sciences*, 1: 3–4.

Patel, V. L., L. A. Gutnik, D. R. Karlin, and M. Pusic. 2008. Calibrating urgency: Triage decision making in a pediatric emergency department. *Advances in Health Sciences Education* 13: 503–520.

Pedro, J. S., F. Burstein, P. Cao, L. Churilov, A. Zaslavsky, and J. Wassertheil. 2004. Mobile decision support for triage in emergency departments. In *Proceedings of the Decision Support in an Uncertain and Complex World: The IFIP TC8/WG8.3 International Conference*, 714–723.

Pitts, S. R., R. W. Niska, J. Xu, and C.W. Burt. 2008. National hospital ambulatory medical care survey: 2006 emergency department summary. National Health Statistics Reports, no. 7, 2008. http://www.cdc.gov/nchs/data/nhsr/nhsr007.pdf (accessed February 03, 2011).

Richardson, L. D., and U. Hwang. 2001a. Access to care: A review of the emergency medicine literature. *Academic Emergency Medicine* 8: 1030–1036.

Richardson, L. D., and U. Hwang. 2001b. America's health care safety net: Intact or unraveling? *Academic Emergency Medicine* 8:1056–1063.

Saaty, T. L. 1980. *The Analytic Hierarchy Process.* New York: McGraw-Hill.

Saaty, T. L. 1994. How to make a decision: The analytic hierarchy process. *Interfaces* 24(6): 19–43.

Singal, B. M., J. R. Hedges, E. W. Rousseau, A. W. Sanders, E. Berstein, R. M. McNamara, and T. M. Hogan. 1992. Geriatric patient emergency visits. Part I: Comparison of visits by geriatric and younger patients. *Annals of Emergency Medicine* 21: 802–807.

Spaite, D. W., F. Bartholomeaux, J. Guisto, E. Lindberg, B. Hull, A. Eyherabide, S. Lanyon, et al. 2002. Rapid process redesign in a university-based emergency department: Decreasing waiting time intervals and improving patient satisfaction. *Annals of Emergency Medicine* 39: 168–177.

Susquehanna Health Official Web Site. http://susquehannahealth.org (accessed November 01, 2010).

Takakuwa, S., and H. Shiozaki. 2004. Functional analysis for operating emergency department of a general hospital. In *Proceedings of the 36th Conference on Winter Simulation*, Washington, DC.

Tanabe P., R. Gimbel, P. Yarnold, and J. Adams. 2004a. The Emergency Severity Index (version 3) 5-level triage system scores predict ED resource consumption. *Journal of Emergency Nursing* 30: 22–29.

Tanabe, P., R. Gimbel, P. Yarnold, D. Kyriacou, and J. Adams. 2004b. Reliability and validity of scores on the Emergency Severity Index version 3. *Academic Emergency Medicine* 11: 59–65.

Tanabe, P., D. Travers, N. Gilboy, A. Rosenau, G. Sierzega, V. Rupp, Z. Matinovoch, and J. G. Adams. 2005. Refining Emergency Severity Index (ESI) triage criteria. *Academic Emergency Medicine* 12: 497–501.

Thevenot, H. J., E. D. Steva, G. E. Okudan, and T. W. Simpson. 2007. A multi-attribute utility theory-based method for product line selection. *Transaction of the ASME. Journal of Mechanical Design* 129 (11): 1179–1184.

Thurston, D. L. 2005. Utility function fundamentals. In *Decision Making in Engineering Design*, ed. K. E. Lewis, W. Chen, and L. C. Schmidt, 15–20. New York: ASME.

Upton, G., and I. Cook. 2002. *A dictionary of statistics*. New York: Oxford University Press.

Weinick, R. M., and H. Burstin. 2001. Monitoring the safety net: data challenges for emergency departments. *Academic Emergency Medicine* 8: 1019–1021.

World Health Organization (WHO). 2000. World Health Organization assesses the world's health systems. http://www.who.int/whr/2000/en/whr00_en.pdf. (accessed February 03, 2011).

Wuerz, R. 2001. Emergency Severity Index triage category is associated with six-month survival. *Academic Emergency Medicine* 8: 61–64.

Wuerz, R. C., L. W. Milne, D. R. Eitel, D. Travers, and N. Gilboy. 2000. Reliability and validity of a new five-level triage instrument. *Academic Emergency Medicine* 7: 236–242.

Wuerz, R., D. Travers, N. Gilboy, D. R. Eitel, A. Rosenau, and R. Yazhari. 2001. Implementation and refinement of the Emergency Severity Index. *Academic Emergency Medicine* 8: 170–176.

Zimmermann, P. G. 2001. The case for a universal, valid, reliable 5-tier triage acuity scale for U.S. emergency departments. *Journal of Emergency Medicine* 27(3): 246–254.

9

Using Process Mapping and Capability Analysis to Improve Room Turnaround Time at a Regional Hospital

John F. Kros and Evelyn C. Brown

CONTENTS

9.1 Introduction

High rates of customer service are expected in the health care industry. At the same time health care managers must plan for highly variable customer demand and face complex and at times customized service requests. In addition, the health care industry faces many challenges regarding service operations due to demand uncertainty, resource allocation, funding constraints, and overall customer access. Although health care is similar in nature to other service operations organizations, health care organizations differ greatly from their counterparts in that they provide services that pertain to human life and well being. The consequence of not meeting these service levels can result in customer death, an outcome that the vast majority of other service providers never face.

The ability to deny or limit demand can have positive effects upon operations (i.e., exclusivity) for some services. However, the health care industry in principle cannot deny or limit service. In fact, slack resources must be carried in quantities large enough to meet increased demand in a timely manner. This, in turn, leads to the emergence of many interesting service operations problems.

The examination of the approach taken by one regional hospital to improve its room turnover process is presented in this work. Aspects of the room turnover process along with data from the existing bed tracking system are analyzed. Specifically, room turnaround time is investigated, as it is the key metric associated with the room turnover process.

This work is organized into sections, with Section 9.2 providing a literature review and Section 9.3 presenting general information on process mapping. A problem definition is given in Section 9.4 with initial conditions and an analysis of relevant data for the system under study are provided in Section 9.5. Section 9.6 discusses the changes implemented. The results of the changes are presented and analyzed in Section 9.7, while Section 9.8 highlights conclusions and future work.

9.2 Literature Review

Service operations literature does contain research in the area of health care. Capacity and demand management dominated much of the early research. For example, the work by Smith-Daniels et al. (1988) discusses three types of resources that are allocated when making health care management decisions. Others have studied performance-based topics such as the impact of operational decisions on clinical performance (Heineke 1995) and the impact of operational decisions on overall hospital performance (Ling et al. 2002). Process analysis and resource allocation has also been studied. Jack and Powers (2004) propose a prescriptive framework that focuses on how health care organizations develop and leverage their resources. More specific process analysis and resource allocation research has been completed in hospital bed allocation (Gaynor and Anderson 1995; Green and Nyugen 2001; Walczak et al. 2003), staffing (Bloom 1997; Grandinetti 2000), outpatient scheduling (Cayirli and Veral 2003), patient arrival and discharge flow (McLendon 2001), and room turnaround (Pellicone and Martocci 2006). Hammer and Champy (1993) provide a good general coverage of business process reengineering.

Process mapping is one approach to improving service that can be applied to the health care industry. Numerous process-mapping approaches have been explored in the literature. Aldowaisan and Gaafar (1999) suggested a generalized assignment linear programming model in an attempt to minimize training requirements for specified categories of employees. Their approach applies observational analysis (Kusiak et al. 1994) and value analysis (Rupp and Russell 1994) in an attempt to reduce the training process by getting rid of some activities and automating others. Bond (1999) applies techniques of process mapping to information systems analysis and develops an approach for modeling data and material flows. The work also explores the application of systems analysis concepts to business process design.

Fulscher and Powell (1999) provide insights on conducting effective process mapping workshops. Like Bond (1999), their case study examines the links between process mapping and the development of an information technology architecture. Biazzo (2002) reveals limitations of process mapping. His work examines "the problems and the limitations of process mapping techniques in the light of sociotechnical experiences in systems analysis" (p. 43). He concludes that process mapping is not all that is needed for process analysis, as mapping may fail to incorporate values and other human characteristics that impact a process. As a note, an alternative to process mapping that is used from time to time is the responsibility assignment matrix (RAM) or the responsibility, accountability, counsel, information matrix (RACI). Many times these techniques are used as they describe in detail the participation by various entities in completing tasks. However, many times process mapping is coupled with this technique to illustrate where physically the participation is applied.

Morrison and Bird (2003) present a methodology, specific to the health care industry, for modeling front office and patient care processes in ambulatory health care. They examine the use of simulation tools and focus on the process modeling and data collection activities and their criticality in the success of process improvement projects. Vergidis et al. (2008) discuss business process modeling and their contribution is "a state-of-the art review in the areas of business process modeling, analysis, and optimization" (p. 69).

The process examined in this chapter is the room turnover process in a health care organization. The work differs from previous research in that a third-party provider is responsible for the process. Specifically, the third-party provider has control over certain aspects of patient and process flows, and allocation of resources such as staff and equipment, and thus can significantly impact the overall efficiency of the room turnover process.

Process analysis and improvement are coupled with data collection in the study of a large regional hospital that services a mostly rural population. The work reports on a process change and documents the effects of that process change. Insights are provided on how a third-party provider functions within a larger health care organization, and demonstrates relevance to and practical implications for health care providers and third-party health care service providers.

9.3 Process Mapping

Process mapping (Boyles 1991) can be used to analyze a process and has four major steps:

1. Process identification—Attaining a full understanding of all the steps of a process.
2. Information gathering—Identifying objectives, risks, and key controls in a process.
3. Interviewing and mapping—Understanding the point of view of individuals in the process and designing actual maps.
4. Analysis—Utilizing approaches and tools to make the process run more effectively and efficiently.

Process identification is the first step in process mapping. Many companies believe they know their processes—manufacturing, sales, accounting, building services. However, this silo mentality can lead to processes that lose their customer-centric approach. Instead of defining processes based on the company's understanding, processes must be defined by the customer's understanding. Literally, walking through customer experiences helps the reviewer identify those trigger points that can make or break success. Trigger points from the walk through form the basis for process identification.

The second step, *information gathering*, begins once the processes are identified. A large volume of information should be obtained before trying to learn the intricacies of a process. Primary among this information is identifying who the true process owners are. In other words, who are the individuals who affect change? Buy-in and agreement throughout the analysis is paramount. Information that should be obtained includes the objectives of the process, risks to the process, key controls over those risks, and measures of success for the process. Information gathering alone is not sufficient to determine all factors influencing the process. Therefore, it is necessary to conduct interviews and create a process map.

To effectively record and maintain this information, some analysts develop worksheets to incorporate into the third step, *interviewing and mapping*. The worksheets are used for gathering information such as the process owner, the trigger events (beginning and ending), inputs, outputs, objectives, risks, key controls, and measures of success.

Analysis is considered the fourth step, but analysis must really occur throughout the review. In defining the processes, the reviewer may determine that objectives are not in line with the processes in place. It may become apparent, in gathering information, that measures of success do not correspond to stated objectives. Specific examples of analysis performed once mapped include identifying unnecessary approvals, removing unnecessary redundant assignments, and investigating decision requirements that lead to no discernable result. No single incident is necessarily wrong, but each must be analyzed in the context of the map to ensure it supports the objectives.

Numerous visits were made to the hospital and interviews of relevant personnel were made. From these visits and interviews, we were able to gain an understanding of the steps involved in the room turnover process. Information gathering and data analysis lead us to a better understanding of response time and cleaning time. The following sections provide details on the analysis.

9.4 Problem Definition

Approximately 500 physicians and 1200 nurses serving over 30,000 inpatients and 265,000 outpatients per year are employed by the hospital analyzed. Room demand has increased and led to analyses of various patient-related systems and processes. The room turnover process is one such process, with a key metric of room turnaround time. Several factors impact room turnaround time and those that are examined are the factors which environmental services (EVS) has control, specifically bed tracking and room cleaning. At this facility, one of EVS' primary responsibilities is room cleaning. Figure 9.1 displays the time frame studied herein. Take note that June 2004 and May 2008 were chosen as representative sampling months based on the knowledge of the hospital's EVS quality manager (Barna 2008).

Room turnaround time is the key metric of the room turnover process. Two components, the response time and the cleaning time, make up the turnaround time for the room of a discharged patient. Response time is calculated based upon the time the dirty room is entered into the bed tracking system and the time the EVS employee logs into the system to indicate he or she has started the cleaning process. Travel time is the main factor impacting response time. The health care facility analyzed is a "horizontal" hospital,

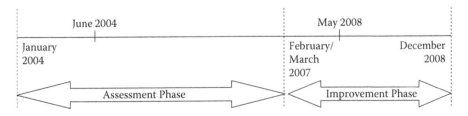

FIGURE 9.1
EVS assessment and improvement phases.

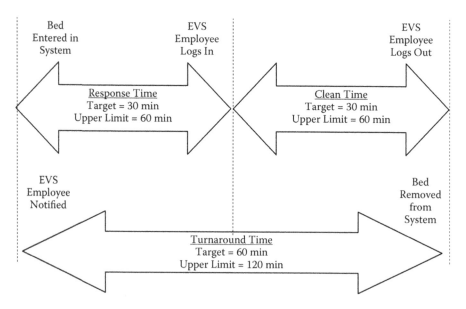

FIGURE 9.2
EVS room turnaround time.

meaning it is spread out over a large area and therefore getting to a dirty room is not simply a matter of getting on an elevator and going up or down a few floors. EVS employees often have to walk more than 10 minutes to get to the room that needs cleaning. Turnaround time for a typical room is depicted in Figure 9.2.

A system for assigning EVS employee to rooms that need to be cleaned is in place. Workers generally are assigned to an area whose rooms are fairly close geographically, and when a room in that area has a discharge, one of those workers is assigned to clean the room. At times, there are situations that arise when a nearby worker cannot be assigned because he or she is currently busy cleaning another room. When this arises, a worker from a more distant area must be assigned to clean the room. Response time is impacted by the increased travel time.

Turnaround time also includes cleaning time. Cleanliness of the room cannot be sacrificed for speed with which the room becomes available. Infection control is another aspect of the work of the EVS team, and infection control is vital to the well-being of all hospital patients, staff, and visitors.

9.4.1 Bed Tracking

The health care facility studied using a bed tracking system (BedTracking by Tele-Tracking Technologies) is in place at numerous hospitals in the United States (Barna 2008). As with any software system, it is only effective if correct information is loaded into it. Visibility of rooms that need to be cleaned is only possible once someone has loaded a discharge (i.e., a discharged patient's room number) into the system. Once identified, the room can be prioritized based upon the knowledge and judgment of EVS bed monitoring staff. In turn, bed-monitoring staff are responsible for prioritizing all beds in the system.

The priority system in place has three levels: "stat," "clean next," and "dirty." A room's status may be changed from dirty to clean next or elevated to stat at any time. A dispatcher monitors the bed tracking system. This position is cornerstone to the ability of the system to fully utilize EVS resources.

A discharge log is also kept as part of the tracking system. This log allows the dispatcher to monitor which employees have cleaned what rooms in what areas. The log is vital because it helps allocate or reallocate an employee if he or she is not busy or if a specific area becomes very busy and needs more resources. Dynamic reallocation of employees is a primary responsibility of the dispatcher, usually by conferring with EVS team leaders. Resources are used more efficiently than if just the main control center allocated them through the dynamic reallocation of employees.

9.4.2 Room Cleaning

Daily clean and a discharge clean make up the two main types of room cleanings. Daily cleaning often occurs with a patient in the room and does not include cleaning activities that would disturb the patient. Discharge cleaning, the focus of this research, is a complete and thorough room cleaning that must occur after one patient has been discharged and before a new patient may be assigned to the room. We were able to determine that the room cleaning times vary little unless the discharged patient leaves items in the room (EVS employee must deal with these items), or new linens to be put on the bed are not clean or have another hospital's logo on them (EVS employee must make another trip to the linens area to get suitable linens), by observing the room cleaning procedure. Although these events do not occur often, they can serve as causes for variation in room cleaning times.

Consistency in the room cleaning process is a must. A veteran employee trains each new employee to ensure quality. As was aforementioned, infection control is a vital aspect of the room cleaning procedure. Mops, rags, disinfectants, and other resources and the procedures used to clean the rooms have been selected based on their ability to maximize infection control.

9.5 Initial Conditions and Data Analysis

The room turnover process was broken down into these logical subprocesses during the initial assessment phase:

1. Discharge order is written and patient is informed he/she can leave.
2. Floor nurse enters room to be cleaned into the electronic bed tracking system when patient physically leaves room.
3. EVS is notified immediately through bed tracking system that a room needs cleaning (i.e., a clean-next notification appears).
4. EVS bed monitoring staff prioritizes clean-next rooms and enters into bed tracking system.
5. EVS employee is notified that a bed needs attention or EVS employee calls into the system to receive his/her next orders.
6. EVS employee responds to orders, incurring travel time to the room (start of response time).
7. EVS employee arrives at room and logs into the system to indicate the room clean has begun (start of clean time and end of response time).
8. EVS employee cleans the room (clean time).
9. EVS employee enters clean bed in bed tracking system to indicate the clean is complete. The bed tracking system records the clean time as the time between the end of response time and the time the EVS employee enters the cleaned room into the bed tracking system.

EVS recognized that discharges were not being handled efficiently and focus was lacking due to conflicting sources of direction that lead to delays in completing room cleans. EVS received on average, during the assessment phase, approximately 163 requests per day for room cleans. The mean number of requests was lower than the median number of requests, suggesting that the distribution is skewed. Skewness can be expected in analysis of service-related data. The data analyzed here came from a time period that did not include emergencies. For example, when demand for hospital services peaks, so do requests for room cleans. The demand for hospital services tends to peak in times of emergencies such as hurricanes, floods, and industrial accidents. These events would skew requests for room cleans higher. Data also indicate that there were days when almost no requests were made (i.e., nonemergency days), driving the mean lower.

Percent of compliant room cleans was a secondary metric measured by EVS. A compliant room clean is defined as a discharge that is logged into the bed tracking system in an appropriate manner, meaning it is properly logged in and properly logged out. EVS achieved a compliant clean rate of approximately 62.7% during the assessment phase. A 62% level of compliant cleans is tolerable; however, compliant clean rates in the 80% range are possible and desired by EVS (Barna 2008). The assessment phase data for compliant cleans are given in Table 9.1.

The metric of interest is room turnaround time as previously noted. However, recall that turnaround time is defined as response time plus clean time. Data for steps 2 through 5 of the room turnover process were

TABLE 9.1

Compliant Clean Data (Assessment Phase)

Assessment Phase	Compliant Cleans
Average	101
SD	35.4
% of total requests	62.74%

TABLE 9.2

Assessment Phase Data

Assessment Phase	Response Time (min)	Clean Time (min)	Turnaround Time (min)	Total Requests
Average	37.41	31.67	69.15	162.46
Median	38	32	70	149
SD	17.93	4.99	18.66	66.87

collected from electronic systems within the hospital. Descriptive statistics for response time, clean time, turnaround time, and total requests during the assessment phase are contained in Table 9.2.

For the EVS data, response times, clean times, and turnaround times all appeared consistent with no outliers skewing the data. This is illustrated in that the means and medians were very close to equal within each data set. It is also evident as response times and clean times were approximately equal but the amount of variation for the two times was quite different.

Time targets and upper specification limits on response time, clean time, and turnaround time are set based on the experience of the EVS managers. Turnaround time targets were set at 60 minutes, with an upper specification set at 120 minutes. EVS determined that when turnaround times fell much below 60 minutes the quality of the room cleaning suffered. In addition, turnaround times of more than 60 minutes were an indication that resources (i.e., EVS employees) were not being used efficiently. In turn, this led to the notion that the process needed to be centered at the target. A capability study was suggested since EVS was concerned with the room turnover process, but as currently administered the process was not capable of achieving these specifications or ever being on target.

Measures of process potential and performance are key to evaluating any process. Capability ratios and indices are widely used to provide that performance measure. C_p, referred to as the capability ratio and C_{pm}, referred to as the Taguchi capability index, are typical process capability measures. While C_p works well as measures of progress for quality improvement and variability reduction, these measures can fail to address the issue of process centering or to distinguish between off-target and on-target processes. In addition, Boyles (1991) shows that C_p is essentially a measure of process yield

TABLE 9.3

C_{pm} Capability Indices for Assessment Phase

	C_p	C_{pm}
Response time	0.56	0.52
Clean time	2.00	1.90
Turnaround time	1.07	0.96

only. In contrast, C_{pm} discerns between the process target value and the process mean.

Larger capability values are more favorable, as they are an indication that a process is operating on target and with low variation. Capability values less than 1.0 indicate that a process is highly variable and not operating at or near target. From Table 9.3, the C_p and C_{pm} for clean time are greater than 1.0, indicating that the process is operating on target and with low variation. However, the C_p and C_{pm} for response time and turnaround time are both less than 1.0. This indicates that the response process as well as the entire room turnover process are not operating on target and are highly variable. To understand the capability of the turnaround process better, each component is examined in detail in the following sections.

Since EVS is interested in a target turnaround time of 60 minutes, the Taguchi capability index, C_{pm}, is appropriate to use. C_{pm} measures process variation from a target and is sensitive to process centering as well as process yield. C_{pm} is calculated using the upper specification limit (USL), lower specification limit (LSL), target (T), and mean and variance of the data. Formulae for C_p and C_{pm} are as follows:

$$C_p = \frac{USL - LSL}{6\sigma} \tag{9.1a}$$

$$C_{pm} = \frac{USL - LSL}{6\sqrt{\sigma^2 + (\mu - T)^2}} \tag{9.1b}$$

Process capability was measured to determine the capability of the room turnaround process in meeting the specifications set forth by EVS. Table 9.3 contains the capability values for response time, clean time, and turnaround time during the assessment phase.

9.5.1 Response Time

For EVS, response time begins when an employee logs into the system and is assigned a new room to clean. Response time is impacted by the distance an employee must travel to get to the room that needs to be cleaned. Since EVS operates in a horizontal hospital travel time can vary greatly. This

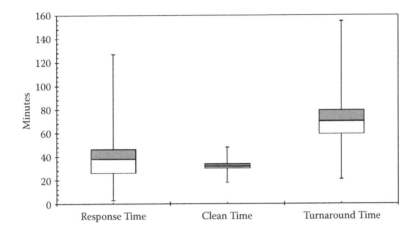

FIGURE 9.3
Boxplot for turnaround time, assessment phase.

variability is reflected by a large relative standard deviation for response time, as seen in Table 9.2. Even though the mean and median of response times are close to each other, the high standard deviation characterizes this distribution as highly volatile. Figure 9.3 contains a boxplot including the response times. The boxplot illustrates the wide range of response times. Each boxplot contains an upper extreme value (top hash mark), a lower extreme value (bottom hash mark), and a box. The box represents the middle 50% of values, 75% of the values are below the top of the box, the line represents the median value, and 25% of values are below the box. The observations translate to inconsistency regarding the deployment of EVS staff or inefficient use of that staff.

Variation in response time could be attributed to a number of different factors including but not limited to the number of workers assigned to various units of the hospital, time of day, height of hospital capacity level alert, and conflicting priorities or lack of priorities within the EVS staff. This variation in response time was a major concern of the EVS staff when it was identified. It quickly became an area for further investigation and possible place for improvement.

9.5.2 Clean Time

The length of time from when an EVS employee logs into the system and identifies he or she is cleaning a specified room until that room is logged out of the system as clean is defined as clean time. Clean times exhibited little variation due to the cleaning process being very consistent across all EVS employees. New employees are trained by veteran employees in an effort to maintain consistency and quality. Through this process, all employees learn the step-by-step procedures used by EVS cleaning staff during the cleaning process.

Consistency in the cleaning process is illustrated by the low relative standard deviation of the clean times (see Table 9.2). Figure 9.3 confirms this notion of consistency in clean times. For example, the box in the boxplot itself is very narrow with the difference between the high point and low point being relatively small.

9.5.3 Turnaround Time

In adding response time to clean time, turnaround time is calculated. Since the mean and the median of turnaround time are almost identical, there appears to be no undue influence by outliers. A normal probability plot of the data indicated an approximately normal distribution. Although normal in nature, the data indicated the range of turnaround times was relatively high. Figure 9.3 displays the boxplot of turnaround times and illustrates this wide range of values. Variation in turnaround time is again relatively high (refer to the standard deviation in Table 9.2) with much of the variation coming from the response time. EVS agreed that this variation and range for turnaround times was too high.

9.6 Implementing Changes

EVS implemented several major changes between February 2007 and March 2007. New discharge protocols were instituted. Duty lists for the area technicians (area techs) were reformatted to reflect the need for reallocation of staff based upon the capacity level alert in existence at the hospital. Capacity level alert ranges on a scale from 1 to 5 with higher numbers indicating a higher probability of the facility reaching maximum capacity. The "alternate" duty lists were created to be implemented when the hospital entered a level 2 or level 3 capacity level alert. Meal times and break times for area techs moved from scheduled time slots to being at the discretion of the manager on duty. The techs' areas of responsibility were also modified to eliminate cleaning of nonpatient rooms during these times. Following the alternate duty list, area techs are assigned to discharges until instructed otherwise by members of EVS management (Barna 2008). Figure 9.4 and Figure 9.5 show the discharge support protocols during the assessment and improvement phases.

Patient placement facilitators were put into place to assist with the prioritization of rooms to be cleaned. The facilitators had direct access to information about incoming patients that allowed the facilitators to place specifically needed beds into the stat category to ensure the beds were ready by the time the incoming patient arrived. Overall, once a discharge is loaded into the bed tracking system, its status is determined by the condition of the patient going into that room. Previously, all that was known was that a set of rooms

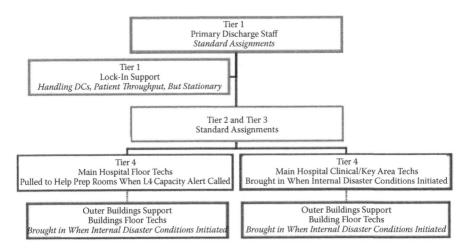

FIGURE 9.4
EVS discharge support protocol, assessment phase.

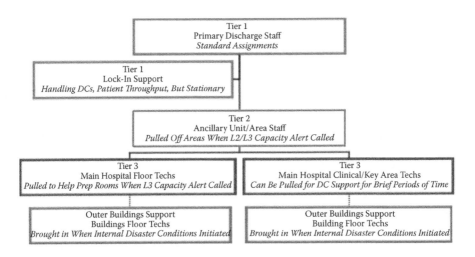

FIGURE 9.5
EVS discharge support protocol, improvement phase.

needed to be cleaned. The cleaning personnel were not told of beds that had higher priority.

Three major results were seen from the impacts of the staffing changes: a decrease in average response times, a decrease in average cleaning times, and a decrease in average turnaround times. The impacts illustrated and highlighted the flexibility of the EVS staff. Results were accomplished as demand for room cleans increased from 163 requests to 225 requests on average per day, an increase of over 38%.

TABLE 9.4

Improvement Phase Data

Assessment Phase	Response Time (min)	Clean Time (min)	Turnaround Time (min)	Total Requests
Average	32.76	30.81	63.19	225.65
Median	33	32	61	250
SD	14.99	6.32	17.87	54.32

9.7 Resulting Conditions and Data Analysis

Following the implemented changes, data were analyzed for the period again. For the new data, the mean number of requests was higher than the median, suggesting that the distribution is skewed. As before, the metric of interest for the process is turnaround time. Table 9.4 contains the descriptive statistics for response time, clean time, turnaround time, and total requests during the improvement phase.

9.7.1 Response Time

Response times improved greatly compared to the assessment phase response times, with the improvements in place. The change is reflected by the decrease in average response times and by the decrease in standard deviation around the mean. Table 9.4 contains the improvement phase data.

Mean and median of response times are very close to each other, indicating the data tend to be symmetric around the mean. The normal probability plot verifies that the data are essentially normal in nature. Improvements in the range of response times is verified in the boxplot contained in Figure 9.6. When the boxplot for response times in Figure 9.6 is compared to the boxplot of response times in Figure 9.3, this improvement can be seen in the fact that the entire plot of response times is narrower/reduced in Figure 9.6. Reductions in the range resulted from the improved consistency regarding the deployment of EVS staff and improved efficiency in staff use.

9.7.2 Clean Time

During the improvement phase of the study, there were few if any modifications to how EVS employees cleaned rooms. Overall, EVS employees were still very consistent in their cleaning regimes. EVS employees again strictly followed the step-by-step procedures of the room cleaning process. Cleaning consistency is illustrated in the low relative standard deviation of the clean times given in Table 9.4. Figure 9.6 confirms this notion of consistency in clean times. It is apparent from the boxplot, as the box is very narrow with the range being relatively low.

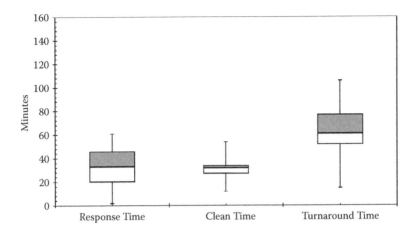

FIGURE 9.6
Boxplot for turnaround time, improvement phase.

TABLE 9.5

Compliant Clean Data (Improvement Phase)

Assessment Phase	Compliant Cleans
Average	141.48
SD	25.89
% of total requests	62.70%

Compliant cleans were at a level of 62.74% (see Table 9.1) in the assessment phase. Changes, implemented by EVS, did not result in an improvement in the percentage of compliant cleans (see Table 9.5). Recall that a compliant clean is defined as a discharge that is logged into the bed tracking system in an appropriate manner, meaning it is properly logged in and properly logged out. After discussions with the EVS quality manager, identification of some reasons that compliant clean numbers did not improve were revealed.

In all, EVS has no authority or control over floor nurses and hospital staff who are responsible for logging discharged patients into the bed tracking system. If these personnel do not enter data for a discharge in a timely manner, the compliant clean percentage is impacted. In addition, it is often the case that soon after a shift change, numerous rooms for a particular floor will appear in the bed tracking system. Many times, the EVS staff have already identified these rooms as dirty and cleaned them, thus the rooms should not have been loaded into the system. Rooms are then deleted from the bed tracking system, but again, the compliant clean percentage is impacted. Only when the personnel responsible for logging discharges into the bed tracking system clearly understand the impacts of their delays will compliant clean percentage improve substantially.

9.7.3 Turnaround Time

Turnaround time not only decreased, as the changes were implemented, the range of turnaround times decreased greatly. This becomes apparent when comparing the boxplots for turnaround times from the assessment and improvement phases. The mean and the median values of the improvement phase turnaround times are both lower than the assessment phase turnaround times (see Figure 9.7). Overall, the mean turnaround time decreased by over 8% while the median turnaround time decreased by more than 12%. This was accompanied by an improvement in the process variability. In comparing the improvement phase to the assessment phase, the standard deviation of turnaround times decreased by over 4%.

Process capability was again measured to determine the capability of the room turnover process after the improvement phase. Table 9.6 contains the C_p and C_{pm} index values for response time, clean time, and turnaround time after the improvement phase.

Larger capability values are more favorable and are indicative of a process that is operating on target and with low variation. From Table 9.6 it can be seen that the C_p and C_{pm} for clean time and turnaround time are both greater than 1.0. As before, this indicates that the cleaning process and the entire turnaround process are operating on target and with low variation. Although response time is still below the 1.0 limit, the C_p and C_{pm} indices have improved from 0.56 to 0.67 and 0.52 to 0.66, respectively, lending credence to

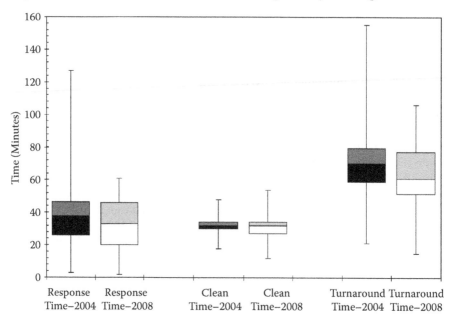

FIGURE 9.7

Boxplot for assessment and improvement phase turnaround time.

TABLE 9.6

C_{pm} Capability Indices for Improvement Phase

	C_p	C_{pm}
Response time	0.67	0.66
Clean time	1.58	1.57
Turnaround time	1.12	1.10

the process improvements EVS instituted. EVS agreed that further reductions in mean response times and variability are needed and made plans for continuous improvement regarding response time.

9.8 Conclusions and Future Work

The work presented here explores a highly significant metric for health care facilities: room turnaround time. Turnaround time improvement leads to reduced waiting times in the emergency department, thus patient satisfaction is improved as well. From a bottom-line perspective, reducing room turnaround times leads to more use of rooms, and thus more profit. Simply put, analyzing the room turnover process to reduce room turnaround time is a topic of interest to many in health care operations.

The analysis here demonstrates how process mapping techniques and heuristic approaches can be utilized along with existing systems to improve the room turnaround process. It is important to note that the changes implemented at this health care facility did not have a negative impact on infection control. Additionally, the solution has not generated any substantial negative feedback by employees to date. At the conclusion of our study the new policy had been in place over 2 years.

Analyses similar to these may be employed by EVS and hospital operations personnel to improve other areas of performance such as emergency department wait time or scheduling of the operating room. For many of the hospital's areas of operations, data are already available for analysis.

References

Aldowaisan, T. A., and L. K. Gaafar. 1999. Business process reengineering: An approach for process mapping. *International Journal of Management Science* 27: 515–524.

Barna, Jeffrey. 2008. Personal communication.

Biazzo, S. 2002. Process mapping techniques and organisational analysis. *Business Process Management* 8(1): 43–52.

Bloom, J. R. 1997. Nurse staffing patterns and hospital efficiency in the United States. *Social Science and Medicine* 346(15): 147–156.

Bond, T. C. 1999. Systems analysis and business process mapping: A symbiosis. *Business Process Management* 5(2): 164–178.

Boyles, R. A. 1991. The Taguchi capability index. *Journal of Quality Technology* 23(1): 17–26.

Cayirli, T., and E. Veral. 2003. Outpatient scheduling in health care: A review of literature. *Productions and Operations Management* 12(4):519–549.

Fulscher, J., and S. G. Powell. 1999. Anatomy of a process mapping workshop. *Business Process Management* 5(3): 208–238.

Gaynor, M., and G. Anderson. 1995. Uncertain demand, the structure of hospital costs, and the cost of empty hospital beds. *Journal of Health Economics* 14(3): 291–318.

Grandinetti, D. 2000. Make the most of your staff. *Medical Economics* 77(8): 56–58.

Green, L., and V. Nyugen. 2001. Strategies for cutting hospital beds: The impact on patient service. *Health Services Research* 36(2): 421–442.

Hammer, M., and J. Champy. 1993. *Reengineering the Corporation: A Manifesto for Business Revolution*. London: Nicholas Brealey.

Heineke, J. 1995. Strategic operations management decisions and professional performance in U.S. HMOs. *Journal of Operations Management* 13: 255–277.

Hunt, V. D. 1996. *Process Mapping: How to Reengineer Your Business Processes*. New York: Wiley.

Jack, E. P., and T. L. Powers. 2004. Volume flexible strategies in health services: A research framework. *Production and Operations Management* 13(3): 230–244.

Kusiak, A., N. T. Larson, and J. Wang. 1994. Reengineering of design and manufacturing processes. *Computers and Industrial Engineering* 26(3): 521–536.

Ling X. L., W. C. Benton, and G. Keong Leong. 2002. The impact of strategic operations management decisions on community hospital performance. *Journal of Operations Management* 20(4): 389–408.

McLendon, M. H. 2001. What causes long waits, diversions, and overcrowding in hospital EDs? *Health Care Strategic Management* 19(5), 12–14.

Morrison, B. P., and B. C. Bird. 2003. A methodology for modeling front office and patient care processes in ambulatory health care. *Proceedings of the 2003 Winter Simulation Conference*, New Orleans, LA.

Pellicone, A., and M. Martocci. 2006. Faster turnaround time. *Quality Progress* 39(3): 31–36.

Rupp, R., and J. Russell. 1994. The golden rules of process redesign. *Quality Progress* 27(12): 85–90.

Smith-Daniels, V., S. Schweikhart, and D. Smith-Daniels. 1988. Capacity management in health care services: Review and future research directions. *Decision Sciences* 19(4): 889–919.

Vergidis, K., A. Tiwari, and B. Majeed. 2008. Business process analysis and optimization: Beyond reengineering. *IEEE Transactions on Systems, Man, and Cybernetics— Part C: Applications and Reviews* 38(1): 69–82.

Walczak, S., W. E. Pofahl, and R. J. Scorpio. 2003. A decision support tool for allocating hospital bed resources and determining required acuity of care. *Decision Support Systems* 34(4): 445–456.

Section 3

Decision Making
in Logistic Services

10

Decision Making in Logistics Outsourcing from the Standpoint of Hiring Companies

Renata Albergaria de Mello Bandeira, Luiz Carlos Brasil de Brito Mello, and Antonio Carlos Gastaud Maçada

CONTENTS

10.1 Introduction

Logistics outsourcing has become a necessity for most organizations due to the fierce competition among companies and can be an important enabler for supply chain success (Schoenherr 2010). Leading organizations outsource many of their logistics processes in an effort to make their supply chains more agile, cost efficient, and competitive (Gunasekaran and Irani 2010). Others adopt outsourcing to focus on the core business (Bot and Neumann 2007) or to compensate for the lack of in-house logistics capacity (Boer et al. 2006).

The logistics industry increased globally, from 1995 to 2007, at a rate of 10% per year (Bot and Neumann 2007). In 2008, despite the economic recession, the third-party logistics (3PL) market increased 6.5% (DiBenedetto 2009), but in 2009 it shrank in response to the reduced trade volumes. The economic revival in 2010 reflected a growth in overall logistics as well as the 3PL market (Berman 2010). The Brazilian market has also experienced significant growth, with revenue up 20% in 2008 (CEL/COPPEAD 2008). Approximately $100 billion U.S. dollars, 11% of Brazilian Gross Domestic Product (GDP), was spent on logistics in 2009, and 63% of this amount was spent on logistics outsourcing, similar to the European (65%) and Asian (62%) outsourcing indexes and higher than the American outsourcing indexes (47%) (Barros 2009).

On the other hand, some companies choose not to outsource their logistics due to disbelief in cost reduction, belief that the service would be better performed internally, fear of knowledge and technical skill loss, and loss of control over logistics activities (Kremic et al. 2006). The unsuccessful logistics outsourcing experiences of some companies can influence the decision as well. About 20% to 25% of logistics agreements failed within 2 years, and 50% failed to succeed in 5 years (Craig and Willmott 2005). Such failures have been attributed mainly to deficiencies in this complex decision process (Isiklar et al. 2007). Thus, it is important to carry out studies focused on the logistics outsourcing decision.

Although the number of publications on logistics outsourcing has expanded rapidly over the past 15 years, much is still to be learned by both academic researchers and practicing decision makers (Schoenherr 2010). Most of the existing studies are concerned with the selection of 3PLs (Jharkharia and Shankar 2007), with little attention paid to the logistics outsourcing decision from the perspective of the hiring company. Harland et al. (2005) point out this topic as a valid research agenda with little support in current literature.

In this context, this chapter seeks to answer the following question: How do organizations that are considering hiring logistics services analyze the decision to outsource? The scope of the present investigation is to identify and analyze decision factors for logistics outsourcing from the standpoint of the hiring organizations in the Brazilian context, although results may also be applied to other countries in the same stage of development as Brazil. This work is an empirical study that uses qualitative and quantitative techniques to propose a set of key decision factors for logistics outsourcing. Logistics managers and decision makers can refer to this set of factors when structuring the decision-making process in their organizations and, therefore, reduce the risk of making wrong decisions.

The chapter proceeds as follows. Section 10.2 describes the definition of logistics outsourcing. Section 10.3 provides a literature review discussing the main theoretical perspectives on which logistics outsourcing has been analyzed, presenting the research framework. Section 10.4 describes the methodology used. Section 10.5 presents the results of the qualitative and quantitative investigations, while Section 10.6 provides assessment of these results. Finally, Section 10.7 concludes and summarizes insights for both academia and industry practices.

10.2 Logistics Outsourcing

There is much debate in the literature defining outsourcing (Harland et al. 2005), due to its many facets (Schoenherr 2010). According to Arnold (2000), outsourcing is a short term for "outside resource using" and is related to the reliance on external resources to add value to the organization. Harland et al. (2005) define outsourcing as turning over to suppliers activities that are not considered core competencies of the firm. For Reeves et al. (2010), outsourcing involves the procurement of physical or service inputs from outside organizations that were originally sourced internally or could have been sourced internally in spite of the decision to go outside. The present research adopts this definition of outsourcing.

The European Commission (2000) defines logistics outsourcing as the

> activities carried out by an external company on behalf of a client and consist of at least the provision of management of multiple logistics services. These activities are offered in an integrated way, not on a standalone basis.

The outsourcing of logistics functions to a 3PL provider has become an increasingly prevalent tendency for most modern companies (Rajesh et al. 2011). In previous eras, only transportation was outsourced, but 3PL providers have broadened their activities to services such as warehousing, distribution, freight forwarding, and manufacturing. Currently, the entire logistics process can be outsourced, so the utility of a 3PL provider is growing at a rapid pace.

10.3 Logistics Outsourcing Decision

The logistics outsourcing decision can be considered a variation of the classical make or buy decision, in which a firm selects to purchase an item or service that was previously made or performed in-house (Harland et al. 2005; Monczka 2005). This decision has been analyzed through an economic or strategic perspective (Schoenherr 2010). For Reeves et al. (2010) and Schoenherr (2010), the most prominent theories in reference to the outsourcing decision are the transaction cost economics (Coase 1937; Williamson 1995; Yang and Huang 2000) and the resource-based view (Barney 1999; Holcomb and Hitt 2006; McIvor 2009; Rodriguez and Robaina 2005; Liu and Lions 2011).

Transaction cost economics (TCE) has been the predominant theoretical approach for the outsourcing decision (Balakrishnan et al. 2008). TCE deals with the limits of firms (Barney and Hansen 1994) and defines transaction costs as the reason why firms choose to adopt market solutions instead of developing them internally. Based on TCE theory, firms seek the sourcing arrangement that minimizes transaction costs (Coase 1937; Gonzalez-Diaz et al. 2000). The decision about governance aims to reduce to a minimum the cost of transaction problems created by limited rationality and opportunism (Williamson 1995).

TCE identifies three critical factors to outsourcing decisions due to their influence on the magnitude of transaction costs and the likelihood of opportunistic behavior (Holcomb and Hitt 2006; Schoenherr 2010): asset specificity, uncertainty, and frequency. Asset specificity, which is defined as investments related to a specific transaction and with limited value when applied on alternative uses (Williamson 1995), is related to greater exposure to risks due to growing opportunism and uncertainty, as well as higher monitoring and bonding costs (Reeves et al. 2010). TCE predicts an increased likelihood of a firm structure as asset specificity or uncertainty increases. Frequency impacts reputation, accrual of trust, and economies of scale, which consequently impacts the likelihood for outsourcing. Holcomb and Hitt (2006) also point out the availability of service providers as a critical factor to the outsourcing decision, since low availability of providers allows them to behave opportunistically and, consequently, increase transaction costs.

Arnold (2000) and Madhok (2002) argue that the exclusive use of economical criteria limits the quality of the decision assessment, so TCE should be used in association with the resource-based view (RBV). Companies should also use aspects intrinsic to their own competence for the outsourcing decision, based on RBV. This theoretical perspective posits that sustainable competitive advantage is possible due to unequal distribution of assets that cannot be easily replicated by another firm (Peteraf 1993). Thus, in regarding the outsourcing decision, companies should own assets that are a source of competitive advantage and purchase those that do not (Reeves et al. 2010).

RBV assesses the organizational resources based on their value, rarity, inimitability, and nonsubstitutability (Barney 1999). Barney (1999) and McIvor (2009) advocate the use of concepts from RBV in the outsourcing decision, since valuable, rare, inimitable, and nonsubstitutable resources can result in competitive advantage and, hence, should not be outsourced.

The concept of core competence can be used to understand why companies choose to outsource. Within the RBV perspective, competitive advantage is derived from capabilities and resources, allowing efforts and investments to be focused on core competencies (Rodriguez and Robaina 2005). As a result, organizations should only invest in activities that constitute their core competencies and outsource the remaining ones.

Nevertheless, Barney and Arikan (2001) and Priem and Butler (2001) criticize the RBV perspective, arguing that the mere possession of rare and valuable resources, which are often unique and difficult to substitute, does not guarantee competitive advantage. Sirmon et al. (2007) emphasize the importance of integrating RBV with theories of the competitive environment, such as the institutional theory and the theory of contingency. Harland et al. (2005) point out common patterns of outsourcing in the same market sector, since some organizations choose to outsource due to the success obtained by other firms that have delegated their logistics processes to providers (Jharkharia and Shankar 2007). DiMaggio and Powell (1983) identify these common patterns of behavior as mimetic isomorphism. According to this concept, which is rooted in institutional theory, standard responses are caused by a set of changing environmental factors. The influence of environmental factors on the outsourcing decision can also be explained by the theory of contingency, which states that environmental factors influence the nature of organizational structure as well as the organizational characteristics (Donaldson 1998).

Schoenherr (2010) affirms that outsourcing can also be looked at with the lens of knowledge-based view (KBV), social exchange theories, or agency theory. KBV provides an incentive for vertical integration, and social exchange theory advocates for outsourcing in cases of high partnership advantage and low dependence (Balakrishnan et al. 2008). From the perspective of agency theory, outsourcing is not recommended in cases with high cost or quality uncertainty or when agents are risk adverse (Balakrishnan et al. 2008). Nonetheless, as a result of their research, Reeves et al. (2010) found inconclusive evidence of possible agency problems with respect to the sourcing of logistics services, but the authors concluded that, despite the complexity of the logistics outsourcing decision-making process, the theoretical constructs of the resource-based view of the firm and transaction cost economics still hold. Therefore, the essential theoretical bases of the present study are the RBV and the TCE approaches. Concepts from contingency and institutional theories are also used to propose a set of decision elements for logistics outsourcing. Table 10.1 presents the decision elements compiled from the literature review as well as their theoretical basis.

The concept of focusing on core competences originates in operations strategy literature (Harland et al. 2005). Core competence is a theoretical concept of RBV that has been used in combination with TCE to address the outsourcing decision (Schoenherr 2010). Based on RBV, products or services that represent a core competence should not be outsourced (Quinn 1999). Additionally,

TABLE 10.1

Decision Elements in Logistics Outsourcing, Theoretical Grounding, and Authors

Factors	Elements	Theoretical Grounding	Supporting Literature
Strategy	Core competence	RBV	Kremic et al. (2006), Mantel et al. (2006), Willcocks and Currie (1997)
	Access to resources	RBV	Boyson et al. (1999), Kremic et al. (2006), Persson and Virum (2001), Willcocks and Currie (1997)
	Strategic risk	RBV	Kremic et al. (2006), Sink and Langley (1997), Willcocks and Currie (1997)
Costs	Logistics costs	TCE	Kremic et al. (2006), Willcocks and Currie (1997), McIvor (2000)
	Investments in assets	TCE	Boyson (1999), Kremic et al. (2006), Sink and Langley (1997)
Characteristics of the logistics process	Complexity	RBV	Cain (2009), Kremic et al. (2006), Sohail and Sohal (2003)
	Specificity	TCE	Holcomb and Hitt (2006), Rodriguez and Robaina (2005)
	Value creation	RBV	Rodriguez and Robaina (2005)
	Difficulty to imitate/substitute	RBV	Rodriguez and Robaina (2005)
	Performance	RBV	Kremic et al. (2006), Persson and Virum (2001), Willcocks and Currie (1997)
	Quality	RBV	Kremic et al. (2006), Ying and Dayong (2005)
	Flexibility	RBV	Kremic et al. (2006), Persson and Virum (2001), Willcocks and Currie (1997)
	Operational risk	RBV	Persson and Virum (2001), Sink and Langley (1997)
Environment	Domestic political environment	Contingency theory	Jharkharia and Shankar (2007), Kremic et al. (2006), Iañes and Cunha (2006)
	Isomorphism	Institutional theory	Jharkharia and Shankar (2007)
	Uncertainty of the internal and external environment	TCE	Kremic et al. (2006), Ivanaj and Franzil (2006)
Logistics providers	Services offered	RBV	Holcomb and Hitt (2006), Kremic et al. (2006), McGinnis et al. (1997)
	Resources offered	RBV	Holcomb and Hitt (2006), Kremic et al. (2006), McGinnis et al. (1997)
	Geographical coverage	RBV	Holcomb and Hitt (2006), Kremic et al. (2006), McGinnis et al. (1997)
	3PL's experience in the client's market	RBV	Holcomb and Hitt (2006), Kremic et al. (2006), McGinnis et al. (1997)
	Image	RBV	Holcomb and Hitt (2006), Kremic et al. (2006), McGinnis et al. (1997)

Note: RBV, resource-based view; TCE, transaction cost economics.

according to RBV theory, companies seek, through outsourcing, external resources and capabilities to improve their performance. The increasing demand for sophisticated technologies, as well as for specialized logistics services, encourages outsourcing to 3PL providers who are more technically skilled and have more resources (Persson and Virum 2001). For Harland et al. (2005), 3PLs can be significantly more advanced in logistics operations, so outsourcing to them allows organizations to utilize their more advanced technologies and resources. However, for the hiring party, outsourcing also represents a series of strategic risks that have been a central variable considered in decision making (Schoenherr 2010). Kremic et al. (2006) list the main strategic risks involved in logistics outsourcing: risk of increased logistics costs; loss of control of outsourced activities; risk of dependence on providers; loss of organizational image due to provider's poor performance; loss of clients due to provider's poor performance; and reduction in employee morale. For Schoenherr (2010), these risks are usually heightened in a more involved outsourcing arrangement, since more control is transferred to the 3PL. Therefore, the decision-making process is based on the assessment of decision factors—such as core competence, resources availability for logistics, and strategic risks involved in the logistics outsourcing process—that are strategic in nature (Kremic et al. 2006). As a result, such discussion led to the following research proposition:

Proposition 1: The logistics outsourcing decision-making process relies on the analysis of the factor strategy.

Besides strategic factors, price and cost advantages are often mentioned in the literature as the primary reasons to outsource (Ross et al. 2005). Top benefits for companies outsourcing are often related to cost savings (Capgemini 2007), since 3PL providers are expected to offer lower logistics costs and better services. As logistics is their core business, these firms can lower costs by being more efficient or by economies of scale (Rajesh et al. 2011). According to TCE, the decision to outsource logistics is taken in order to reduce to a minimum the cost of transaction problems created by limited rationality and opportunism (Williamson 1995). Thus, the decision-making process of logistics outsourcing should include cost analysis, though the assessment of logistics costs and investments is needed for logistics assets. Outsourcing aims to reduce logistics costs and release resources for other activities. Consequently, high-cost processes that require high investments are more likely to be outsourced (Schoenherr 2010). Stated in a hypotheses form, we have:

Proposition 2: The logistics outsourcing decision-making process relies on cost analyses.

In addition, according to RBV, logistics outsourcing is an opportunity to improve the quality, performance, and flexibility of processes (Persson and

Virum 2001). The ability of a 3PL provider to offer quality service and performance creates higher value for clients (Rajesh et al. 2011). Moreover, outsourcing of noncore business, such as logistics, has positive impacts on flexibility (Lau and Zhang 2006; Power et al. 2006) in adapting to market changes, such as demand for products, services, and technologies (Greaver 1999).

Based on RBV theory, value created, specificity, difficulty in imitating, and difficulty in substituting are important concepts used to classify processes according to the generation of core competences, and, therefore, influence the outsourcing decision (Rodriguez and Robaina 2005). According to TCE theory, specificity of assets is considered a major cause of increased transaction costs (Holcomb and Hitt 2006). Razzaque and Sheng (1998) also suggests that the growing complexity of the supply chain leads to a greater propensity toward logistics outsourcing. Supply chain complexity can be defined as the level and type of interactions present in the supply chain (Milgate 2001). Hence, supply chain complexity is higher when the numbers of suppliers or customers increase. They become globally spread, or the variety of products and uncertainty increases (Hsiao et al. 2010). Logistics operations have become so complex that some companies face difficulties in managing them and, therefore, opt for outsourcing (Sohail and Sohal 2003). Moreover, in the case of outsourcing, operational risks of the logistics process are shared between the provider and hiring company, mitigating the operational risk (Persson and Virum 2001).

Such discussion shows that outsourcing practices are influenced by process type (Harland et al. 2005). The characteristics of the process—specificity, ability to create value, performance, quality, flexibility, difficulty in substituting, and imitating the logistics process, complexity, and operational risk of logistics processes—may interfere in the decision-making process for logistics outsourcing (Kremic et al. 2006), as stated in the following research proposition:

Proposition 3: The logistics outsourcing decision-making process relies on the analysis of the factor characteristics of the logistics process.

Moreover, the organization's political environment, represented by senior management's commitment and the engagement of the remaining collaborators, is a crucial element for the success of outsourcing (Ren et al. 2010). Too often though, outsourcing does not receive sufficient support within the organization due to the unwillingness to rely on external companies and the perceived threat to internal employment. Another relevant concept to the analysis of logistics outsourcing is mimetic isomorphism (DiMaggio and Powell 1983), since organizations choose to outsource because other firms were successful when delegating logistics processes to providers (Jharkharia and Shankar 2007). Additionally, uncertainty—coupled with the difficulty in estimating future needs (Ivanaj and Franzil 2006) or external uncertainties, as defined as the degree of volatility and unpredictability in the supply market—also influences the outsourcing decision, especially in long-term agreements (Kremic et al. 2006). For Gilley and Rasheed (2000), environmental uncertainty has a negative impact on outsourcing due to greater transaction costs associated with negotiating, monitoring, and executing arrangements. In uncertain situations, it seems more prudent to conduct the activity in-house (Schoenherr 2010). Therefore, the logistics outsourcing decision-making process should involve analysis of the environmental elements—the organization's political environment, success of organizations that outsourced logistics processes (isomorphism), and environmental uncertainty—such as stated in the following research proposition:

Proposition 4: The logistics outsourcing decision-making process relies on the analysis of the factor environment.

Finally, the availability of 3PL providers that fulfill the needs of the hiring organization is a primary motivator for outsourcing (McGinnis et al. 1997), because a limited availability of logistics providers allows for opportunistic behavior, increasing transaction costs (Holcomb and Hitt 2006). However, it is necessary that providers meet the high standards and demands of the hiring company. The service offerings of 3PL providers and the expected requirements of their clients influence the improvement of performance indicators from clients (Rajesh et al. 2011). Consequently, the decision-making process for logistics outsourcing should include the analysis of the logistics provider market, checking the availability of providers who offer the required services, resources, and geographic coverage and have experience in the market of the hiring company. This factor is conceptually grounded in contingency theory, since the characteristics of the logistics provider market are a dimension of the external environment (Donaldson 1998), as well as in TCE, due to the relation with 3PL availability. Such discussion led to the following research proposition:

Proposition 5: The logistics outsourcing decision-making process relies on the availability of logistics providers.

Figure 10.1 illustrates the preliminary research model compiled from the literature review, which represents the decision factors in logistics outsourcing.

10.4 Research Method

This investigation uses both qualitative and quantitative techniques to identify key factors in the logistics outsourcing decision-making process. The research methods applied were structured interviews, multiple case studies, and a survey.

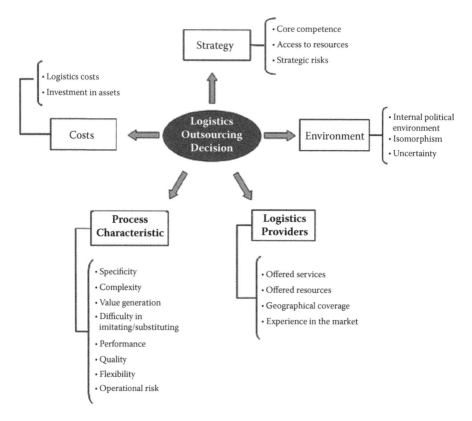

FIGURE 10.1
Research model.

In the first stage of the research, exploratory interviews were conducted according to a script based on the literature and validated by three logistics experts. The interviewees were four executives who participated in the logistics outsourcing decision-making process. The executives have, on average, 10 years experience in the logistics departments of leading companies with similar revenues ($1.5 billion) in different sectors (metal, petrochemical, and industrial gases). At this stage, the research model (Figure 10.1) was discussed and assessed by the interviewees, who agreed on the decision factors and elements proposed. The executives concluded that the preliminary model (Figure 10.1) is suitably composed and represents the multidimensionality of the logistics outsourcing decision-making process.

In the second phase, multiple case studies were conducted to examine the research propositions and generate hypotheses. The same procedures and research protocol were applied in three large companies that outsource logistics activities to 3PL. The interviewees were logistics executives who participated in the logistics outsourcing decision-making process. Table 10.2 shows

TABLE 10.2

Characteristics of the Interviewees in the Case Study

Assessment Unit	Justification for Selection	Position	Time in the Hiring Company	Time Working in Logistics
Industrial gas company	The logistics cost is considered the second largest value in product cost structure. Transport logistics is important due to the risk to the product. The products are commodities, so it is necessary to add value through logistics services.	Logistics manager	26 years	10 years
		Logistics analyst	10 years	6 years
		Quality manager for logistics operations	20 years	10 years
Automaker company	Logistics is strategic to competitiveness and performance in the auto industry. The automotive industry is complex because of its breadth, the level of global competition, and penetration.	Supply chain coordinator	15 years	8 years
		Logistics coordinator	10 years	7 years
		Logistics analyst	5 years	5 years
Engine manufacturer	The company participates directly in the automotive supply chain, given the demand for engines. The selection of the company is justified because it is a major supplier in the chain.	Logistics analyst	2 years	9 years
		Logistics supervisor	9 years	20 years
		Logistics manager	15 years	12 years

the interviewees' characteristics as well as the justification for the selection of the assessment units. The content analysis technique was adopted in the interview assessment.

In the third stage, the hypotheses, generated during the qualitative research, were tested by a survey. The sample consisted of 160 logistics executives from 100 Brazilian organizations that act in different sectors, such as automotive, metalworking, industrial gas, pharmaceutical, construction, food, beverages, agribusiness, electro-electronic, steel, and chemistry industries. These are mostly large companies that have outsourced their logistics. Elements of the research model were operationalized in a Likert scale of 7 points. The SPSS and AMOS (Analysis of Moment Structures) software packages were applied in the data analysis. The process of validation and refinement of the instrument followed the steps proposed by Koufteros et al. (2009), presented in Table 10.3.

TABLE 10.3

Steps in the Process of Refining and Validating the Research Instrument

Stage	Technique
Preparation	Development of the instrument based on theory and qualitative research.
Pretest	The following statistics techniques were applied: Corrected Item—Total Correlation Index (CITC), Convergent Exploratory Factor Analysis, and Cronbach's alpha.
Factor identification	The factor relationship was analyzed within the instrument. Instrument reliability was tested using Cronbach's alpha.
Confirmatory factor analysis	(1) Theoretical and conceptual analysis of the model and the relationship among decision factors and elements; (2) Analysis of the individual factor's validity; (3) Confirmatory factor analysis (CFA) for measurement Model A: one first-order latent factor; (4) CFA for measurement Model B: five first-order uncorrelated latent factors (orthogonal); (5) CFA for measurement Model C: five first-order correlated latent factors; (6) CFA for measurement Model D: second-order factor and five first-order latent factors; (7) comparison analysis among models and model selection.
Reliability analysis	New verification of Cronbach's alpha for the instrument and factors, considering only the results of the previous stage.

Source: Adapted from Koufteros, X., Babbar, S., and Kaighobadi, M., A Paradigm for Examining Second-Order Factor Models Employing Structural Equation Modeling, *International Journal of Production Economics*, 112(2009): 2–39.

10.5 Assessment of the Multiple Cases Study

The following section presents the characteristics of each assessment unit and the results crossing for each decision factor.

10.5.1 Case 1: Industrial Gas Company

The company operates in the industrial gas market, having 4000 employees in Brazil and another 1000 in nine South American countries. In 2009, the organization held 62% percent of the Brazilian market and presented a revenue of US$1.4 billion. The research investigated the outsourcing process of the outbound logistics activities related to liquid gas products. The logistics department for these products is formed by a national distribution center (responsible for planning the supply chain) and nine regional centers (responsible for delivery and routing). Transportation, loading, picking/packing, information technology (IT), billing, and vehicle maintenance are the responsibility of six hired logistics providers. Such activities have been outsourced for 10 years and represent the second-largest portion of the organizational cost structure.

10.5.2 Case 2: Automaker Company

The organization employs 266,000 employees to manufacture cars and commercial vehicles in 35 countries. The Brazilian subsidiary, consisting of three industrial areas and a distribution center, is the second largest operation outside the United States. The plant analyzed in this study, opened in 2000, employs 5200 workers, besides 18 modular suppliers, and forms a gated community where the involved parts keep their identities and share costs and expenses. The plant uses the services of five providers, who play the role of milk-run distribution of components of nonmodular providers; full load transportation haulage; express delivery and aerial transportation and contingency logistics; international logistics in Mercosur (Southern Common Market); and remaining international logistics activities.

10.5.3 Case 3: Engine Manufacturer

The organization participates directly in the automotive supply chain, being the leading engine manufacturer in Mercosur. The company has 2500 collaborators acting in four plants and finished 2009 with a net turnover of US$790 million. The logistics department is composed of three sectors: internal logistics, expedition, and external logistics. Most of the outsourced logistics activities are the responsibility of external logistics: consolidation and packaging activities, transportation, milk-run system, storage, load consolidation, fleet managing operations, reverse logistics, and information technology. The reasons that led to such activities being outsourced vary according to their characteristics. The company currently works with 20 3PLs.

10.5.4 Decision Factors for Logistics Outsourcing: A Comparative Assessment of the Cases

Table 10.4 presents a comparative analysis of the multiple case studies for each decision factor. By the content analysis of the qualitative study, the decision elements that were assessed by the analyzed companies that were not originally on the theoretical model (Figure 10.1) were identified. Therefore, the need to alter the original model consisting of 5 factors and 20 elements became clear, and the number of elements was increased to 26. Changes were made in 5 elements initially proposed within the theoretical model: (1) the element Access to Resources (Factor Strategy) was partitioned in Access to Equipment and Assets for Logistics Activities and Access to IT for Logistics; (2) the element Strategic Risks (Factor Strategy) was divided into Risk of Loss of Control of Outsourced Activities, Risk of Dependence on Providers, and Loss of Organizational Image due to Poor Performance of the Provider; (3) the element Logistic Costs (Factor Costs) was partitioned into Transport Cost, Warehousing Cost, and Inventory Cost; (4) the element

TABLE 10.4

Comparative Assessment of the Cases

Factor	Considerations of Comparative Assessment of the Cases
Strategy	• Companies outsource logistics because they consider it a complementary competence and so they hope to focus on their core competences. In the case of the automaker, the insource of logistics operations is considered impractical due to the high concentration of skills, resources, and assets it would require. The case studies are aligned to the RBV perspective, which holds that only processes that do not generate core competencies are apt for outsourcing (Rodriguez and Robaina 2005).
	• The organizations chose to outsource to gain access to resources specific to the logistics process (such as equipment that meet the characteristics of a particular activity and IT) without the need for investments. Boyson et al. (1999) and Persson and Virum (2001) state that, according to the RBV, organizations seek, through outsourcing, external expertise to increase IT availability, as well as access to the latest products and technologies.
	• The strategic risks related to logistics outsourcing that most concern the decision makers are associated to the organizational dependence on the providers. These are risk of loss of control of outsourced activities, risk of dependence on providers, and loss of organizational image due to poor performance of the provider. Businesses are aware of the risks involved in logistics outsourcing and try to manage them by a careful selection and monitoring of providers.
Costs	• Most of the expected benefits of logistics outsourcing involve financial aspects such as reduction of logistics costs, reduction of investments, provision of resources for other activities, and replacement of fixed costs by variable costs. The industrial gas company, for example, achieved a 4% reduction in distribution costs after outsourcing. In the view of TCE, logistics outsourcing provides the opportunity to minimize the costs of transactional problems (Williamson 1995).
	• In the assessment of case studies, logistics costs were analyzed as a single item, but such costs are composed of various elements such as inventory costs, transportation and storage costs (Bowersox and Closs 2001). The analysis of the case studies highlighted the importance of different logistics costs. In one of the companies studied, the storage cost is more representative in the composition of logistics costs and, consequently, exerts a greater influence on the logistics outsourcing decision.
	• The assessed companies expect that the outsourced process will require less investment and consequently it would release resources to core competencies. In the case of the engine manufacturer, most of the outsourced processes, such as transport and distribution, demanded high investments in assets.
Environment	• The organizations received support from top management to outsource their logistics processes, a practice in line with literature's recommendations (Ren et al. 2010). But the support of other employees is not usually considered in the decision-making process, because workers feel threatened by the possibility of job loss. As a result of the assessment of the case studies, it was identified that the internal environment becomes more favorable as outsourcing becomes a widespread practice.

TABLE 10.4 *(Continued)*

Comparative Assessment of the Cases

Factor	Considerations of Comparative Assessment of the Cases
Environment	• In the case of the industrial gas company, logistics outsourcing was influenced by success rates of other companies. The outsourcing feasibility was analyzed because of the successful cases of other companies and benchmarking studies were developed. Therefore, mimetic isomorphism can occur due to the success obtained by leading companies or by competitors.
	• The analyzed units opted to outsource logistics processes that are susceptible to the uncertainty of the external environment. For instance, the automotive market (case of the automaker and engine manufacture) is highly influenced by the economic environment and is subject to uncertainties and contingencies of the external environment such as variability in demand for these services. These results are contrary to the arguments from Kremic et al. (2006), who argue that the propensity to outsource is lower in cases with high uncertainty.
Logistics providers	• The availability of 3PLs capable of efficiently performing the activities outsourced is an essential factor to be considered in the decision-making process.
	• Logistics outsourcing would be impossible without the availability of logistics providers who offer the services and resources demanded, meet coverage requirements, and have experience in the hiring company's market.
Characteristics of the logistics process	• The decision to outsource a certain process is directly related to its characteristics and circumstances. The assessment of the case study developed at the engine manufacturer shows that the reasons that led to the outsource of certain activities vary according to their own characteristics: (a) the outsource of the storage process was encouraged by the lack of space to keep stocks internally, and (b) the milk run system was outsourced to a specialized provider because it requires specific knowledge.
	• The studied units are market leaders who had logistic services with high standards of quality, flexibility, and performance, but even so they chose to outsource their logistics processes to seek higher earnings due to the experience of 3PLs. For Ying and Dayong (2005), the success of outsourcing depends on maintaining or improving the performance and quality standards.
	• According to the results obtained in the multiple case studies, there is a satisfactory availability of logistics providers in the market that can perform the activities outsourced by the hiring companies. Therefore, one can conclude that such activities are not difficult to imitate or substitute. The RBV perspective states that processes with specific capabilities that are difficult to substitute or imitate should not be outsourced (Rodriguez and Robaina 2005).
	• The analysis of the case studies shows that the logistics processes add value to the product of the hiring organizations. For instance, in the case of the industrial gas company, distribution services are essential to differentiate their products that are considered commodities. Even so, the companies opted for outsourcing this process due to the fact that the value generated does not make it a core competency. Rodriguez and Robaina (2005) point out that the ability to generate value is an important element to process classification according to the generation of core competencies.

TABLE 10.4 *(Continued)*

Comparative Assessment of the Cases

Factor	Considerations of Comparative Assessment of the Cases
Characteristics of the logistics process	• The cases reinforce the hypothesis that the process complexity encourages outsourcing. The logistical process of the automobile company, for example, is so complex that it becomes impossible to perform it internally. • The organizations require certain resources from their logistics providers that can also be employed in alternative applications. For instance, when the engine manufacturer was deciding whether to outsource the process of cargo consolidation, the company searched for a provider who already owned a distribution center with the required characteristics and did not need to construct a new warehouse. Thus, these outsourced processes do not have high specificity, as proposed transaction cost economics and resource-based view (Holcomb and Hitt 2006). • Through outsourcing, the assessed units try to share the operational risks of their logistics processes with the 3PL. Sink and Langley (1997) emphasize that companies outsource logistics activities to mitigate operational risks.

Domestic Political Environment (Factor Environment) was substituted by the element Commitment of Top Management; (5) and the element Mimetic Isomorphism (Factor Environment) was partitioned into Success Rate with Logistics Outsourcing by Competitors and Success Rate with Logistics Outsourcing by Market Leaders. Such changes do not aggregate new concepts to the model but enrich the assessment with greater detail. The validity of the model is tested through the statistical analysis developed in the survey.

10.6 Assessment of the Survey Study

The results of the assessment of the pretest and the survey are presented as follows.

10.6.1 Pretest Survey

The instrument applied in the pretest survey consisted of 26 items and was developed based on the results of the bibliographical assessment and qualitative research. The questionnaire's face and content validity was carried out by three academics experts on logistics and five logistics managers of different industrial organizations. Seventy-five questionnaires were answered by 18 logistics managers and 57 executive graduate students in logistics, who are also professionals in the field. The database was purified and one questionnaire was eliminated. Reliability analysis was performed. As a result of Corrected Item–Total Correlation Index (CITC) analysis, the element

Difficulty in Substituting or Imitating the Logistics Process was eliminated from the factor Characteristics of the Logistics Process. The unidimensionality of the factors was verified by convergent exploratory factor analysis, applying the method of orthogonal rotation varimax and principal component analysis. Finally, a new reliability analysis was applied to validate the instrument that resulted from the pretest, composed of 5 factors and 25 elements. The resultant Cronbach's α index of the final instrument is equivalent to 0.87 and values for the factors range from 0.71 to 0.75.

10.6.2 Complete Survey Study

Data collection occurred in two distinct ways. Fifty-five percent of the questionnaires were delivered personally to the respondents, and a return rate of 95% was obtained. The remaining questionnaires were sent by e-mail, and postnotification was forwarded a month afterward. Seventy-five questionnaires were collected with a return rate of 30%. Forty-five questionnaires were collected before the postnotification dispatch. In total, the final sample consisted of 160 respondents.

The database purification resulted in the elimination of seven questionnaires. None of the elements showed a significant number of missing data points, and atypical observations from the perspective univariate (Z scores of standardized variables) or multivariate (D² Mahalanobis) were not detected. Thus, the sample came to be composed of 153 executives who work in 100 large companies that keep contracts with 3PL providers in distinct markets such as presented in Section 10.3. Of the respondents, 77% work in large organizations (over 1500 employees) and 60% are employed by companies with annual sales exceeding $600 million. Seventy percent of respondents are top management executives.

Data analyses provided evidence of normality, multicollinearity, and heteroscedasticity. Sampling adequacy for the factor analysis was confirmed by a KMO test (0.83) and a Bartlett's test of sphericity (0.00) (Meyers et al. 2006). The ratio of respondents per item equals 6.12 questions (153 respondents to 25 items), exceeding the limit of five questionnaires to each item recommended by Bentler (1990). The study employed confirmatory factor analysis to test the significance of the research hypothesis. According to the research hypotheses, the logistics outsourcing decision-making process must be evaluated in part due to its complexity. The decision-making process relies on the analysis of five latent factors—Strategy, Characteristics of the Logistics Process, Logistic Operators, Costs, and Environment—that are subdivided into multiple elements. Therefore, a second-order measurement model was developed for testing the research hypotheses and to verify if the five first-order factors are dimensions of a broader and more general construct: the Logistics Outsourcing Decision. The five factors are reflective indicators of the Logistics Outsourcing Decision since the removal of the elements does not change the factor nature. Moreover, the elements were placed under the same factor due to their similar contents. These factors also share a more abstract construct, which is the second-order factor Logistics Outsourcing Decision.

Model validation followed a two-step procedure. Initially, latent factors were individually validated, and, subsequently, the hypothesized relationships were validated with the use of integrated models (Garver and Mentzer 1999; Byrne 2001). The process of validating the hypothesized relationships followed the paradigm steps proposed by Koufteros et al. (2009) for higher-order structural equation modeling. Four types of models were analyzed (Table 10.3). To compare the four posited models and assess whether a second-order model is plausible, various fit indices were compared (Baumgartner and Homburg 1996; Hair et al. 2005; Harrington 2008). However, the final selection of a measurement model rests upon soundness that goes beyond the comparison of fit indices. The models under comparison must be supported theoretically, so the measurement model that generates the best fit indices cannot automatically secure its position as the leading model (Koufteros et al. 2009).

10.6.2.1 Individual Validation of the Factors

The validation of each individual factor is an important step to ensure the correct evaluation of the measurement model (Brasil 2005). Five factors and 25 observed elements, which form the initial instrument for the complete survey, were considered in the analysis. Each factor was evaluated in terms of its unidimensionality, reliability, extracted variance, convergent validity, and discriminant validity. Table 10.5 summarizes the main results of this analysis. The resultant research model is composed of 5 factors and 16 elements.

10.6.2.2 Measurement Model Analysis and Comparison: Models A, B, C, and D

This section presents the validation of the first- and second-order models, as well as their comparison, as proposed by Koufteros et al. (2009). Four measurement models were analyzed: (1) Model A, a first-order measurement model that specifies that all 16 elements are reflective of one latent variable—Logistics Outsourcing Decision; (2) Model B, a first-order measurement model that posits all five uncorrelated latent variables (orthogonal specification)—Strategy, Costs, Characteristic of the Logistics Process, Environment, and Logistics Providers—are related to their respective elements of the 16 observed variables; (3) Model C, a first-order measurement model that posits all five latent variables are free to correlate; and (4) Model D, a measurement model that specifies a second-order factor (i.e., Logistics Outsourcing Decision) that is related to five first-order factors that, in turn, are related to the respective observed variables. Table 10.6 presents the fit indices for all four models along with the respective criteria (Hair et al. 2005; Harrington 2008).

TABLE 10.5

Assessment of the Validation of the Individual Factors

Factor	Factor's Individual Validation
Strategy	Three items with standardized factor loadings below 0.5 were eliminated (Harrington 2008): Core Competence, Access to Logistics Equipment, and Access to Human Resources Skilled in Logistics. The values of composite reliability (0.63) and average variance extracted (0.3) are lower than recommended, so Cronbach's α index was evaluated (Baumgartner and Homburg 1996). The Cronbach's α index equals to 0.68, confirming factor reliability. Standardized factor loadings are significant (*t*-test) with values higher than 0.5 (0.52 to 0.59). Residuals' covariance were smaller than 2.58, confirming the factor unidimensionality.
Costs	The element Need for Investments in assets was eliminated due to its standardized factor loading of 0.3. Satisfactory values for composite reliability (0.71) and average variance extracted (0.51). Standardized factor loadings were higher than 0.5 (0.55 to 0.77) and significant (*t*-test). The largest residuals' covariance equals to 0.2, confirming factor unidimensionality.
Characteristics of the logistics process	Four elements were removed due to standardized factor loadings smaller than 0.4: Flexibility, Complexity, Specificity, and Value Creation. The resulting factor loadings proved high (0.57 to 0.73) and significant (*t*-test). Composite reliability (0.68) and average variance extracted (0.47) were close to satisfactory values. The Cronbach's α index equals to 0.67, confirming factor reliability. The largest value for residuals' covariance equals to 0.13, confirming factor unidimensionality.
Environment	Two elements were removed due to standardized factor loadings smaller than 0.4: Support from Organization's Employees and Environmental Uncertainty. Standardized factor loadings were high (0.55 to 0.85) and significant (*t*-test). Composite reliability (0.76) and average variance extracted (0.52) are adequate. The largest residuals' covariance equals to 0.11, indicating factor's unidimensionality.
Logistics providers	The element Availability of providers offering the required Resources was removed due to standardized factor loadings smaller than 0.4. Average variance extracted (0.68) and composite reliability (0.47) are almost satisfactory, but Cronbach's α index equals to 0.64, confirming factor reliability. The element Availability of operators offering the required Geographic Coverage (0.44) presented standardized factor loading smaller than 0.5. However, the element was held because it was significant at a 0.01 level and presented standardized factor loading higher than the minimum threshold of 0.3 (Hair et al. 2005). Without its presence, the factor would be composed of only two items, making it unidentified. The factor is unidimensional, with the highest residual covariance equal to 0.32.

To validate Model A, eight observed variables were eliminated so fit indices met the respective criteria: Risk of Loss of Control of Outsourced Activities; Transportation, Warehouse, and Inventory Costs; Quality of the Logistics Process; Availability of Logistics Providers with Experience in the Hiring Company's Market; Availability of Logistics Providers That Offer the Required Services; Risk of Loss of Organizational Image Due to Provider's Poor Performance; and Availability of Logistics Providers That

TABLE 10.6

Alternative Measurement Model Structures (Models A, B, C, and D)

Fit indices	Criteria	Final Model A	Final Model B	Final Model C	Final Model D
Chi-square/df (χ^2/df)	= 3.00	2.15	3.27	1.45	1.54
Goodness-of-fit index (G FI)	= 0.90	0.95	0.76	0.93	0.91
Normed-fit index (NFI)	= 0.90	0.90	0.55	0.92	0.90
Non-normed fit index (NNFI)	= 0.90	0.91	0.63	0.93	0.92
Comparative fit index (CFI)	= 0.90	0.94	0.65	0.95	0.92
Standardized root mean square residual (RMSEA)	= 0.10	0.087	0.122	0.05	0.06

Attend the Required Geographic Coverage. All remaining elements have statistically significant relationships with the factor Logistics Outsourcing Decision (*t*-value test). Composed reliability and average variance extracted (AVE) equals to 0.81 and 0.4, respectively. AVE is smaller than the recommended value (0.5), but it is still accepted due to the exploratory nature of the research (Harrington 2008). However, the necessity to eliminate eight elements (observable variables) to achieve an acceptable model with fit indices that met the respective criteria strengthens the adverse consequences of combining manifest variables from various content domains into one first-order latent variable (Koufteros et al. 2009).

Model B posits five uncorrelated latent variables (first-order factors) related to their respective observed variables of the 16 elements. The fit indices are indicative of a poor model (Table 10.6). This result can be explained by the high correlation among the first-order factors, as can be seen from the analysis of the factor loadings obtained for Model C. According to Koufteros et al. (2009), where there is a strong correlation between latent variables, an orthogonal specification for the relationships between latent variables would be expected to produce a poorly fitting model.

Model C was similar to Model B, except that the latent variables are free to correlate. In Model C, discriminant validity was not confirmed for the factor Strategy. The result does not indicate that the elements of this factor are less important to the logistics outsourcing decision, but it points out that such elements are also correlated to other factors. Thus, the factor Strategy was removed from the final measurement model. Koufteros et al. (2009) state that discriminant validity emerges as a critical concern in cases where first-order factors exhibit high correlations. The final measurement Model C posits four first-order factors related to twelve observed elements. Model fit was acceptable as all indices and *t*-values met respective criteria, providing evidence of convergent validity (Table 10.6). The significance of the *t*-value, associated with factor-to-item loadings, exceeds the critical value at the significance level of 0.01. Composite reliability (0.71–0.74) and AVE (0.51–0.6) were higher than the recommended values.

Model D presents a total disaggregation second-order factor model, in which the second-order factor (Logistics Outsourcing Decision) is related to five first-order factors (Strategy, Cost, Characteristics of the Logistics Process, Environment and Logistics Providers) that, in turn, are related to 16 observed variables. To validate Model D, the element Risk of Loss of Control of Outsourced Activities was eliminated from the factor Strategy, because its standardized factor loading (0.43) was less than the recommended value by Harrington (2008). The model fit was acceptable since all fit indices met respective criteria. The fit indices along with the *t*-values provide evidence of convergent validity. The significance of the *t*-values, associated with factor-to-item loadings and the first-order to the second-order factor, exceeds critical value at the 0.01 significance level. Discriminant validity was not analyzed, since it is not relevant to second-order models (Koufteros et al. 2009).

Models A, B, C, and D were compared in order to select the final measurement model. The evaluation through fit indices was considered a first cut, so models with poor fit indices ought not to advance to the next stage of scrutiny (Koufteros et al. 2009). Thus, Model B is not considered since its fit indices do not meet criteria. Although Model A is acceptable, eight elements whose relevance to logistics outsourcing was supported by the literature (Table 10.1) were eliminated. Model C is also acceptable as all fit indices met the respective criteria. Nonetheless, a first-order factor (Strategy) and three observed elements were eliminated from Model C. Since this factor presented the highest factor loadings, it should be analyzed in the logistics outsourcing decision-making process. Model D does not present fit indices as high as those obtained for Model C. This result is justified by the additional restrictions in cases of transition from a first-order measurement model to a second-order measurement model (Koufteros et al. 2009). Nevertheless, given the need for a higher-order measurement model according to the conceptual perspective, Model D is the best representing measurement model. Therefore, the logistics outsourcing decision-making process takes place through the analysis of 5 factors—Strategy, Characteristics of the Logistics Process, Costs, Environment, and Logistic Providers—that give rise to 15 elements. Model D and standardized factor loadings are presented in Figure 10.2. Values above 0.7 were obtained for the Cronbach's α index for all factors and the index of the final instrument (5 factors and 15 items) equals 0.85.

All factor loadings are above 0.53 and a majority of them is above 0.7. The significance of the *t*-values (Table 10.7) associated with factor-to-item loading exceeds the critical value at a 0.01 significance level. The loadings of the first-order factors to the second-order factor vary from 0.6 to 0.97 and *t*-values are statistically significant, attesting convergent validity of the second-order model. As such, the second-order model is quite effective in representing the data. It is important to notice that the Logistics Outsourcing Decision-Making process contributes the most to Strategy (std. coefficient = 0.97) and the least to Costs (std. coefficient = 0.6).

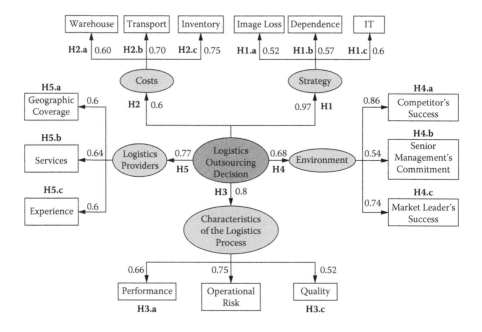

FIGURE 10.2
Second-order structural model (Model D).

All 20 hypothesized causal relationships in the final model were statistically supported. As a result, according to the perceptions of logistics executives acting in the Brazilian market, the Logistics Outsourcing Decision-Making Process, from the perspective of the hiring company, occurs through the analysis of five latent decision factors—Strategy, Costs, Characteristics of the Logistics Process, Environment, and Logistics Providers—that relate to their perspective observed variables (decision elements). The applicability of the final research model was confirmed by three executives, each of who works in one of the case study's assessment units. Interviewees considered the second-order structural equation model representative, remarking that the model reveals the multidimensional nature of the logistics outsourcing decision.

10.6.2.3 Analysis for Nonresponse Bias

The extrapolation method was applied to evaluate the existence of non-response bias, comparing the responses between two waves of respondents—the first wave after sending the questionnaire and second after the postnotification. The first group consisted of 123 respondents and the second of 30 respondents. As a result, t-values were significant at a 1% significance level, indicating lack of differences between the responses of both groups.

TABLE 10.7

Statistics Analysis of the Second-Order Structural Model (Model D)

	Factors	Nonstandardized Factor Loading	Standardized Factor Loading	*t*-Value	Hypothesis
Logistics outsourcing decision-making process	Characteristics of the Process	1	0.8		H3
	Strategy	1.19	0.97	4.703	H1
	Cost	0.76	0.60	4.018	H2
	Environment	1.247	0.68	5.175	H4
	Logistics Providers	0.97	0.77	4.551	H5
Strategy	Risk of Organizational Images Loss Due Provider's Poor Performance	1	0.50		H1.a
	Risk of Organizational Dependence to Provider	1.258	0.57	5.009	H1.b
	Access to IT Resources Applied to Logistics	1.249	0.60	5.181	H1.c
Costs	Warehouse Costs	1	0.6		H2.a
	Inventory Costs	1.419	0.74	5.681	H2.b
	Transport Costs	1.126	0.68	5.593	H2.c
Characteristics of the logistics process	Process Performance	1	0.66		H3.a
	Operational Risk of the Logistics Process	1.213	0.74	6.391	H3.b
	Process Quality	0.925	0.52	5.042	H3.c
Environment	Competitor's Success with Logistics Outsourcing	1	0.86		H4.a
	Senior Management Commitment	0.636	0.54	6,127	H4.b
	Competitor's Success with Logistics Outsourcing	0.906	0.74	7.972	H4.c
Logistics providers	Provider's Geographic Coverage	1	0.60		H5.a
	Services Offered by Available Providers	0.98	0.63	5.176	H5.b
	Provider's Experience on the Hiring Company's Market	0.987	0.60	5.02	H5.c

10.6.2.4 Analysis of Executives' Perception Regarding the Logistics Outsourcing Decision

In this section, we evaluate the differences of perception about the logistics outsourcing decision between different groups of respondents. As presented in Section 10.3, the survey respondents were 153 Brazilian executives who act in 100 organizations that maintain contracts with 3PLs and operate in different industries. These companies were classified into five categories according to the total number of employees: up to 100 employees; 101–500 employees; 501–1500 employees; 1501–5000 employees; and 5001–50,000 employees. Additionally, the companies were classified into categories based upon their annual revenues: below $10 million; $10 million–$60 million; $60 million–$500 million; $500 million–$5 billion; and $5 billion–$50 billion. The categories for number of employees and revenues were divided so that there was an equal distribution of companies between each category. ANOVA was run to check for differences of perception on the logistics outsourcing decision between executives who work in organizations of different industries and sizes, represented by the independent variables of number of employees and annual revenues. The dependent variables of the analysis were the observed variables, as well as the first- and second-order factors, presented in Figure 10.2. The results did not detect any influence of the business segment, annual revenue, or number of employees in the executives' perception of logistics outsourcing decision (p-values were over 0.05 for all analyzed items). The analysis was complemented by the post hoc test (LSD), and these results corroborate to those obtained in ANOVA test. Thus, for the study sample, there is no significant difference in perception between managers given the size of the organization or their business segment.

10.7 Conclusions

This research aims to discover which factors and elements must be considered by contracting organizations in the logistics outsourcing decision-making process. The research question is justified by the number of unsuccessful cases of logistics outsourcing, as well as a great amount of this failure is attributed to deficiencies in the decision-making process (Khan and Schroder 2009). For Harland et al. (2005), these deficiencies are caused by lack of knowledge relating to the decision-making process on logistics outsourcing.

In the bibliographical research, it was found that an error in the outsourcing decision may be crucial to the company due to a possible increase in logistics cost and process flaws (performance and quality), affecting organizational competitiveness and image (Kremic et al. 2006). The errors are a consequence of the complexity of the decision, which involves a large amount and variety of both quantitative and qualitative factors that are interdependent on each other (Water and Peet 2007). Adding to the complexity, the logistics outsourcing decision becomes even more difficult due to the growing competition, fast technological change, and dispersal of knowledge and markets (Holcomb and Hitt 2006).

Qualitative research shows that hiring companies have not structured a logical process and criteria to conduct the logistics outsourcing decision-making process, even though such decision influences the organization's competitive position (Ren et al. 2010). The results of the qualitative study also show that Brazilian executives have difficulty defining which elements should be analyzed in the decision-making process.

The combination of qualitative and quantitative research methods allowed identification of a second-order structural equation model that represents the logistics outsourcing decision-making process. The Logistics Outsourcing Decision, as a latent second-order factor, is an abstract concept, which is related to five decision factors—Strategy, Cost, Characteristics of the Logistics Process, Environment, and Logistics Providers. In the Brazilian context, the decision-making process of logistics outsourcing, according to the perspective of the contracting companies, occurs through the analysis of such factors.

Quantitative research highlighted Strategy as the most important factor in the decision-making process of outsourcing logistics. This was a surprising result, since price and cost advantages are often mentioned as the primary reason to outsource in the literature (Ross et al. 2005; Schoenherr 2010). However, the case studies analysis revealed that, in reality, strategic factors are discussed in the decision-making process but are not considered in a detailed evaluation. Consequently, even though executives consider the logistics outsourcing decision a strategic one, all the perspectives of this factor are rarely evaluated during the decision-making process. Such contradiction between the results obtained by the survey and the case studies can be explained by the difficulty to predict and measure the strategic risks involved in logistics outsourcing.

The Strategy factor, as perceived by executives, should be evaluated through the analysis of the elements Access to IT Resources for Logistics, Risk of Loss of the Organizational Image Due to Operator's Poor Performance, and Risk of Organizational Dependence on Logistics Providers. This result reveals important points that drive the logistics outsourcing decision:

- With outsourcing, companies seek access to resources for complementary activities without the need for investments. According to Schoenherr (2010), the unavailability of resources is a compelling reason to outsource.

- Information technology is a relevant resource to the outsourcing decision. Wu et al. (2005), in previous studies, had identified IT as a vital resource to the success of logistics processes. However, this study concludes that access to IT is an essential strategic element in the logistics outsourcing decision-making process.

- Decision makers should be aware that outsourcing leads to loss control of logistics activities, causing dependence on 3PLs' performance. For that reason, despite the eventual short-term balance sheet benefits, there are strategic risks to organizations (McIvor 2000) related to such dependence, which should be analyzed during the decision-making process, such as Risk of Organizational Dependence on Logistics Providers and the Risk of Loss of Organizational Image Due to Provider's Poor Performance. Therefore, corporations are advised to formulate strategies for outsourcing to minimize these long-term strategic risks from the cumulative impact of outsource decisions (Harland et al. 2005).

The decision-making process of logistics outsourcing should also include the evaluation of the economic criteria. Qualitative research revealed that the reduction in logistics costs is an outsourcing benefit that allows the release of resources for core activities. Logistics costs, according to the executives' perception, should be analyzed through the assessment of the following elements: Cost of Transport, Storage Cost, and Inventory Cost. These are the most important logistic costs, since almost 90% of a company's logistics costs correspond to transport, storage, and inventory (Bowersox and Closs 2001).

Quantitative research showed that the analysis of characteristics of the logistics process is crucial to the outsourcing decision. Elements of this factor considered relevant to the decision-process are Quality, Performance, and Operational Risks. Executives consider outsourcing as an opportunity to improve processes with low performance and quality standards, in addition to the providing, sharing, and mitigating of logistics' operational risks. This result resonates with Power et al. (2006) and Rajesh et al. (2011), whose studies pointed out a positive influence on performance and quality indicators for organizations in association with a 3PL provider. Nonetheless, the results of a survey in the Taiwanese food processing industry, developed by Hsiao et al. (2010), indicated that the effect of outsourcing on performance is contingent on the characteristics of the organizational supply chain and logistics' environment. Thus, decision makers should carefully analyze the characteristics of their logistics process in order to make a valid decision regarding outsourcing.

Other important points that drive the decision to outsource logistics that were revealed by the survey are:

- Top management's commitment is crucial to the success of the logistics outsourcing decision-making process. Ren et al. (2010) argue that outsourcing success depends on top management commitment.
- Success achieved by other companies (competitors or market leaders), through logistics outsourcing, encourages the outsourcing decision by organizations. The importance of elements related to isomorphism was corroborated by the survey results and by the case

of the industrial gas company, since its CEO decided to examine the feasibility of logistics outsourcing due to cases of success.

Quantitative research also highlighted the importance of analyzing the logistics provider market in the decision-making process. The feasibility of outsourcing depends on the availability of logistics providers capable of meeting the needs of the hiring company. This result resonates with Rajesh et al. (2011) whose research indicated that the service offerings of 3PL providers contribute largely to the logistics process performance indicators measures. Therefore, one should consider the availability of providers capable of offering required services, of meeting required geographic coverage, and that have experience in the contractor's market when in the decision-making process.

It should be stressed that the five decision factors remained constant throughout the entire study, with changes only with respect to the elements that compose them. Such changes do not add new concepts to the model. Moreover, the fact that the final model contains fewer parameters than the theoretical model originally proposed does not mean it is less adjusted. Instead, these changes have significantly improved the model, making it more parsimonious. This difference can also be explained by the exploratory nature of the study: the authors identified as many decision elements as possible through the bibliographical and qualitative research to be validated by the quantitative research. The final model was analyzed by three executives who confirmed its applicability.

In conclusion, the final research model consists of 15 elements structured in 5 factors—Strategy, Costs, Environment, Logistic Providers, and Characteristics of the Logistics Process—that should be analyzed in the logistics outsourcing decision-making process. The contributions of the literature are empirical results that lend support for the applicability of resource-based view constructs and transaction cost economics constructs to the logistics outsourcing decision. These results are quite useful to researchers and professionals in the logistics and supply chain field, as they contribute towards a better understanding of the logistics outsourcing decision and of the decision-making process structure.

The proposed second-order structural equation model defines which factors and elements should be analyzed in the decision-making process when contemplating outsourcing logistics activities. Nevertheless, it requires the assessment of a large amount (15 on total) and variety of quantitative and qualitative factors that are interdependent on each other. To support such decision-making process, the use of multicriteria decision analysis (MCDA) techniques is recommended. These techniques take into account that the decision-making processes are, to a certain extent, subjective and complex in nature and try to eliminate the inconsistencies in the decisions made (Water and Peet 2007). Although the analytic hierarchy process (AHP) is one of the most widely applied approaches to handle the multicriteria decision-making problem (Saaty 1980), we recommend the use of the analytic network process

(ANP) for the studied problem because it also allows for the inclusion of all relevant criteria (tangible or intangible, objective or subjective) without making assumptions about independency among them (Jharkharia and Shankar 2007). We suggest, for future research, the use of the proposed second-order structural equation model as an ANP-based model for the decision-making process of logistics outsourcing. The opinion of logistics executives of the hiring company should be sought in the formation of pairwise comparison matrices, using a ratio scale of 1 to 9 to compare any two elements or factors. The application of software and decision support systems may also reduce the complexities in implementing ANP technique.

Moreover, this study is relevant not only to contracting firms but also to logistics providers. Through the assessment of the proposed decision factors, providers can better understand how their clients structure their decision-making process and, thus, meet their customers' needs.

Furthermore, no evidence was found supporting the influence of the business segment, annual revenue, or number of employees in the executives' perception of the logistics outsourcing decision. It is suggested that this research instrument be applied in different countries to study the influence of other cultural, economic, social, and business factors and scenarios in the logistics outsourcing decision. It is possible that the research results can be applied to other countries that are in the same stage of development of Brazil as long as more studies are developed to validate their usage.

References

Arnold, U. 2000. New dimensions of outsourcing: A combination of transaction cost economics and the core competencies concept. *European Journal of Purchasing and Supply Management* 6:23–29.

Balakrishnan, K., Mohan, U., and Seshadri, S. 2008. Outsourcing of front-end business processes: Quality, information, and customer contact. *Journal of Operations Management* 26(2):288–302.

Barney, J. 1999. How a firm's capabilities affect boundary decisions. *Sloan Management Review* Spring:137–45.

Barney, J., and Arikan, A. 2001. The resource-based view: Origins and implications. In *Handbook of Strategic Management*, ed. M. A. Hitt, R. E. Freeman, and J. S. Harrison, 124–188. Oxford: Blackwell.

Barney, J., and Hansen, M. 1994. Trustworthiness as a source of competitive advantage. *Strategic Management Journal* 15:175–190.

Barros, M. 2009. Terceirização Logística no Brasil. Working paper Ilos–Instituto de Logística e Supply Chain. http://www.ilos.com.br/site/index.php?option=com_content&task=view&id=738&Itemid=279 (accessed March 1, 2011).

Baumgartner, H., and Homburg, O. 1996. Applications of structural equation models in marketing and consumer research: A review. *International Journal of Research in Marketing* 13(6):139–161.

Bentler, P. 1990. Comparative fit indexes in structural models. *Psychological Bulletin* 10(7):238–246.

Berman, J. 2010. 3PL news: Annual 3PL study indicates industry is showing post-recession growth signs. *Logistics Management*, September 30. http://www. logisticsmgmt.com/article/3pl_news_annual_3pl_study_shows_industry_is_ showing_post-recession_growth_s/ (accessed March 2, 2011).

Boer, L., Gayatan, J., and Arroyo, P. 2006. A satisfying model of outsourcing. *Supply Chain Management: An International Journal* 11(5):444–455.

Bot, B., and Neumann, C. 2007. Growing pains for logistics outsourcers. *The McKinsey Quarterly* 2:24–32.

Bowersox, D., and Closs, D. 2001. *Logística Empresarial: O Processo de Integração da Cadeia de Suprimento*. São Paulo: Atlas.

Boyson, S., Corsi, T., Dresner, M., and Rabinovich, E. 2009. Managing effective third-party logistics relationships: What does it take? *Journal of Business Logistics* 20(1):73–100.

Brasil, V. 2005. Análise das Variáveis Antecedentes e das Conseqüências do Uso de Diferentes Sistemas de Entrega de Serviços. Ph.D. Thesis, UFRGS.

Byrne, B. 2001. *Structural Equation Modeling with AMOS*. Mahwah, NJ: Lawrence Erlbaum Associates.

Cain, R. 2009, January. Outsourcing without fear. *World Trade,* available at: www. worldtrademag.com.

Capgemini. 2007. Third-party logistics. http://www.capgemini.com (accessed October 11, 2010).

CEL/COPPEAD. 2008. Indicadores sobre Prestadores de Serviços Logísticos. http:// www.ilos.com.br/site/index.php?option=com_deeppockets (accessed October 18, 2009).

Coase, R. 1937. The nature of the firm. *Economica, New Series* 4(16):386–405.

Craig, D., and Willmott, P. 2005. Outsourcing grows up. *The McKinsey Quarterly* 1:13–26.

DiBenedetto, B. 2009. Global 3PLs grew 6.5%. *The Journal of Commerce*, January 29. www.joc.com (accessed February 20, 2010).

DiMaggio, P., and Powell, W. 1983. The iron cage revisited: Institutional isomorphism and collective rationality in organizational fields. *American Sociological Review* 47:147–160.

Donaldson, L. 1998. Teoria da contingência estrutural. In *Handbook de Estudos Organizacionais*, ed. S. Clegg, C. Hardy, and W. Nord, 105–113. São Paulo: Atlas.

European Commission. 2000. *Protrans: Analysis of third-party logistics market*. Deliverable No. 1, October, Competitive and Sustainable Growth Programme of the 5th Framework Programme.

Garver, M., and Mentzer, J. 1999. Logistics research methods: Employing structural equation modeling to test for construct validity. *Journal of Business Logistics* 20(1):33–57.

Gilley, K., and Rasheed, A. 2000. Making more by doing less: An analysis of outsourcing and its effects on firm performance. *Journal of Management* 26(4):763–790.

Gonzalez-Diaz, M., Arrunada, B., and Fernandez, A. 2000. Causes of subcontracting: Evidence from panel data on construction firms. *Journal of Economic Behavior and Organization* 42(2):167–187.

Greaver, M. 1999. *Strategic Outsourcing: A Structured Approach to Outsourcing Decisions and Initiatives*. New York: American Management Association.

Gunasekaran, A., and Irani, Z. 2010. Modeling and analysis of outsourcing decisions in global supply chains. *International Journal of Production Research* 48(2):301–304.

Hair, J., Anderson, R., Tatham, R., and Black, W. 2005. *Análise Multivariada de Dados*. New York: Bookman.
Harland, C., Knight, L., Lamming, R., and Walker, H. 2005. Outsourcing: Assessing the risks and benefits for organizations, sectors and nations. *International Journal of Operations and Production Management* 25(9):831–850.
Harrington, D. 2008. *Confirmatory Factor Analysis*. Oxford: Oxford University Press.
Holcomb, T., and Hitt, M. 2006. Toward a model of strategic outsourcing. *Journal of Operations Management* 10:42–65.
Hsiao, H., Kemp, R., and Van der Vorst, J. 2010. A classification of logistic outsourcing levels and their impact on service performance: Evidence from the food processing industry. *International Journal of Production Economics* 124:75–86.
Iañes, M., and Cunha, C. 2006. Oma metoddogia para a selecto de un provedor de servicor logisticos. *Produsgar* 16(3): 394–412.
Isiklar, G., Alptekin, E., and Büyüközkan, G. 2007. Application of hybrid intelligent decision support model in logistics outsourcing. *Computers and Operations Research* 34:3701–3715.
Ivanaj, V., and Franzil, Y. 2006. Outsourcing logistics activities: A transaction cost economics perspective. Paper presented at XVème Conférence Internationale de Management Stratégique, Annecy (AIMS), Geneva.
Jharkharia, S., and Shankar, R. 2007. Selection of logistic service provider: An analytic network process (ANP) approach. *Omega: The International Journal of Management Science* 35:274–289.
Khan, S., and Schroder, B. 2009. Use of rules in decision-making in government outsourcing. *Industrial Marketing Management* 38(4):379–386.
Koufteros, X., Babbar, S., and Kaighobadi, M. 2009. A paradigm for examining second-order factor models employing structural equation modeling. *International Journal of Production Economics* 112:2–39.
Kremic, T., Tukel, O., and Rom, W. 2006. Outsourcing decision support: A survey of benefits, risks, and decision factors. *Supply Chain Management: An International Journal* 11(6):462–482.
Lau, K., and Zhang, J. 2006. Drivers and obstacles of outsourcing practices in China. *International Journal of Physical Distribution and Logistics Management* 36(10):776–792.
Liu, C., and Lions, A. 2011. An analysis of third-party logistics performance and service provision. *Transportation Research Part E* 47:547–570.
Madhok, A. 2002. Reassessing the fundamentals and beyond Ronald Coase, the transaction cost and resource-based theories of the firm and institutional structure of production. *Strategic Management Journal* 23:535–550.
Mantel, S., Mohan, T., and Lizo, Y. 2006. A behavioral study of supply manager decision making: Factors influencing making versus buy evaluation. *Journal and Operations Management* 24:822–838.
McGinnis, M., Kochunny, C., and Ackerman, K. 1997. Third party logistics choice. *The International Journal of Logistics Management* 6(2):93–102.
McIvor, R. 2000. A practical framework for understanding the outsourcing process. *Supply Chain Management: An International Journal* 5(1):22–36.
McIvor, R. 2009. How the transaction cost and resource-based theories of the firm inform outsourcing evaluation. *Journal of Operations Management* 27(1):45–63.
Meyers, L., Gamst, G., and Guarino, A. 2006. *Applied Multivariate Research: Design and Interpretation*. New York: Sage.

Milgate, M. 2001. Supply chain complexity and delivery performance: An international exploratory study. *Supply Chain Management: An International Journal* 6(3):106–118.

Monczka, R. 2005. *Outsourcing strategically for sustainable competitive advantage.* CAPS Research Report.

Persson, G., and Virum, H. 2001. Growth strategies for logistics service providers: A case study. *The International Journal of Logistics Management* 12(1):53–64.

Peteraf, M. 1993. The cornerstones of competitive advantage—A resource-based view. *Strategic Management Journal* 14(3):179–191.

Power, D., Sharafali, M., and Bhakoo, V. 2006. Adding value through outsourcing contribution of 3PL services to customer performance. *Management Research News* 30(3):228–235.

Priem, R., and Butler, J. 2001. Is the resource-based view a useful perspective for strategic management research? *Academy of Management Review* 26:22–40.

Quinn, J. 1999. Strategic outsourcing: Leveraging knowledge capabilities. *Sloan Management Review* 68:9–21.

Rajesh, R., Pugazhendhi, S., Ganesh, K., Murlidharan, C., and Sathiamoorthy, R. 2011. Influence of 3PL service offerings on client performance in India. *Transportation Research Part E* 47:149–165.

Razzaque, M., and Sheng, C. 1998. Outsourcing of logistics functions: A literature survey. *International Journal of Physical Distribution and Logistics Management* 28(2):89–107.

Reeves, K., Caliskan, F., and Ozcan, O. 2010. Outsourcing distribution and logistics services within the automotive supplier industry. *Transportation Research Part E* 46:459–468.

Ren, S., Ngai, E., and Cho, V. 2010. Examining the determinants of outsourcing partnership quality in Chinese small- and medium-sized enterprises. *International Journal of Production Research* 48(2):453–475.

Rodriguez, T., and Robaina, V. 2005. A resource-based view of outsourcing and its implications for organizational performance in the hotel sector. *Tourism Management* 26:707–721.

Ross, W., Dalsace, F., and Anderson, E. 2005. Should you set up your own sales force or should you outsource it? Pitfalls in the standard analysis. *Business Horizons* 48(1):23–36.

Saaty, T. 1980. *The Analytic Hierarchy Process.* New York: McGraw-Hill.

Schoenherr, T. 2010. Outsourcing decisions in global supply chains: An exploratory multi-country survey. *International Journal of Production Research* 48(2):343–378.

Sink, H., and Langley, C. 1997. A managerial framework for the acquisition of third party logistics services. *Journal of Business Logistics* 18(2):163–188.

Sirmon, D., Hitt, M., and Ireland, R. 2007. Managing firm resources in dynamic environments to create value: Looking inside the black box. *Academy of Management Review* 32(1):273–292.

Sohail, M., and Sohal, A. 2003. The use of third party logistics services: A Malaysian perspective. *Technovation* 25:401–408.

Water, H., and Peet, H. 2007. A decision support model based on the analytic hierarchy process for the make or buy decision in manufacturing. *Journal of Purchasing and Supply Chain Management* 12:258–271.

Willcocks, L., and Currie, W. 1997. Information technology in public services: Towards the contractual organization? *British Journal of Management* 8(1):107–120.

Williamson, O. 1995. Hierarchies, markets and power in the economy: An economic perspective. *Industrial and Corporate Change* 4(1):21–49.

Wu, F., Yeniyurt, S., Kim, D., and Cavusgil, S. 2005. The impact of information technology on supply chain capabilities and firm performance: A resource-based view. *Industrial Marketing Management* 35:493–504.

Yang, C., and Huang, J. 2000. A decision model for IS outsourcing. *International Journal of Information Management* 39(1):225–239.

Ying, J., and Dayong, P. 2005. Multi-agent framework for third party logistics in e-commerce. *Expert Systems with Applications* 29:431–436.

11

Revenue Management of Transportation Infrastructure during the Service Life Using Real Options

Hongyan Chen and Ruwen Qin

CONTENTS

11.1 Introduction

Transportation infrastructures, such as highways, railways, and airports, require a huge initial capital investment. They are often planned and constructed to serve the public for decades. Operating and maintenance (O&M) costs during the long service life of the infrastructure are another important cost component to be considered for effective revenue management. These two types of costs are relevant since superior construction quality requires higher capital investment to promote lower O&M costs during the service life. A trade-off between these has to be considered by service operations. Because traffic demand may change significantly during the service life of transportation infrastructure, service revenue may significantly deviate from an original prediction. Facing the deep uncertainty during a project's long life, service providers (usually in the public sector) may confront great challenges to ensure desired services for travelers and a self-liquidating investment in the project (i.e., producing sufficient returns to repay the total investment in the project). Therefore, the public sector favors the involvement of private participants in financing and operating these high-risk, expensive projects. Public–private partnerships (PPPs) provide an ideal way to promote and realize this collaboration.

Although PPPs have provided guidelines for collaborations between public and private sectors, they are far from sufficient to effectively manage infrastructure services. For example, existing literature on build–operate–transfer (BOT) primarily focuses on the concession, which is often shorter than the service life of transportation infrastructure. After an infrastructure is transferred back to the public sector, there is a risk that toll revenues to be collected will not be sufficient to operate and maintain the infrastructure if the actual travel demand is lower than expected. Also, the concession may prevent the public sector from generating timely revenue to alleviate potential traffic congestion by capacity expansions when future travel demand is unexpectedly high. This chapter introduces an incentive scheme of using real options (termed RO incentives) to resolve these problems. The RO scheme (introduced in the following sections) can promote an effective collaboration between the private and public sectors during the entire service life, and help investors and service providers better plan the construction, operation, and maintenance of transportation infrastructure to respond to stochastic travel demand.

RO applies the financial option theory to the valuation of real assets (Hull 2008). An RO is value-added because it allows its owner to take asymmetric actions to favorable conditions and unfavorable ones. It has been used to model flexibility-type incentives such as guarantees, subsidies, and rights of abandonment for the service operations of transportation infrastructure. The public sector is the option writer if it offers the private sector flexibility in operating a PPPs project. The flexibility (e.g., an option to abandon the operation before the concession expires) either mitigates the private sector's concern with project risks or provides greater profits, thereby motivating it to collaborate with the public sector under a PPP agreement. The private sector pays a premium and become the option owner if it decides to take the option. The connections between incentives and ROs have been discussed in the literature. For instance, Mason and Baldwin (1988) modeled subsidies as put options for energy infrastructure development. Huang and Chou (2006) formulated the minimum revenue guarantee (MRG) as European call options. Cheah and Liu (2006) modeled the MRG to the private sector as a put option and the government's right as a call option for a bridge project.

Alonso-Conde et al. (2007) evaluated timing flexibilities for a toll road project. Lara Galera and Sánchez Soliño (2010) used RO to value the minimum traffic guarantee (MTG) for a BOT highway project. Previous studies focused on modeling existing types of incentives as real options and usually valued these solely from the viewpoint of an option owner. This chapter considers the behavioral changes of both parties in the design of an RO incentive contract. It provides insightful guidelines for investors and service providers to correctly value the incentives and use these to support their decisions on investment in, and operations and maintenance of, transportation infrastructure.

Traditional RO valuation has limitations when the options value depends on the decisions of multiple parties. An approach to correct the problem is to embed RO models in a game framework to capture the interaction between them (for example, Smets 1991; Lambrecht and Perraudin 1994; Grenadier 2002). This approach is named option game and has been used to manage infrastructure projects. The joint analysis of RO and game theory is also suitable for deriving RO strategies for collaborations or service operations. The problem discussed in this chapter belongs to this category.

This chapter models the investment in, and operation and maintenance of, transportation infrastructure during its service life as a portfolio of ROs. Game theory is used to model the decision changes, prompted by the RO incentive, between the option owner and option writer. The option game framework developed in this chapter can be generalized and applied to the management of collaborative relationships for other industries, such as other infrastructure projects or the outsourcing of manufacturing.

11.2 Decisions under Build–Operate–Transfer (BOT)

This chapter describes the development of an option game framework using an example of a highway build–operate–transfer (BOT) project. BOT is a PPP agreement often used in transportation infrastructure projects. Figure 11.1 describes a typical BOT project. The private concessionaire (PRI) constructs, owns, and operates the infrastructure during the concession period. After

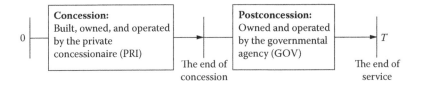

FIGURE 11.1
A BOT project without options.

the concession ends, the governmental agency (GOV) will own and operate the infrastructure until the end of its service life, T.

The entire design of a BOT contract is a complex process. This chapter only focuses on problems related to the design of RO incentive and, thus, assumes that some issues have already been settled between the GOV and the PRI in their initial negotiations. These include the length of the concession period, T_c, the toll price, P, the minimum requirement for construction quality, h, and the PRI's minimum required rate of return, r_c. Now, the GOV is planning the highway capacity to provide the best service for the society, as indicated by the maximum social welfare offered by this project. The PRI is the appropriate party to control the construction quality because it is the builder and owner of the highway during the concession period. The total construction cost, I, is assumed to increase linearly with highway capacity and construction quality; that is,

$$I(k,C) = (h+k)C \qquad (11.1)$$

In Equation (11.1) h is the minimum requirement for construction quality and k is the unit cost of quality improvement. This assumption is made based on the empirical finding in Levinson and Karamalaputi (2003), which estimated the highway construction cost function using data on adding new links and expanding existing capacity in the Twin Cities (Minneapolis and St. Paul, Minnesota) from 1978 to 1998. The construction cost is not deterministic in practice. Cost overruns commonly occur in construction projects because of unpredictable factors, such as the scope or design changes, unforeseen site conditions, and material price changes (Attalla and Hegazy 2003). Since this chapter mainly focuses on the revenue management during the service life of transportation infrastructure, the construction cost is not a primary stochastic process during that long time horizon. This chapter simplifies the uncertainty pertaining to construction cost and yet should be studied further.

The annual O&M cost is assumed to be proportional to a highway's capacity. The annual O&M cost per unit of capacity is a decreasing convex function of the construction quality and an increasing convex function of highway age (Chen and Qin 2012). That is,

$$M(k,C,t) = mk^{-\theta}e^{\delta t}C, \quad m,\theta,\lambda > 0 \qquad (11.2)$$

In Equation (11.2) m is the capacity coefficient, θ is the quality improvement factor, and δ is the aging factor.

11.2.1 Decisions under BOT without Options

To measure and demonstrate the effects of RO incentives, a highway BOT project without options is discussed first. The decision process for this

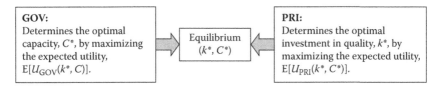

FIGURE 11.2
Contracting process for the BOT without options.

project is a two-stage game with complete and perfect information, as illustrated in Figure 11.2.

The GOV first decides a level of initial capacity, C; after observing the GOV's decision, the PRI determines the level of investment in quality improvement, k. The game can be solved backward to find the equilibrium of the game. The PRI maximizes the expected profit from the concession by optimizing k, for any given C

$$E\left[U_{PRI}(k^*,C)\right] = \max_k \left\{ -I(k,C) + \sum_{t=0}^{T_c-1} e^{-rt}\left[R(\hat{Q}_t) - M(k,C,t)\right] \right\} \quad (11.3)$$

where r is the discount rate, $\hat{Q}_t$ is the expected travel demand during $[t, t+1)$, and $R(\hat{Q}_t) = P\hat{Q}_t$ is the expected toll revenue during that period. The PRI determines the optimal investment in quality improvement,

$$k^* = \left[\theta m \sum_{t=0}^{T_c-1} e^{(\delta-r)t} \right]^{\frac{1}{\theta+1}} \quad (11.4)$$

The GOV's target is the social welfare created by the project, which is measured by the sum of travelers' utility and the service provider's net profit (Xiao et al. 2007),

$$W(k,C) = -I(k,C) + \sum_{t=0}^{T-1} e^{-rt}\left[B(\hat{Q}_t) - T(\hat{Q}_t,C) - M(k,C,t)\right] \quad (11.5)$$

where $B(\hat{Q}_t)$ represents the expected benefit to travelers who use the highway during $[t, t+1)$. $T(\hat{Q}_t,C)$ measures the travel time cost,

$$T(\hat{Q}_t,C) = \beta\hat{Q}_t t^0[1 + a(\hat{Q}_t/C)^b] \quad (11.6)$$

where β represents the average time value per traveler per unit of time. t^0 is the travel time under free flow conditions, and is the traditional BPR

(Bureau of Public Roads) travel time function measuring the time needed to go through a specific route (Tampère et al. 2009).

The GOV must consider the profitability of the PRI when it plans the highway capacity (Yang and Meng 2000). Therefore, the GOV solves the following problem to optimize the highway capacity,

$$E[U_{GOV}(k^*, C^*)] = \max_{C} W(k^*, C)$$

$$\text{s.t.} \quad E[U_{PRI}(k^*, C)] \geq \pi_c$$

(11.7)

Consequently, the payoffs of the GOV and PRI depend on the decisions of both, which is the equilibrium of the game as shown in Figure 11.2. The detailed solution process is available on the authors' Web sites.

Annual travel demand, Q_t, may deviate significantly from the original expectation during the service life of the highway. The two-stage game for determining the BOT contract has limitations because it bases the decisions on the expected annual travel demand. If travel demand during the post-concession period is lower than expected, the GOV may not generate sufficient toll revenues to pay off the initial investment and cover ongoing O&M costs. It has to find additional funds to cover the shortfall. If travel demand is higher than expected, travelers suffer from heavy traffic congestion although the GOV probably generates revenue in excess of the required O&M costs. Excess toll revenue cannot be used to alleviate traffic congestion or improve services in a timely manner considering the lead time of construction.

11.2.2 A BOT Contract with Options

The aforementioned problems may be addressed by adding an RO incentive to the BOT contract. The RO incentive includes two options that are compounded sequentially: an option to continue operating the project after the concession expires and an option to terminate operation of the highway at any time during the postconcession period if the continuation option is exercised. The first option is a European-style continuation option because it can be exercised only at the expiration of concession; the second option is an American-style abandonment option since the PRI can exercise it anytime on or before the end of the highway's service life. This abandonment option is compounded with the continuation option because the former is valid only if the PRI has exercised the latter. Clearly, the options are value-added to the PRI. The GOV prices the options to determine the option premium, which can be used to cover the possible shortage in O&M costs or to add service capacity. The options have the potential to benefit both PRI and GOV, by improving the basic BOT scheme.

The ownership of the highway will be transferred from one party to the other when the PRI exercises either of the two options, as Table 11.1 describes.

TABLE 11.1

Change in Highway Ownership in the BOT with Options

Phase of Service	Ownerships and Transfers
Concession	Solely owned by PRI
Postconcession	Owned by GOV if PRI does not exercise the continuation option; otherwise, owned by PRI until the exercise of abandonment option

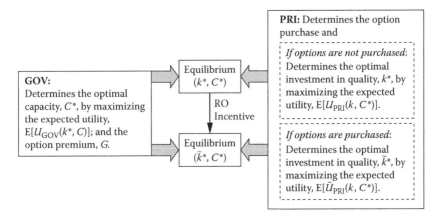

FIGURE 11.3
Contracting process for the BOT with the RO incentive.

The ownership of the option is different from that of highway: the PRI is the owner of the continuation and the abandonment options, and the GOV is the writer of these options.

If the options are sold to the PRI, the GOV must price the options and develop a plan to convert the option premium to travel services. Offered the RO incentive to continue the operation, the PRI faces decisions on option acquisition and exercise. Given the benefits associated with the options, previous decisions on highway capacity and quality improvement may no longer be optimal. New decisions are associated with behavioral changes and may also change the value of options.

The RO incentive is designed as an add-on to the BOT contract. Therefore, the GOV's decision on capacity, C^*, has to keep unchanged (although it may no longer be optimal) and yet the PRI itself can determine the level of its investment in quality (beyond the minimum). This is due to the fact that the BOT contract without options defines just the minimum requirement for construction quality. Consequently, this chapter discusses only the behavioral change of the PRI in the investment, which is stimulated by the RO incentive. Figure 11.3 describes the potential change in the contracting process after the RO incentive is added to the BOT contract. The GOV determines the premium of the RO incentive first. Then, the PRI decides the options purchase by valuing the potential return from the premium. If the PRI decides not to buy the options, the decision regarding quality improvement remains the same as that for the BOT without options. Otherwise, it will determine a new investment in quality improvement, $\tilde{k}^*$, that maximizes the expected profit from the entire project.

After the options for the postconcession period are added to the BOT contract, the PRI's valuation function for the project changes to

$$E\left[\tilde{U}_{PRI}\left(\tilde{k}^*, C^*\right)\right] = \max_k \left\{ \begin{aligned} &-I\left(k, C^*\right) \\ &+\sum_{t=0}^{T_c-1} e^{-r_{rf}t}\left[R(\hat{Q}_t) - M\left(k, C^*, t\right)\right] \\ &tV_{PRI}\left(k, C^*\right) - G \end{aligned} \right\} \tag{11.8}$$

where V_{PRI} is the options value to the PRI, and G is the premium determined by the GOV. The value of G should be no greater than $V_{PRI}(\tilde{k}^*, C^*)$ to attract the PRI to purchase the options. A comparison of Equation (11.8) to Equation (11.3) clearly shows that the option incentive may cause the PRI to change its investment decision which, in turn, can also change the option value.

11.2.3 Valuation of the Options

The PRI's new valuation function in Equation (11.8) indicates that the valuation of the options is critical to the derivation of both parties' optimal decisions. This section presents a discrete-time method to value the options.

Lara Galera and Sánchez Soliño (2010) tested the traffic volume data from 11 toll highway stretches and proved that the hypothesis of the geometric Brownian motion (GBM) process for travel demand is valid (i.e., the growth rate of traffic volume follows a normal distribution). Based on their work, this chapter assumes that travel demand follows a GBM process during the service life of transportation infrastructure:

$$\frac{dQ_t}{Q_t} = \mu dt + \sigma dW_t \tag{11.9}$$

In Equation (11.9), the annual growth rate of travel demand has a normal distribution, $N(\mu, \sigma^2)$. The term σ is commonly known as volatility, which measures the scale of travel demand uncertainty. The stochastic movement of Q_t is modeled by W_t, a standard Wiener process (Hull 2008). Travel demand, following a GBM, indicates that it is independent of travel costs (toll price and travel time cost). The assumption of rigid traffic demand is not universally valid. Demand could be elastic, decreasing with increased travel costs

(Yang and Meng 2000; Subprasom and Chen 2007; Tan et al. 2010). However, rigid travel demand can often be observed. For example, such demand is commonly seen in developing regions where the construction of transportation infrastructure lags behind an exploding economy; and in urban areas, especially during rush hours and when only one route is available to reach a particular destination. For these cases, travel demand is not affected by toll price or congestion.

This chapter values the RO incentive in a risk-neutral world where all investors are indifferent to risks (Hull 2008). A problem can be transformed from the real world to the risk-neutral world by measuring the uncertainty in the underlying asset using a risk-neutral probability, p, and discounting cash flows using the risk-free rate, r_{rf}. Decision outcomes in a risk-neutral world remain the same as those in the real world. However, the risk-neutral decision makes the process easier for operations managers because no risk-adjusted rate needs to be estimated for each party or at various times in the life of the project. A real asset, depending on the variable S, can be valued in the risk-neutral world by reducing the growth rate of S from μ to $\mu-\lambda\sigma$, where λ is the market price of the risk of this asset, and discounting cash flows at the risk-free rate (Hull 2008).

Considering that decisions in infrastructure development and operations are often made at discrete time, we use the binomial lattice method to value the options. First, the lattice of the underlying process (i.e., the annual travel demand Q_t) needs to be built. During the service life of the highway, Q_t can either move up to a higher level, uQ_t, or move down to a lower level, dQ_t, at a time interval, Δt. As in Cox, Ross, and Rubinstein (1979), the increase and decrease ratios are set to be

$$u = e^{\sigma\sqrt{\Delta t}} \quad \text{and} \quad d = 1/u \tag{11.10}$$

where the term σ is the volatility rate of Q_t. The risk-neutral probability of the up movement, p, is

$$p = \frac{e^{(\mu-\lambda\sigma)\Delta t} - d}{u - d} \tag{11.11}$$

Starting with the initial traffic volume, Q_0, the evolution of the Q_t throughout the highway's service life is discretely approximated by the binomial lattice shown in Figure 11.4.

The binomial lattice lays out the annual traffic volume at discrete time, $t = 0, 1, \ldots, T$. At anytime t the lattice has $t + 1$ nodes (t, i) $(i = 0, 1, \ldots, t)$. Each node represents a unique level of annual travel demand, Q_{ti}, and

$$Q_{ti} = Q_0 u^i d^{t-i} \tag{11.12}$$

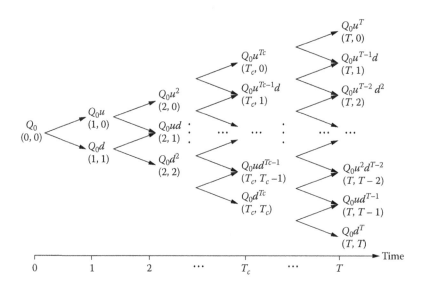

FIGURE 11.4
Binomial lattice approximating annual travel demand process.

The RO value measures the value of flexibilities that the option owner has under uncertain environments. With the arrival of new information, the option owner is able to have asymmetric responses to opportunities of gaining profits and risks of losing profits. The flexibilities can then be valued by an option valuation lattice that is derived from the underlying variable lattice. In this chapter, the flexibilities are an abandonment option compounded with a continuation option. Therefore, the abandonment option is evaluated first.

Denote A_t as the PRI's action at t (for $t = T_c$, $T_c + 1$, ..., $T-1$). At the beginning of each year during the postconcession period, the PRI needs to decide whether effectively abandoning the operation the next year will maximize the expected value-to-go,

$$Z(Q_{ti}, A_t) = \max_{A_{t+1}} \left\{ \begin{aligned} &R(Q_{ti}, A_t) - M(k, C^*, t, A_t) \\ &+ e^{-r_{rf}\Delta t}\left[pZ(Q_{(t+1)i}, A_{t+1}) + (1-p)Z(Q_{(t+1)(i+1)}, A_{t+1}) \right] \end{aligned} \right\} \quad (11.13)$$

where actions A_{t+1} can be either "operate" or "abandon." $Z(Q_{ti}, A_t)$ is the expected value-to-go at the node (t, i) of the option valuation lattice. $R(Q_{ti}, A_t) - M(k, C^*, t, A_t)$ is the net cash flow of year t at travel demand Q_{ti} and action A_t. The decision in A_{t+1} is made only if A_t is operate because the PRI can no longer operate the highway after it has abandoned the right to continue operating the infrastructure. Therefore, $Z(Q_{ti}, \text{abandon})$ is equal to 0. The backward recursion process stops at T_c when the PRI must decide whether to exercise the continuation option.

The continuation option is European style and can only be exercised at the end of the concession period, T_c. The PRI values the expected value-to-go associated with each possible action to find the best strategy. That is,

$$Z\left(Q_{T_ci}\right) = \max_{A_{T_c}}\left\{Z\left(Q_{T_ci}, A_{T_c}\right)\right\} \tag{11.14}$$

where action A_{T_c} is either "continue" or "expire." If the PRI decides to exercise the continuation option, it obtains the abandonment option for the postconcession period, and $Z(Q_{T_ci}) = Z(Q_{T_ci}$, continue). Otherwise, the continuation option expires and $Z(Q_{T_ci}) = Z(Q_{T_ci}$, expire) = 0.

The expected present value of $Z(Q_{T_ci})$ is the value of options to the PRI:

$$V_{\text{PRI}}\left(k,C^*\right) = e^{-r_{rf}T_c}\sum_{i=0}^{T_c} Z(Q_{T_ci})p(Q_{T_ci}) \tag{11.15}$$

Here, $V_{\text{PRI}}\left(k,C^*\right)$ is presented as a function of both k and C^* because it is also relevant to the GOV and PRI's investment decisions, according to Equation (11.13). p in Equation (11.15) is the risk-neutral probability that Q_{ti} appears.

11.2.4 Determination of Options Premium

Determination of the premium for the continuation and abandonment options is critical because it directly affects the implementation of the RO incentive scheme. The GOV takes a variety of objectives into account in this decision. It particularly considers the needs to realize the desired service, to increase the possibility that the project will be self-liquidating, and to ensure the profit required by the private sector.

Attractive profits may motivate the PRI to purchase the option to continue the project. Therefore, the option premium, G, must be bounded above by the net profit of PRI, $V_{PRI}(\tilde{k}^*, C^*) - \pi_o$, where π_o is the minimum required return of the PRI for the postconcession period. The GOV hopes that the option premium is an effective financing resource for possible highway maintenance and service improvement, such as capacity expansions. Hence, G is bounded below by $F_{OS}(\tilde{k}^*, C^*) + F_E$, where $F_{OS}(\tilde{k}^*, C^*)$ is the total potential O&M expense shortage with the RO incentive, and F_E is the expected fund needed for the service improvement. Considering all possible relationships between $V_{PRI}(\tilde{k}^*, C^*) - \pi_o$ and $F_{OS}(\tilde{k}^*, C^*) + F_E$, the option premium, G, is set to

$$\min\left[F_{OS}(\tilde{k}^*, C^*) + F_E, V_{PRI}(\tilde{k}^*, C^*) - \pi_O\right] \le G \le V_{PRI}(\tilde{k}^*, C^*) - \pi_O \tag{11.16}$$

11.3 Case Study of a Highway Project

11.3.1 Case Description

This chapter presents a case study to illustrate the implementation of the RO incentive. The case originally discussed in Chen and Qin (2012) is about a highway project in China. To support the fast development of economy, city I is developing a new industrial park in its exurb. City II is near the new park and has an old provincial highway passing by the park. To meet the growing travel demand, the Department of Transportation of City II (GOV) would like to construct a 30-km highway connecting the industrial park and the urban area of city II. The highway is designed to have a 30-year useful life. This highway will reduce the travel distance by 25 km and the travel time by 30 minutes, from 50 to 20 minutes. Facing a large investment in construction and uncertainty over the lengthy service life, the GOV let a private company (PRI) develop the highway under a BOT agreement. They have decided an average toll rate (P) of 10 yuan per vehicle, a 20-year concession period (T_c), and a minimum construction quality requirement (h) of 6.2 yuan per unit of capacity.

The GOV anticipates that the travel demand in the first year of operation is approximately 20 million vehicles. The growth rate of the travel demand is 4% per year, with a standard deviation of 15%. The high volatility in travel demand promises great uncertainty in toll revenue. Therefore, besides the basic BOT contract, the GOV is considering offering the PRI an RO incentive, which includes an option to continue operating the highway after the concession period and an option to terminate the operation early with a 1-year notice. The GOV will use the upfront option premium to expand the capacity of the highway when needed or cover the possible shortage in O&M expenses. Two additional lanes can be added, which will increase the capacity by 17.4 million vehicles per year. Demand model parameters and contract model parameters are summarized in Table 11.2.

TABLE 11.2

Parameter Values for the Numerical Example

Demand Model Parameters	
$\sigma = 0.15$	$\lambda = 0.05$
$r_{rf} = 0.05$	$Q_0 = 20$ [million vehicle/year]
$\mu = 0.04$	$T(\hat{Q}_t, C) = 5\hat{Q}_t\left[1 + 0.15(Q_t/C)^4\right]$
Contract Model Parameters	
$P = 10$ [yuan/vehicle]	$C = \{C^*, C^* + 17.4\}$ [million vehicle/year]
$T = 30$ [year]	$M = 15k^{-1}e^{0.06t}C$ [yuan/year]
$T_c = 20$ [year]	$I = (20 + k)C$ [yuan]

The calculation of minimum profit required by the PRI during the concession period, π_c, is based on the fact that the PRI needs a equivalent annual rate of return at 10% during the concession period, that is

$$\pi_c = (e^{(0.1-r_{rf})T_c} - 1)I \tag{11.17}$$

The option will yield no return until the concession period expires in 20 years; however, the PRI passes the option premium, G, to the GOV when the concession just starts. Besides a reasonable profit from the options (for example, a 10% annual rate of return from the premium during the postconcession period), the PRI requests an 8% annual rate of return as the opportunity cost of waiting. The expected net profit from the options investment, π_O, is determined by

$$G = V_{\mathrm{PRI}} e^{-\left[0.08T_c + 0.1(T-T_c) - r_{rf}T\right]}$$

$$\pi_O = V_{\mathrm{PRI}} - G \tag{11.18}$$

11.3.2 Option Valuation

To illustrate the calculation of the RO incentive, we perform the valuation procedure described in Section 11.2.3 using the data pertaining to this case. First, the lattice of the annual travel demand can be built. The initial travel demand is $Q_0 = 20$ million vehicles/year, the up/down movement factors are $u = 1.1618$ and $d = 0.8607$, and the risk neutral probability of the up movement is $p = 0.5723$, respectively, according to Equation (11.10) and Equation (11.11). Then, the option value lattice is generated based on the traffic demand lattice, as shown in Figure 11.5. The number in a bubble is the value of $Z(Q_{ti}, A_t)$ (see Equation 11.13 and Equation 11.14) at node (Q_{ti}, A_t). The action, *operate* or *abandon*, in any bubble during $[T_c + 1, T - 1]$ is the decision to abandon for the next time step, A_{t+1}, made on this node. The action of continue or expire in the bubbles, at time T_c, is the decision to continue at the current node. The A and B highlight bubbles in Figure 11.5 are two examples that illustrate the valuation of options and the decision-making process.

Node A is at time $T - 1$ when the abandonment option may be alive. The value is the expected value-to-go, which is estimated at node A, given A_{T-1}. At node A if the abandonment option is still alive (i.e., A_{T-1} = operate), the PRI needs to decide whether to exercise the abandonment option at time T, based on the expected remaining value under each action at time T. If the action for time T, A_T, made at node A is operate, the expected remaining at time T would be either −326.52 million yuan, with a probability of 0.5723, or −327.52 million yuan, with a probability of 0.4277. Thus the risk neutral expectation from operating the project on year T is $e^{-0.5 \times 1} \times (-326.52 \times 0.5723 - 327.52 \times 0.4277) = -326.83$. If the A_T made at node A is abandon, the expected remaining value at time T is 0. The annual net cash flow at node A is independent to A_T, as it is decided by A_{T-1}, which is made at time $T - 2$. At node A, $Q_A = 0.40$ million

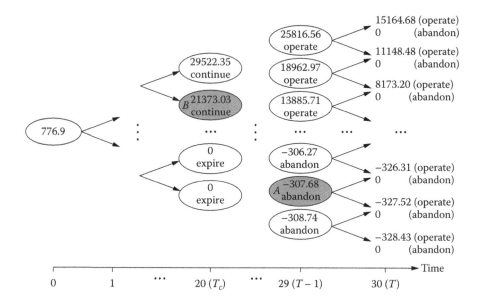

FIGURE 11.5
Option valuation lattice.

vehicles per year and $M(\tilde{k}^*, C^*, T-1) = 311.68$. If A_{T-1} for node A is *operate*, the annual net cash flow at node A is $R(Q_A, operate) - M(\tilde{k}^*, C^*, T-1, operate) = 10 \times 0.4 - 311.68 = -307.68$ million yuan. Therefore, according to Equation (11.12), if A_{T-1} is operate, the value function on node A is

$$Z(Q_A, operation) = \max_{A_T}\left\{\frac{-307.68 - 326.83}{\text{if } A_T \text{ is "operate"}}, \frac{-307.68 + 0}{\text{if } A_T \text{ is "abandon"}}\right\} \quad (11.19)$$

$$= -307.68 \text{ million yuan}$$

Apparently, the decision on A_T at node A would be *abandon*. If A_{T-1} for node A is *abandon*, the PRI's operation will be terminated at node A; thus, $Z(Q_A, abandon)$ will be 0 and no decision on A_T is needed.

Node B is at time T_c when the decision on whether to exercise the continuation option needs to be made. The expected remaining profits during the years that follow would be 21,373.03 million yuan, if the PRI chooses to continue the operations. It would be 0 if the PRI chooses to not exercise the option. According to Equation (11.14), the value function at node B is

$$Z(Q_B) = \max_{A_{Tc}}\left\{\frac{21373.03}{\text{if } A_{T_c} \text{ is "operate"}}, \frac{0}{\text{if } A_{T_c} \text{ is "abandon"}}\right\} \quad (11.20)$$

$$= 21{,}373.03 \text{ million yuan}$$

Therefore, the PRI's decision on node B is to exercise the continuation option. When all of the values on the nodes at time T_c are obtained, we discount them along the lattice to node $(0, 0)$ using Equation (11.15). The value of the RO incentive yields 776.9 million yuan.

The computational complexity of a binomial tree with n steps is $O(n^2)$. We code the options valuation using MATLAB, a popular technical computing language. The computational time for solving this 30 time steps binomial tree is 0.06 of a second. Commercial software and open sources, such as DerivaGem (Hull 2008), are available for valuating standard options.

11.4 Numerical Results

The numerical results of this case are summarized in Table 11.3.

11.4.1 Decision Change Motivated by the RO Incentive

We can prove that the optimal investment in quality improvement when the RO incentive is offered, $\tilde{k}^*$, is no less than that when no RO incentive is offered; that is, $\tilde{k}^* \geq k^*$. By offering the options, the GOV motivates the PRI to increase the investment in quality improvement, thereby reducing maintenance costs. This is further demonstrated by the results in Table 11.3. Without an RO incentive, the optimal capacity is 32.1 million vehicles per year (about four lanes, assuming the average capacity per lane is 1000 vehicles per hour), and the optimal quality factor is 6.2 yuan per capacity unit. Offered the RO incentive, the PRI increases the quality factor from 6.2 to 8.8 yuan per unit of capacity. Consequently, the initial investment is increased by 10%, from 840 to 925 million yuan.

TABLE 11.3

Results for the Example

	Governmental Agency (GOV)		Private Concessionaire (PRI)	
	Decision	Outcomes	Decision	Outcomes
No RO incentive	$C^* = 32.1 \times 10^6$	$F_{BS} = 49.4$	$k^* = 6.2$	$I = 840$
				$E[U] = 1514$
RO incentive added	$C^* = 32.1 \times 10^6$	$F_{OS} = 35.7$	$\tilde{k}^* = 8.8$	$I = 925$
		$F_E = 287.8$		$E[U] = 2028$
Changes	$G = 285.6$	$\Delta F_S = 17.3 \ (28\%)$	$\Delta k = 2.6 \ (42\ \%)$	$V_{PRI} = 776.9$
		$\Delta W_k = 28.4$		$\Delta I = 85 \ (10\%)$
		$\Delta W_c = 3.5 \times 10^4$		$\Delta E[U] = 514 \ (34\%)$

Note: The unit of capacity is vehicle per year, the unit of quality investment is yuan per vehicle per year, and the unit of all other monetary measurements is million yuan.

The increased investment in quality improvement reduces the expected shortfall in O&M expenses facing the GOV, that is, $F_{OS}(\tilde{k}^*, C^*) \leq F_{BS}(k^*, C^*)$. In this case, the expected shortfall in O&M expenses drops by 28%, from 49 to 36 million yuan. The increased investment in quality also makes the project produce greater social welfare, that is, $W(\tilde{k}^*, C^*) \geq W(k^*, C^*)$. In this case, the social welfare increases 28 million yuan because of improved quality, designated by ΔW_k in Table 11.3.

11.4.2 Benefits of Option Premium

The RO incentive, V_{PRI}, shown in Table 11.3 is worth 777 million yuan to the PRI and the GOV receives a premium, G, equal to 259 million yuan. The premium can effectively meet the need for O&M expenses if the PRI abandons operation of the expressway during the postconcession period (36 million yuan in this case).

The benefits of the option premium to the GOV are not limited to the provision of a financial resource when toll revenue is insufficient to pay for highway operation. The GOV may consider using the premium to further improve the highway service. In this case, the GOV uses the premium as a resource to finance expansions of highway capacity and to maintain the added capacity. Using the optimal expansion strategy*, the premium can provide 91% of the expected expansion-related expenses, when two lanes are added, and effectively add 3.50×10^4 million yuan of social welfare (ΔW_C) over the 30-year lifetime of the project (near 40 yuan per vehicle).

11.4.3 Total Social Welfare Improved

The total social welfare added by the RO incentive is the sum of the social welfare added by the improved construction quality, ΔW_k, and that added by increased service capacity, ΔW_C. The values of ΔW_k and ΔW_C are both nonnegative. Thus, the RO incentive can always improve the social welfare produced by the project. Table 11.3 demonstrates that ΔW_C is significantly higher than ΔW_k; therefore ΔW_k can be seen as a favorable side effect of the RO incentive.

11.4.4 Effectiveness of the RO Incentive

The effectiveness of the RO incentive may vary depending on project conditions. This section extends the case to a wider range of conditions and examines the impact of volatility and the length of the concession on the decisions and outcomes.

---

* The optimal expansion strategy can be realized using dynamic programming. Please refer to Puterman (1994).

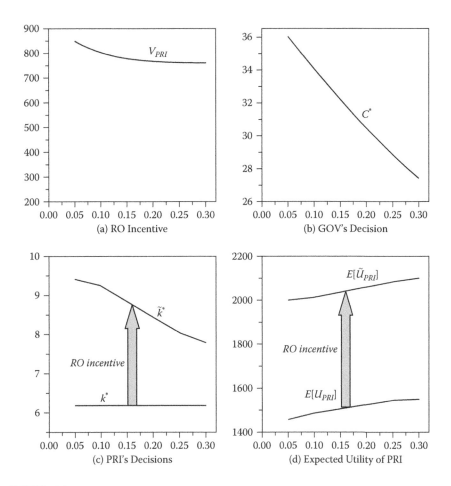

FIGURE 11.6
Impacts of demand volatility on the effectiveness of the RO incentive.

The uncertainty in annual traffic volume is measured by the volatility, σ. The impact of volatility on the effectiveness of the RO incentive is illustrated in Figure 11.6. In a standard RO valuation, options value increases with the growth of volatility. Figure 11.6a shows that the value of the RO incentive decreases as volatility increases. This unusual observation is related to the behavioral changes of the GOV and the PRI. Figure 11.6b,c indicate that they tend to be conservative in their investments: the GOV will reduce the initial highway capacity if future traffic demand is highly uncertain and, similarly, the PRI will also reduce its investment in quality improvement. These changes will increase O&M costs, which thereby reduces the value of the RO incentive. Therefore, the behavioral dynamics of the option issuer and option owner can affect the option value, and a standard RO valuation has its limitations for conditions that multiple parties' decisions are dependent and

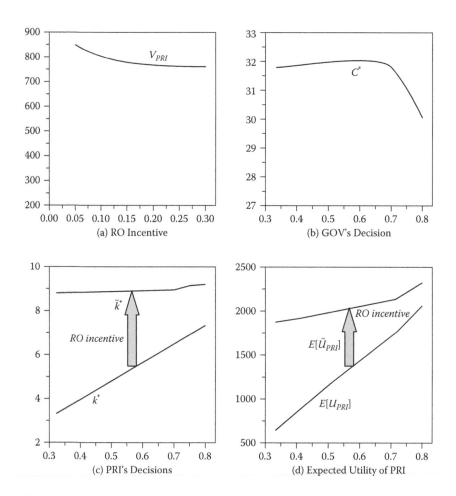

FIGURE 11.7
Impact of the concession period on the effectiveness of the RO incentive.

sensitive to uncertainty. Nevertheless, the decrease in the value of the RO incentive is not large, as Figure 11.6a shows. The RO incentive can still attract the PRI to participate in PPPs projects, as Figure 11.6d indicates. Therefore, as the writer of the RO incentive, the GOV is capable of managing the investment and services of the highway under high volatility.

The PRI is discouraged from participating in a BOT if only a short period of ownership is given. Figure 11.7 illustrates how the RO incentive improves a BOT scheme with a short concession period. Although the PRI can reduce investment in quality, k^*, as the concession length decreases (shown in Figure 11.7c), it may still not be able to maintain the expected utility, $E[U_{PRI}]$. Figure 11.7d shows that $E[U_{PRI}]$ decreases quickly as the concession length decreases. If the GOV offers a longer concession period, in order to encourage the PRI to participate, it may have to lower the capacity requirement to meet the profitability constraint (see Figure 11.7b). Offering a longer concession period is not a good solution for both parties in the BOT contract.

The RO incentive is able to address this dilemma. The options value will increase if the postconcession period becomes longer (i.e., the concession period decreases), as Figure 11.7a shows. The increase in the options value compensates for the loss in the PRI's profit due to the shortened concession period. Therefore, Figure 11.7d shows that the PRI's expected profit with options, $E[\tilde{U}_{PRI}]$, decreases slowly with the decreased concession period than $E[U_{PRI}]$ does. Moreover, when a shorter concession period is selected for the BOT contract, the PRI can generate a larger portion of profit from the postconcession period, which is less risky than that from the concession period. Therefore, the PRI will not significantly reduce its investment in quality improvement even though the concession period is short, as Figure 11.7c shows. The GOV can ask a higher premium although it offers a shorter concession period. This is because it gives the PRI a longer period with options. The higher premium will produce greater social welfare. The RO incentive added to the basic BOT contract improves the positions of both parties.

11.5 Conclusions and Future Research

Transportation infrastructure projects usually require huge capital investments and face high uncertainty in travel demand during lengthy service lifetimes. Investors and service providers of transportation infrastructure thereby confront significant revenue and service management challenges. The public sector often partners with the private sector to develop such projects under PPP agreements such as BOT projects to alleviate financial pressure and transfer risk. Revenue management for PPP projects frequently focuses on the concession period, which conflicts with the growing concern with sustainability. This chapter has presented an RO incentive scheme of managing revenues and services under BOT. The RO incentive gives the private sector a right to continue the operation after the concession period expires (i.e., a continuation option) and a right to terminate the continued operation early (i.e., an abandonment option). Due to the involvement of multiple parties in BOT, the optimal design of the RO incentive cannot be derived from the standard RO valuation. Consequently, this chapter analyzes the problem in an option game framework.

The work presented in this chapter also initiates discussions of important future research. Most RO studies for transportation consider rigid travel demand, that is, traffic volume is independent of the toll rate and congestion level. Although rigid demand assumption is valid for many situations, elastic demand is the other important situation often seen in the real world. Relaxation of the rigid demand assumption is an important extension of this research and will broaden the application of RO in transportation service operations. This chapter uses a linear model for the construction cost. Nonlinear models can be inevitable (e.g., increasing to decreasing economies of scale). As a new concept for the investment and service management of transportation infrastructure, the proposed work will need some time to test, improve, and enrich its implementation in practice. An important future work is to further verify the proposed methodology with real data. Besides designing options for the postconcession period to promote sustainability and effective service management, other aspects of service management can be improved by option designs as well.

Acknowledgments

The authors are grateful to the editors and anonymous referees for their valuable comments and suggestions in this chapter.

Appendix

Table of Notations

Notation	Description
B	Travelers benefits
C^*	Optimal initial highway capacity
C_t	Highway capacity at year t
U_{GOV}	GOV's utility
U_{PRI}	PRI's utility without RO incentive
$\tilde{U}_{PRI}$	PRI's utility with RO incentive
F_{BS}	Expected shortage of maintenance expenses if GOV does not offer the RO incentive
F_{OS}	Expected shortage of maintenance expenses if GOV offers the RO incentive
F_E	Expected expansion-related costs
G	RO incentive premium
I	Construction investment
M	Maintenance costs
P	Toll price
$\hat{Q}_t$	Average traffic volume during $[t, t+1)$
Q_t	Traffic volume during $[t, t+1)$
R	Toll revenue
T	Highway service life
T_c	Concession length
V_{PRI}	RO incentive value
V^*	Expansion option value

W	Expected social welfare produced by the project
ΔW	Expected social welfare added by the RO incentive
ΔW_k	Expected social welfare associated with the change in quality investment
ΔW_C	Expected social welfare produced using the option premium
h	Unit construction cost that meets the minimum quality requirement
k	Unit investment in quality improvement
k^*	Optimal investment in quality improvement without RO incentive
$\hat{k}^*$	Optimal investment in quality improvement with RO incentive
p	Risk-neutral probability
r	Discount rate
r_{rf}	Risk-free rate
t^0	Travel time through the highway under free flow condition
β	Average time value per traveler per unit time
λ	Market price of risk for Q_t
μ	Expected annual growth rate of travel volume
π_c	Minimum profit request of the PRI from concession
π_o	Minimum profit request of the PRI from purchasing the RO incentive
σ	Volatility of annual travel demand

References

Alonso-Conde, A. B., C. Brown, and J. Rojo-Suarez. 2007. Public Private Partnerships: Incentives, Risk Transfer and Real Options. *Review of Financial Economics* 16: 335–349.

Attalla, M., and T. Hegazy. 2003. Predicting Cost Deviation in Reconstruction Projects: Artificial Neural Networks versus Regression. *Journal of Construction Engineering and Management* 129: 405–411.

Cheah, C. Y. J., and J. Liu. 2006. Valuating Governmental Support in Infrastructure Projects and Real Option Using Monte Carlo Simulation. *Construction Management and Economics* 24: 545–554.

Chen, H., and R. Qin. 2012. Real Options as an Incentive Scheme for Managing Revenues in Transportation Infrastructure Projects. *International Journal of Revenue Management* 6: 77–101.

Cox, J. C., S. A. Ross, and M. Rubinstein. 1979. Option Pricing: A Simplified Approach. *Journal of Financial Economics* 7: 229–263.

Grenadier, S. R. 2002. Option Exercise Games: An Application to the Equilibrium Investment Strategies of Firms. *The Review of Financial Studies* 15:691–721.

Huang, Y.-L., and S.-P. Chou. 2006. Valuation of the Minimum Revenue Guarantee and the Option to Abandon in BOT Infrastructure Projects. *Construction Management and Economics* 24: 379–389.

Hull, J. C. 2008. *Options, Futures, and Other Derivatives*, 7th ed. Upper Saddle River, NJ: Prentice Hall.

Lambrecht, B., and W. Perraudin. 1994. *Option Games*. Working paper 9414, University of Cambridge.

Lara Galera, A. L., and A. Sánchez Soliño. 2010. A Real Options Approach for the Valuation of Highway Concessions. *Transportation Science* 44: 416–427.

Levinson, D., and R. Karamalaputi. 2003. Induce Supply: A Model of Highway Network Expansion at the Microscopic Level. *Journal of Transport Economics and Policy* 37:297–318.

Mason, S. P., and C. Y. Baldwin. 1988. Evaluation of Government Subsidies to Large-Scale Energy Projects. *Advances in Futures and Options Research* 3: 169–181.

Puterman, M. L. 1994. *Markov Decision Processes: Discrete Stochastic Dynamic Programming*. New York: John Wiley & Sons.

Smets, F. 1991. *Exporting versus FDI: The Effect of Uncertainty, Irreversibilities and Strategic Interactions*. Working paper, Yale University.

Subprasom, K., and A. Chen. 2007. Effects of Regulation on Highway Pricing and Capacity Choice of a Build–Operate–Transfer Scheme. *Journal of Construction Engineering and Management* 133: 64–71.

Tampère, C., J. Stada, and B. Immers. 2009. Calculation of Welfare of Road Pricing on a Large-Scale Road Network. *Technological and Economic Development of Economy* 15: 102–121.

Tan, Z. J., H. Yang, and X. L. Guo. 2010. Properties of Pareto-Efficient Contracts and Regulations for Road Franchising. *Transportation Research Part B* 44: 415–433.

Xiao, F., H. Yang, and D. Han. 2007. Competition and Efficiency of Private Toll Roads. *Transportation Research Part B* 41: 292–308.

Yang, H., and Q. Meng. 2000. Highway Pricing and Capacity Choice in a Road Network under a Build–Operate–Transfer Scheme. *Transportation Research Part A* 34: 207–222.

12

Evaluating Different Cost-Benefit Analysis Methods for Port Security Operations

Galina Sherman, Peer-Olaf Siebers, David Menachof, and Uwe Aickelin

CONTENTS

12.1 Introduction

Businesses are interested in the trade-off between the cost of risk mitigation and the expected losses of disruptions (Kleindorfer and Saad 2005). Port operations, due to their complexity, have a large impact on the economy, for example, a paralyzed port will have an enormous effect on an area or country.

Airports and seaports have an additional complexity when conducting such risk analysis, because there are two key stakeholders with different interests involved in the decision processes concerning the port operation: port operators and security operators (Bichou 2004). In this chapter we focus on the tangible costs and benefits for port stakeholders.

Airports and seaports as well as other service industries are interested in smooth traffic flow, short service times at different stations, quality, and efficiency. In ports we have port operators who are service providers and as such interested in all the performance parameters mentioned above, and we have the border agency that represents national security interests that need to be considered. However, the interests of the border agency are different from ones of the service providers. Detection of threats such as weapons, smuggling, and sometimes even stowaways is of high importance. Nevertheless, security checks that are too long can compromise the service-standard targets that port operators are trying to maintain.

Besides these two conflicting interests there is also the cost factor for security that needs to be kept in mind. Security checks require expensive equipment and well-trained staff. However, the consequences for the public of undetected threats crossing the border can be severe. Therefore, it is in the interest of all parties to find the right balance between service, security, and costs.

How can we decide the level of security required to guarantee a certain detection threshold for threats while maintaining economic viability and avoiding severe disruptions to the process flow? A tool frequently used by business and government officials to support such investigations is cost-benefit analysis (CBA) (Hanley and Spash 1993). Compared to other methods, such as cost effectiveness analysis or cost utility analysis, it provides a monetary value quantifying the expected benefits and costs. In order to use CBA, certain input parameters need to be known or, in absence of real data, estimated.

Methods frequently used to estimate these input parameters are scenario analysis, decision trees, and Monte Carlo simulation (Damodaran 2007). Discrete-event simulation is less frequently used in risk analysis but often used in operational research (Turner and Williams 2005; Wilson 2005) and, in particular, in service industries (Laughery et al. 1998) (e.g., banks, medical, transportation) to investigate different operational practices. When reviewing the relevant literature on risk analysis there appears to be a gap comparing the efficiency of all these methods using a single case study (Virta et al. 2003; Jacobson et al. 2006).

In this chapter, we present such a comparison of different probabilistic techniques that are used for conducting CBA. In particular, we focus on scenario analysis, decision tree, and simulation. We demonstrate step by step the application of these methods, while conducting a CBA of different cargo screening policies. Our main aim is to demonstrate how CBA can be applied to service industry problems. Furthermore, we want to show what the data requirements are for the different modeling methods and what additional decision support intelligence an analyst can obtain by applying each of the modeling methods mentioned earlier. This information is intended to support analysts in the decision-making process of which method to use for their analysis.

Section 12.2 comprises a brief review of the existing literature. First, we focus on the advantages and disadvantages of CBA compared to alternative methods. Then we look at case study examples from the service industry where the different modeling methods mentioned earlier have been applied. In Section 12.3, we introduce our case study (cargo screening process) and show step by step how to conduct a CBA and how to apply the different modeling methods to the case study scenario. In Section 12.4, we summarize our findings and conclude.

12.2 Literature Review

12.2.1 Cost-Benefit Analysis and Alternative Analysis Methods

The use of CBA is described in the literature from the early 19th century. Already at that time this approach was used by U.S. governmental agencies in environmental management (Hanley and Spash 1993). According to Guess and Farnham (2000), since 1960 the application of CBA was expanded to "human beings" and "physical investment programs."

There are some alternatives to CBA such as cost effectiveness analysis and cost utility analysis. The economics literature compares these three methods. The difference between these three approaches is that CBA allows a comparison of a wider range of scenarios, because the costs and the results have a monetary expression unlike the two other techniques that are observed by using a single result every time. Thus the advantage of CBA over the other two methods is that it has a wider scope of possibilities and is more relevant for the service sector. However, sometimes it is impossible to give a monetary value to the costs or to the benefits (Guess and Farnham 2000).

A related method is risk analysis. "Risk analysis consists of the repeated random extraction of a set of values for the critical variables, taken within the respective defined intervals, and then calculating the performance indices for the project resulting from each set of extracted values" (European Commission Directorate General Regional Policy 2008). It does not help to reduce the risk itself, but to identify and manage it. Our work has commonalities with both risk analysis and CBA, and in fact is a form of CBA that assesses risks.

A competitive method to CBA that we find in the literature is multicriteria analysis. Our method allows a comparison between a mixture of inputs, monetary and nonmonetary. It can use the results of a standard CBA as monetary inputs and service quality estimators as nonmonetary inputs. From these it produces results to show the relationship between costs and benefits of different options (Department for Communities and Local Government 2009). In this chapter we conduct a relatively simple CBA study based on a single case of port operation; this can be extended in the future using multicriteria analysis.

De Langen and Pallis (2006) conduct a qualitative study to analyze the benefits of intraport competition. We learn that the use of costs versus benefits can be valuable at many aspects of the port performance. In their research the authors use a qualitative methodology (e.g., survey), which we find less beneficial for our study. Jacobson et al. (2006) and Virta et al. (2003) suggest CBA in evaluation of trade-offs and cost effectiveness while screening 100% of the cargo and using single or double screening devices. The authors suggest a cost model that contains direct costs and indirect costs associated with costs of failure.

Bichou (2004) finds that CBA is a model used to evaluate optimal policy decisions. According to him this model is useful when port authorities are interested to add value, for example, a program that is nonmandatory as long as the benefits of the new behavior can be quantified. According to Bichou (2004), there are, however, some difficulties to applying cost analysis to port performance and one of them is the difficulty estimating efficiency measures. Moreover, Bichou et al. (2009) argue that an additional concern in using a CBA while dealing with security issues arises as a result of uncertainty, which means that the benefits will also be uncertain or difficult to quantify and evaluate, for example, in case of a terrorist threat. They find that it is very difficult to judge the effectiveness of the security policy. On the other hand they argue that an additional difficulty lies in the fact that one cannot know what would happen if the policy did not exist.

While we acknowledge Bichou et al.'s (2009) argument that conducting a CBA can be limited when dealing with rare security issues such as terrorist attacks, we tend to agree with other authors such as Jacobson et al. (2006) and Virta et al. (2003) who suggest using CBA to compare different screening policies for better service and performance routines.

In conclusion, we find that CBA is a powerful tool for comparing costs versus benefits of different practices in the service sector. It allows the user to compare a wide number of variables and provides a monetary value to the comparison. We find CBA suitable for our research purpose.

12.2.2　Modeling Methods Applied in Service Industry Cost-Benefit Analysis

12.2.2.1　Scenario Analysis

Scenario analysis is a version of sensitivity analysis that is used to study the possible impact on outcomes while different variables are tested (European

Commission Directorate General Regional Policy 2008). It is often used to analyze possible future scenarios by considering possible best, worst, and average outcomes. According to Damodaran (2007), this technique is suitable for single events. Daellenbach and McNickle (2005) state that any business faces uncertainty and as a result creates an unlimited number of possible futures to be considered. However, the number of possible scenarios to be considered is limited to three or four in scenario analysis. Bruzzone et al. (1999) use this method to verify the suitability of simulators of container terminals.

12.2.2.2 Decision Trees

A decision tree is a decision-support tool (diagram) used in operational research. It can be helpful in deciding about strategies and dealing with conditional probabilities. According to Anderson et al. (2008), decision trees are particularly useful when dealing with relatively few possible solutions. Decision trees are diagrams that can be used to represent decision problems so that their structure is made clearer. Unlike decision tables, decision trees can be used to represent problems involving sequences of decisions, where decisions are made at different stages in the problem. For instance, Kim et al. (2000) use decision trees to decide about storage policies for transshipments.

12.2.2.3 Simulation

Simulation is widely used in logistics. Turner and Williams (2005) specify that simulating the behavior of complicated systems gives the ability to experiment with different scenarios and has the power of generalization of the insight on the performance of the complicated systems. Also discrete-event simulation is often used in modeling of traffic movement (Robinson 2004; Laughery et al. 1998) and traffic and transport management (Davidsson 2005). Furthermore, Wilson (2005) confirms the usefulness of simulation for investigating security issues and states that "simulation modeling allows the analysis or prediction of operational effectiveness, efficiency, and detection rates (performance) of existing or proposed security systems under different configurations or operating policies before the existing systems are actually changed or a new system is built, eliminating the risk of unforeseen bottlenecks, under- or over-utilization of resources, or failure to meet specified security system requirements."

In the medical literature, we find a successful use of simulation for CBA. For example, Habbema et al. (1987) suggest simulation as an appropriate technique to conduct the CBA to compare two different cancer screening policies. The authors use a micro simulation approach to explore different scenarios and their outputs. Also Pilgrim et al. (2009) suggest conducting a cost effectiveness analysis for cancer screening policies while using discrete-event simulation. The authors support their choice of methodology with previous research using the same research strategy.

Simulation is widely used in different service industries, including medical and transportation (Laughery et al. 1998). It is a powerful tool that can help an analyst to learn about the system under research. In addition simulation allows the user to compare different scenarios before implementing them in real world.

There are different types of simulation and here we focus on static discrete stochastic simulation (i.e., Monte Carlo simulation) and dynamic discrete stochastic simulation (i.e., discrete-event simulation). The difference between them is that in Monte Carlo simulation time does not play a natural role, but in discrete-event simulation it does (Kelton et al. 2010). Whereas the first is often used in risk assessment, the second is used when further investigation into the system behavior on the operational level is required for the decision making.

12.3 Case Study

In this section we apply the aforementioned approaches in a real-world situation. Furthermore, Farrow and Shapiro (2009) identify that the literature dealing with the CBA is often based on a case study methodology while focusing on a sensitive or strategic topic. We find this approach applicable for the service industry as well.

12.3.1 The Case Study System

In our research we chose to make the comparison of scenario analysis, decision tree, and simulation for a CBA based on the same real-world system, which is different from the practices we find in the literature where we find different examples for each approach (Damodaran 2007). We find that the users can benefit and learn more if a comparison of methods is conducted, based on the same data. We chose a case study approach as our research methodology, because it will supply us the data needed for all the methods. Our case study involves the cargo screening facilities of the ferry port of Calais (France). This case study was conducted in collaboration with the UK Border Agency (UKBA).

At Calais we find two main stakeholders related to service provision: port operators, who are service providers and as such interested in a smooth flow of port operations to provide certain service standards (e.g., service times), and the border agency, which represents national security interests that need to be considered. Checks have to be conducted to detect threats such as weapons, smuggling, and stowaways. If the security checks take too long they can compromise the service standard targets to be achieved by the port operators. We have chosen Calais as our case study system for two reasons: First, it has limited number of links; Calais operates only with Dover leading to a simple cargo flow. Second, there is only one major threat of interest to the

TABLE 12.1

Statistics from Calais (April 2007–April 2008)

Statistic	Value
Total number of lorries entering Calais harbor	900000
Total number of positive lorries found	3474
Total number of positive lorries screened on French site	900000
Total number of positive lorries found on French site	1800
Total number of positive lorries screened on UK site	296406
Total number of positive lorries found on UK site	1674
... In UK sheds	890
... In UK berth	784

British government: clandestines, who are people trying to enter the United Kingdom illegally, that is, without having proper papers and documents.

In Calais there are two security areas: one is operated by French authorities and one is operated by UKBA. According to the data collected between April 2007 and April 2008 about 900,000 lorries passed the border and approximately 0.4% of these lorries have been found to have an additional human freight (UK Border Agency 2008). More details can be found in Table 12.1.

The search for clandestines is organized in three major steps, one by France and two by the UKBA. On the French side, after passing the passport check (referred to as Passport in the decision tree) all arriving lorries are screened, using passive millimeter wave scanners (PMMW) for soft-sided lorries and heartbeat detectors (HB) for hard-sided lorries. If lorries are classified as suspicious after the screening, further investigations are undertaken. For soft-sided lorries there is a second test with CO_2 probes (CO2) and if the result is positive the respective lorry is opened. For hard-sided lorries there is no second test and they are opened immediately. The cleared lorries proceed to purchase a ticket for the ferry and then to the UK passport check and screening.

On the British side only a certain percentage of lorries (currently 33%) is searched at the British sheds. Here a mixture of measures is used for the inspection, including CO_2 probes, dogs, and opening lorries. Once the lorries pass the British sheds they will park in the berth to wait for the ferry. In the berth there are mobile units operating that search as many of the parked lorries as possible before the ferry arrives, using the same mixture of measures as in the sheds. As shown in Table 12.1 only about 50% of the clandestines detected were found by the French, about 30% in the sheds and 20% by the mobile units in the berth. The overall number of clandestines that are not found by the authorities is of course unknown.

12.3.2 Analysis Framework

We consider two factors with three scenarios each in our analysis: traffic growth (TG) and clandestine growth (CG) (Table 12.2). For each factor and

TABLE 12.2

Two Factors with Three Scenarios and One Decision Variable
with Three Options

Traffic Growth (TG)	p (TG)	Clandestine Growth (CG)	p (CG)	Search Growth (SG)
+0%	0.25	−50%	0.33	+0%
+10%	0.5	+0%	0.33	+10%
+20%	0.25	+25%	0.33	+20%

scenario combination, we have estimated the probability of it happening, as described in the following paragraphs. The question we are trying to answer is how the UKBA should respond to these scenarios. We assume that there are three possible responses: increasing the searches by either 0%, 10%, or 20%.

Our TG scenarios are based on estimates by the port authorities who are planning to build a new terminal in Calais in 2020 to cope with the anticipated additional traffic. According to Dover Harbour Board (2008), between 2010 and 2020 the traffic in the Port of Dover is expected to double. Due to their direct connection, one can assume that this is also applicable to the Port of Calais. Thus an annual traffic growth of 0% to 20% is a realistic factor range, with an increase of 10% most likely, whereas the other two are equally likely. It is assumed that any increase in traffic is proportional, that is, the ratio of soft- to hard-sided lorries remains the same. It is an important assumption because different types of lorries are screened by different technologies. If the ratio between them changes it might cause bottlenecks at some places, and empty and waiting staff at others.

The second factor under consideration is CG. This is the most unpredictable of the three factors, as clandestine numbers greatly vary from year to year based largely on external factors such as the economic attractiveness of the United Kingdom, the number and intensity of wars and other conflicts worldwide, and other political initiatives.

Local aspects also play a role, for example, an increase in searches in Calais can displace clandestines to other nearby ports and vice versa. Due to the uncertainty attached to this factor, a range of +25% to −50% is considered, with all scenarios being equally likely. A higher maximum decrease than increase is assumed to the recent clearing in late 2009 of the Calais "jungle" (illegal encampment of clandestines near the port). We will assume in the following that any changes in clandestine numbers will proportionally affect successful and unsuccessful clandestines.

Search growth (SG) describes the percentage increase in search activity by the UKBA. Currently, UKBA searches 33% of traffic. To keep this proportion stable, UKBA will need to respond to a growth in traffic by increasing the number of lorries it searches. At the same time there is political pressure to search more vehicles, although budget pressures limit the number of vehicles that can be inspected. Searching 33% of the traffic represents a trade-off

TABLE 12.3

Combined Probabilities Assuming Independence of Probabilities

	–50% CG	0% CG	+25% CG
0% TG	0.083	0.083	0.083
10% TG	0.167	0.167	0.167
20% TG	0.083	0.083	0.083

TABLE 12.4

Cost of Extra Searches (As Mentioned Before)

TG vs. SG	SG 0%	SG +10%	SG +20%
TG 0%	£0	£5,000,000	£10,000,000
TG 10%	£0	£5,000,000	£10,000,000
TG 20%	£0	£5,000,000	£10,000,000

between the service levels and the investments. Thus we assume that search growth may also vary between 0% and 20% and not higher because of costs. As before, we assume that any increase in search activity is proportional to hard- and soft-sided lorries.

Combining this information, we arrive at the following combined probabilities of each scenario to occur: $p(TG,CG) = p(TG) \times p(CG)$. The results can be seen in Table 12.3.

It is estimated by the UKBA that each clandestine that reaches the United Kingdom costs the government approximately £20,000 per year. Moreover, it is estimated that the average duration of a stay of a clandestine in the United Kingdom is 5 years, so the total cost of each clandestine slipping through the search in Calais is £100,000. In addition, the average number of clandestines on a lorry is four, which means £400,000 per positive lorry that is missed.

The cost for increasing the search capacity in Calais is more difficult to estimate, as there is a mixture of fixed and variable costs, and operations are often jointly performed by French, British, and private contractors. However, if we concentrate on UKBA's costs, we can arrive at some reasonable estimates, if we assume that any increase in searches would result in a percentage increase in staff and infrastructure cost. Thus we estimate that a 10% increase in search activity (10% SG) would cost £5M and a 20% increase £10M (20% SG) (Table 12.4).

As 33% of vehicles are searched by UKBA, we can calculate resultant percentages of vehicles searched by combining the aforementioned two factors (assuming linear relationships). For example, if traffic increase is matched by search increase, this will remain the same. Or if there is a +10% TG and +0% SG results in 33% × (100%/110%) = 30% of vehicles searched (Table 12.5).

A key question is the relationship between the percentage of vehicles searched versus the number of clandestines found, or more important, the number of clandestines not found. Searching 33%, UKBA finds approximately 1674 lorries in Calais with additional cargo. A best estimate of "successful"

TABLE 12.5

Proportion of Vehicles Searched

TG vs. SG	SG 0%	SG +10%	SG +20%
TG 0%	0.3300	0.3630	0.3960
TG 10%	0.3000	0.3300	0.3600
TG 20%	0.2750	0.3025	0.3300

clandestines is approximately 50 per month (600 per year) or 150 lorries per year. Establishing a clear relationship between these 150 and the figure of 1674 is difficult, as 1674 lorries do not represent unique attempts by the clandestines. Unsuccessful clandestines will try again, that is, finding one more positive lorry does not remove four clandestines from the system.

Based on advice by the UKBA, we estimate the effect of increased searches as follows: a search increase of 10% yields an extra 167 positive lorries found. Ten percent more positive lorries detected is estimated to reduce the number of successful clandestines by 10%. Thus the number of nondetected positive lorries reduces from 150 lorries by 15 to 135 lorries. Please note that these numbers vary for the more detailed simulation scenarios due to effects such as queue jumping, that is, a 10% increase in searches will yield less than a 10% reduction in undetected positive lorries.

It is probably a fair assumption that an increase in searches will yield a decrease in the number of successful clandestines and vice versa. In absence of further information and considering that the variation of percentage of searches is in a relatively limited range of 27.5% to 39.6%, we will make the same assumption here as in the rest of the scenario analysis: the relationship between both parameters is linear. Based on this, we obtain the number of clandestines missed as given in Table 12.5, for example, searching only 30% of traffic results in 600 × (33%/30%) = 660 missed clandestines.

For our case study system we conduct a scenario analysis, build a decision tree that fully represents the traffic flow inside the system boundaries, and build a simulation model of the system using the same input data as for the decision tree. Comparing the results of all these methods allowed us to validate the models. While we use Microsoft Excel for scenario analysis and modeling decision tree, our simulation models are implemented in AnyLogic (XJ Technologies 2009), a multiparadigm simulation package.

We use different methods for estimating the "adjusted" number of positive lorries found if there is no growth of positive lorries. Once we have the matrix (iteration over our two factors) we conduct some data analysis to estimate the costs. The data analysis is the same for all the different approaches we will present. We demonstrate this for scenario analysis, and for all others we will only report on the key outputs (adjusted number of positive lorries found for CG = 0, total expected costs). Finally, we will compare the expected costs that resulted from the different models, and demonstrate the inputs and outputs of each one of the tools.

TABLE 12.6

Adjusted Number of Positive Lorries Found If CG = 0%*

TG vs. SG	SG 0%	SG +10%	SG +20%
TG 0%	1674.0	1841.4	2008.8
TG 10%	1521.8	1674.0	1826.2
TG 20%	1395.0	1534.5	1674.0

* Calculated based on Table 12.5 where 0.3300 represents 1674 lorries.

TABLE 12.7

Relative Number of Positive Lorries Found When Compared to Base Scenario If CG = 0%

TG vs. SG	SG 0%	SG +10%	SG +20%
TG 0%	1	1.1	1.2
TG 10%	0.909091	1	1.090909
TG 20%	0.833333	0.916667	1

TABLE 12.8

Number of Positive Lorries Missed If CG = 0%*

TG vs. SG	SG 0%	SG +10%	SG +20%
TG 0%	150.0	136.4	125.0
TG 10%	165.0	150.0	137.5
TG 20%	180.0	163.6	150.0

* Calculated based on the probabilities in Table 12.5, where 0.3300 results in 150 missed positive lorries. Similar tables can be computed for the other CG values.

12.3.3 Scenario Analysis

Table 12.6 is the result from the scenario analysis: number of positive lorries found. The cost estimation that follows next is the same for all methods discussed in this book chapter and will only be shown once in detail. From Table 12.6 if the TG is 20% but the SG stays the same, the number of positive lorries detected will be 1395, and the other way around if the traffic does not grow (TG 0%) but the SG 20% the number of detected lorries will grow as well.

From Table 12.7 comparing the number of positive lorries from Table 12.6 to the factor of CG = 0%, the 1395 represents 83.3333% of the positive detected lorries.

Table 12.8 presents the number of positive lorries missed for the variable CG = 0%. Similar results were calculated for CG = 25% and CG = –50%. Table 12.9 represents an optional scenario number of positive lorries missed (e.g., 180) divided by base scenario (150 positive lorries missed) for TG 20%. From here we can learn that if the traffic will grow at 20% but search percentage will not grow, an extra of 20% of the positive lorries will be missed.

TABLE 12.9

Relative Number of Positive Lorries Missed Compared
to the Base Scenario

TG vs. SG	SG 0%	SG +10%	SG +20%
TG 0%	1.00	0.91	0.83
TG 10%	1.10	1.00	0.92
TG 20%	1.20	1.09	1.00

Note: 1 means 150 lorries.

TABLE 12.10

Expected Costs Including SG Costs for CG = 0%*

TG vs. SG	SG 0%	SG +10%	SG +20%
TG 0%	£60,000,000	£59,545,455	£60,000,000
TG 10%	£66,000,000	£65,000,000	£65,000,000
TG 20%	£72,000,000	£70,454,545	£70,000,000

* Calculated by combining the information of Table 12.4 with that of
Table 12.9, where the cost of one positive lorry is £400,000 each.

TABLE 12.11

Expected Costs

SG = 0%	SG = 10%	SG = 20%
£60,500,000	£60,000,000	£60,416,667

Table 12.10 was calculated by using ratios from Table 12.9 multiplied by number of positive lorries missed (150 not detected) multiplied by cost per missed positive lorry (£400,000) plus cost of extra searches (Table 12.4).

Table 12.10 can be used to calculate the tables for CG = −50% and CG = 25% by multiplying each value in the table with $1 + CG(x)$ where $CG(x)$ is the proportion of CG. The expected costs for each policy according to scenario analysis are as follows from Table 12.11. It was calculated by multiplying the probabilities of each scenario to occur with its costs, for example, $C(SG) = \sum(EC(SG,TG,CG) \times p(TG,CG))$.

Overall, we can conclude from the scenario analysis that the best option seems to be to change SG by 10%. In following decision tree and simulation we use the same assumptions, costs, and scenarios as described earlier. When we use additional data for an approach we will state it in the relevant section.

12.3.4 Decision Tree

In order to conduct a comparison between the different techniques, we used for the decision tree the same scenarios we used for the scenario analysis.

Also, here we considered the same two factors: TG and CG and as a reaction SG. The aim is to compare the inputs and outputs of different tools and the level of details needed, as well as to investigate what is an appropriate policy for the UKBA to adopt according to the decision tree. For our case study we have built a decision tree that fully represents the flow inside our system (see Figure 12.1). On basis of the decision tree we also built a Monte Carlo

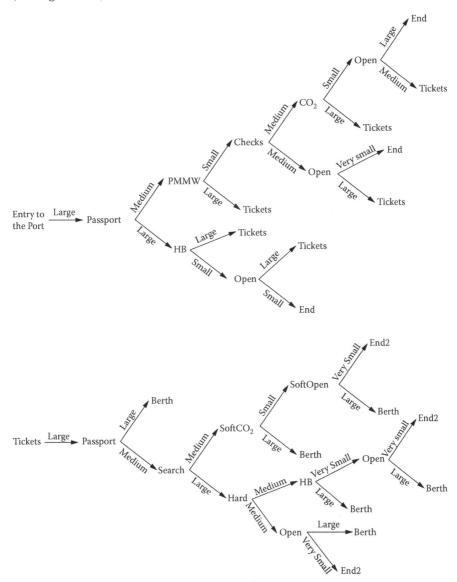

FIGURE 12.1
Decision tree.

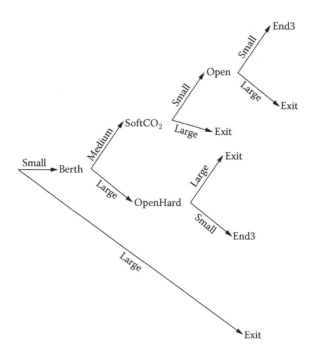

FIGURE 12.1 *(Continued)*

TABLE 12.12

Probabilities That Are Used in the Decision Tree and Their Equivalent in Words

Probabilities	p > 50%	10% < p ≤ 50%	1% < p ≤ 10%	p ≤ 1%
Equivalence	Large	Medium	Small	Very small

simulation. Building both allows us to check the results and validate the models, as the results should be identical.

Building the decision tree using probabilities demands more data than scenario analysis on the one hand; however it allows us to receive more precise outcomes on the other hand. Due to the sensitivity of the data we change the numerical probabilities in the decision tree to their equivalent in words, for example, a probability that is referred to as small at the figure means 1% < p ≤ 10% (see Table 12.12).

From the decision tree we can calculate the following results (Table 12.13), which are identical to the scenario analysis, apart from small rounding errors (due to spreadsheet calculations).

Using the same combined probabilities for each scenario as we have used in scenario analysis we find that the results of the two approaches are almost identical in the monetary value and according to both we should adopt the same search strategy SG by 10%. One of the downsides of the decision tree

TABLE 12.13

Decision Tree Results: Number of Positive Lorries Found If CG = 0%

TG vs. SG	SG 0%	SG +10%	SG +20%
TG 0%	1674	1841	2008
TG 10%	1522	1674	1826
TG 20%	1395	1534	1674

that we have discovered during our work is its linearity. It is an easy task to build a linear decision tree when the data is linear, however, it is more complex if the data is distributed exponentially and requires some manipulations. The decision tree helps the user to better understand the layout of the system.

12.3.5 Simulation

In addition to the data already mentioned, we have collected data on operation times of the activities in the shed and the berth and from ferry operation manuals. We have used the decision tree graphical representation as a conceptual model basis for the Monte Carlo simulation and discrete-event simulation implementation. As we mentioned earlier, the process flow representation in the simulation is equivalent to the decision tree layout. However, the simulation uses probabilities and frequency distributions, for example, exponential arrival times of lorries. This randomness (e.g., slightly different number of arrivals each time) requires us to undertake several replications of each simulation scenario and to calculate the means of the replication outputs. These means we compare with the results of the decision tree to validate our model. We then translate the results to the monetary value and compare the outcomes to other methods.

As we have mentioned already there are different types of simulation and here we use Monte Carlo simulation and discrete-event simulation. We use the same simulation model for both types. The parameter set loaded in the initialization phase of the simulation determines the type of simulation run. The first parameter set sets all delay times to zero and all queue sizes to 1,000,000. This allows emulating a Monte Carlo simulation where all events are executed in the correct sequence but the time to execute them is zero.

The second parameter set defines triangular frequency distributions for the delay times (based on our collected case study data), sets all queue sizes to 1,000,000, and defines the resources that are available (based on our collected case study data). Queue sizes can later be restricted by defining routing rules. For both, Monte Carlo simulation and discrete-event simulation, routing decisions (e.g., which lorries to inspect and how to inspect them) are derived from uniform probability distributions (given probabilities are based on our collected case study data).

Having stochastic inputs means that we also have stochastic outputs. Therefore, we have to do multiple runs. We have conducted some tests to

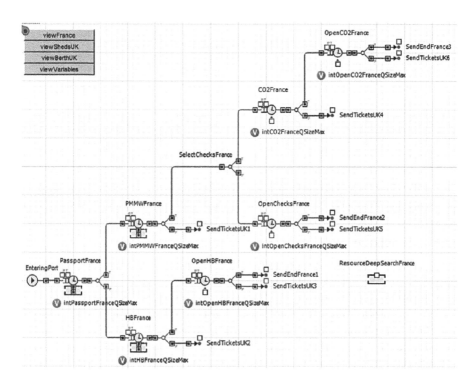

FIGURE 12.2
The French site (AnyLogic Simulation).

determine the number of replications required using confidence intervals (95%) following the guidelines in Robinson (2004). The test results suggest that we run at least eight replications. To be on the safe side we decided to do 10 replications for each iteration of our experiments. To report our experimental results we use mean value as a point estimator and standard deviation as an estimator of the variability of the results.

12.3.6 Monte Carlo (MC) Simulation

We have run all scenarios as defined in the previous sections. The results are as follows in Table 12.14. Comparing the results with the ones from decision tree are shown in Table 12.15. We find that the differences are relatively small and can be attributed to the fact that we use a stochastic method.

We then extend this Monte Carlo simulation model to a discrete-event simulation model by adding elements that are linked to time (e.g., arrival rates, delays, queues) and constraints that are linked to time (e.g., queue size restrictions) and new performance measures linked to time (e.g., utilization, time in system, max queue length) in order to demonstrate the full power

TABLE 12.14

Monte Carlo Simulation Results: Number of Positive Lorries Found If CG = 0%

TG vs. SG	SG 0%	SG +10%	SG +20%
TG 0%	1678.75	1846.25	2027.75
TG 10%	1531.30	1674.15	1827.50
TG 20%	1404.25	1540.90	1670.70

TABLE 12.15

Comparing Decision Tree Results with Monte Carlo Simulation Results (Errors)

TG vs. SG	SG 0%	SG +10%	SG +20%
TG 0%	−5.0	−5.2	−19.4
TG 10%	−9.6	−0.4	−1.7
TG 20%	−9.4	−6.6	3.0

of simulation as a tool in the decision-making process. These modifications reflect the real world and make our simulation model more realistic than the decision tree. Also Monte Carlo simulation allows us to learn more about system variability, an ability of running many replications and having a range of results, can help to learn about sensitivity inside the system.

As we mentioned before we use the same combined probabilities for each scenario as we have used in scenario analysis and decision tree. We find that the simulation results are similar to the other two approaches. Although here, according to the simulation outcomes we should adopt the same search strategy, for example, SG 10%. The difference in the monetary value between Monte Carlo simulation and the other two approaches varies as a result of calculating the costs according to the averages (decision tree and scenario analysis) versus according to the output for the simulation.

12.3.7 Discrete-Event Simulation 0 (Discrete-Event Simulation)

To expand our model to discrete-event simulation we have added additional data that represents the real-world situation. The extra data that we use is service times and resources. We run our simulation again to check if this additional data makes an effect on the results. We find that the impact is very small and as a result we would choose the same strategy as before, SG 10%.

The cycle times are based on data that we collected through observations and from interviews with security staff. In order to represent the variability that occurs in the real system we use different triangular distributions for each sensor types. Triangular distributions are continuous distributions bounded on both sides. In absence of a large sample of empirical data,

TABLE 12.16

Standard Discrete Event Simulation Results: Number of Positive
Lorries Found If CG = 0%

TG vs. SG	SG 0%	SG +10%	SG +20%
TG 0%	1674.50	1833.00	2013.90
TG 10%	1512.00	1667.20	1818.15
TG 20%	1387.65	1527.90	1694.20

TABLE 12.17

Discrete Event Simulation with Variable Arrival Rates Results:
Number of Positive Lorries Found If CG = 0%

TG vs. SG	SG 0%	SG +10%	SG +20%
TG 0%	1681.55	1843.00	2008.70
TG 10%	1519.25	1687.20	1852.85
TG 20%	1385.05	1534.85	1658.85

a triangular distribution is commonly used as a first approximation for the
real distribution (XJ Technologies 2005). Every time a lorry arrives at a shed a
value is drawn from a distribution (depending on the device that will be used
for screening) that determines the time the lorry will spend in the shed (see
Table 12.16).

To show the additional possibilities of discrete-event simulation we have
added some additional features we want to demonstrate here. They are
(1) variable arrival rates, (2) queue size restrictions at sheds in the United
Kingdom, and (3) a combination of the first two features.

12.3.8 Discrete-Event Simulation 1

In order to be able to emulate the real arrival process of lorries in Calais we
created hourly arrival rate distributions for every day of the week from a year
worth of hourly arrival records that we received from the UKBA. These distributions allow us to represent the real arrival process, including quiet and
peak times. In cases where this level of detail is not relevant we use an exponential distribution for modeling the arrival process and the average arrival
time calculated from the data we collected as a parameter for this distribution.

After running the discrete-event simulation with actual arrival rates we
find that also here the results are very similar to previous runs and scenarios
and the appropriate strategy to adopt is SG 10%. The monetary difference is
very small. The results are logical if there is no queue capacity involved and
the only change is at the arrival rates (see Table 12.17).

TABLE 12.18

Discrete Event Simulation with Queue Size Restriction Results:
Number of Positive Lorries Found If CG = 0%

TG vs. SG	SG 0%	SG +10%	SG +20%
TG 0%	1666.95	1833.30	1998.90
TG 10%	1507.80	1657.85	1795.35
TG 20%	1383.90	1526.70	1653.95

TABLE 12.19

Discrete Event Simulation with Variable Arrival Rates and Queue Size
Restriction Results: Number of Positive Lorries Found If CG = 0%

TG vs. SG	SG 0%	SG +10%	SG +20%
TG 0%	1626.35	1788.10	1891.50
TG 10%	1500.75	1619.15	1731.25
TG 20%	1374.8	1492	1600.9

12.3.9 Discrete-Event Simulation 2

At this stage we run the discrete-event simulation while queue capacity is set. We find that here the results are different than in previous runs and scenarios, but the appropriate strategy to adopt is still SG 10%. Although we find that these results are logical the queue capacity only has its impact on the system, but this effect is not big enough to make change in the strategy (see Table 12.18).

12.3.10 Discrete-Event Simulation 3

As a final stage for this book chapter we have combined the three previous scenarios for discrete-event simulation: resources, arrival rates, and queue capacity (see Table 12.19). From the results at this stage we find that the most appropriate strategy to adopt is to keep SG 0%. The reason for the different outcome lies in the combination of arrival rates and queue capacity, when the effort to search more is wasted by queue jumping during the peak times.

12.4 Conclusions

We obtained similar results for three of the four methods we examined: scenario analysis, decision tree, and MC. The discrete-event simulation results

TABLE 12.20

Overall Cost Comparisons of All Methodologies

Option	Total Expected Costs			Cheapest Option
	1: SG = 0%	2: SG = 10%	3: SG = 20%	
Scenario analysis	£60,500,000	£60,000,000	£60,416,667	2
Decision tree	£60,497,446	£60,000,000	£60,418,795	2
Monte Carlo simulation	£60,335,818	£60,058,184	£60,461,341	2
Discrete-event simulation 0	£60,797,873	£60,250,740	£60,350,102	2
Discrete-event simulation 1	£60,881,284	£60,017,602	£60,406,308	2
Discrete-event simulation 2	£60,714,953	£60,166,442	£60,857,915	2
Discrete-event simulation 3	£59,817,382	£60,116,618	£61,624,835	1

TABLE 12.21

Relative Cost Comparisons of All Methodologies

Option	Relative Difference from Lowest Costs			Cheapest Option
	1: SG = 0%	2: SG = 10%	3: SG = 20%	
Scenario analysis	£500,000	£0	£416,667	2
Decision tree	£497,446	£0	£418,795	2
Monte Carlo simulation	£277,633	£0	£403,156	2
Discrete-event simulation 0	£547,133	£0	£99,362	2
Discrete-event simulation 1	£863,682	£0	£388,706	2
Discrete-event simulation 2	£548,511	£0	£691,473	2
Discrete-event simulation 3	£0	£299,236	£1,807,453	1

are different from the others as we explained in the relevant sections. One could argue that the data that was used for the decision tree is more detailed than the one for the scenario analysis, and the data that was used for the simulation is more detailed than decision tree. However, as in our case, full data are often collected by the stakeholders anyway, thus no "extra data" are required by simulation. The summary of the results are presented in Table 12.20. Table 12.21 presents the relative cost to the lowest cost scenario for all methods: option 3 will be always rejected as the most expensive one and the more detailed simulation results show that the best policy is the current policy, for example, SG = 0%.

The main purpose of this chapter has been to demonstrate the difference between the tools, and illustrate their data requirements when applied to the same case study (Table 12.22). Also, we need to remind the reader that these results are based on a mixture of assumptions and real-world data. In addition to our limitations we find that we cannot capture the intangible costs or benefits, such as the benefits of the society as a result of a policy adopted. In our further work (Siebers et al. 2011) we will compare

TABLE 12.22

Factors to Take into Consideration before Making Decisions

		Scenario Analysis	Decision Tree	Monte Carlo	Discrete-Event Simulation
Risk type	Discrete/continuous	D	D	C	C
	Correlated/independent	C	I	Both	Both
	Sequential/concurrent	C	S	Both	Both
Decision process	Strategic/operational	S	S	S	O
	Broad/detailed	B	B	B	D
Model characteristics	Difficulty	L	M	H	H
(low, medium, high)	Data requirements	L	L	M	H
	Tool costs	L	L	M	H
	Training costs	L	L	H	H
	Assumptions	H	M	L	L

different discrete-event simulation techniques, for example, process oriented and object oriented, and suggest using MCA to evaluate nonmonetary benefits.

When dealing with continuous types of risks (i.e., illegal immigration) the simulation approach will be more suitable and it does not depend if the risk type is correlated or independent, sequential or concurrent. However, for scenario analysis and decision tree the situation is different. Scenario analysis can be used for correlated, concurrent, and discrete types of risk, whereas decision tree can be used for independent, sequential, and discrete types. The decision process level and the information required for the three basic approaches will be similarly applied at the strategic level and require broad information. On the other hand, discrete-event simulation requires detailed information and can be applied at the operational level. Total costs of the approaches (including training costs) vary from low for scenario analysis (that require just pen and paper) and decision tree to high for simulation.

Table 12.23 summarizes real-world data that was collected during our case study and the results and outputs that can be obtained from the model (in italics) and help in decision making (resources might be collected as part of a real world data, whereas resource utilization can be a part of the output from the model). As we can conclude from Table 12.23, discrete-event simulation requires the largest amount and greatest detail of data. However, it also provides many different outputs such as peak times, bottlenecks, resource utilization, system throughput, and service quality (e.g., waiting times in the system). This can be so useful for the decision makers in services and other industries. This output can be very valuable for an analyst or a decision maker. It is important to clarify that the tools mentioned in Table 12.23 (i.e., simulation or decision tree) will not directly provide the total expected costs. The user will need to conduct some extra mathematical calculations on the outputs to obtain these costs as well as making a decision about the policy alternatives.

TABLE 12.23

Real-World Data That Can Be Considered in the Model and Decision Support Output That Can Be Collected from the Model (in *Italics*)

Scenario Analysis	Decision Trees	Monte-Carlo Simulation	Process Oriented DES
Scenarios	Scenarios	Scenarios	Scenarios
Total expected costs	*Total expected costs*	*Total expected costs*	*Total expected costs*
Positive lorries detected	*Positive lorries detected*	*Positive lorries detected*	*Positive lorries detected*
	System structure	System layout	System layout
	Existing resources	Existing resources	Existing resources
		System variability	*System variability*
			Resource utilization
			Dynamic system constraints
			Throughput (capacity)
			Waiting time distributions
			Bottleneck analysis
			Dynamic system decisions

Acknowledgments

This project is supported by the EPSRC (EP/G004234/1) and the UK Border Agency.

References

Anderson, D. R., Sweeney, D.J., Williams, T. A., and Martin, K. 2008. *An Introduction to Management Science: Quantitative Approaches to Decision Making*, 12th ed. Mason, OH: Thomson Higher Education.

Bichou, K. 2004. The ISPS code and the cost of port compliance: An initial logistics and supply chain framework for port security assessment and management. *Maritime Economics and Logistics*, 6(4), 322–348.

Bichou, K., Gordon, P., Moore, J. E., Richardson, H. W., de Palma, A., and Poole, R. 2009. Security, risk perception and cost benefit analysis, discussion paper No. 2009-6, Joint Transport Research Centre, Paris.

Bruzzone, A.G., Giribone, P., Revetria, R. 1999. Operative requirements and advances for the new generation simulators in multimodal container terminals. *Winter Simulation Conference Proceedings*, 1243–1252.

Daellenbach, H. G., and McNickle, D. C. 2005. *Management Science Decision Making through Systems Thinking*. New York: Palgrave Macmillan.

Damodaran, A. 2007. *Strategic Risk Taking: A Framework for Risk Management*. Upper Saddle River, NJ: Wharton School Publishing.

Davidsson, P., Henesey, L., Ramstedt, L., Tornquist, J., and Wernstedt, F. 2005. An analysis of agent based approaches to transport logistics. *Transportation Research: Part C: Emerging Technologies*, 13(4), 255–271.

De Langen, P. W., and Pallis, A. A. 2006. Analysis of the benefits of intra-port competition. *International Journal of Transport Economics*, 33(1), 69–86.

Department for Communities and Local Government. 2009. *Multi-criteria analysis: A manual*. London: DCLG. Available from http://eprints.lse.ac.uk/12761/1/Multi-criteria_Analysis.pdf (accessed July 19, 2011).

Dover Harbour Board. 2008. *Planning for the next generation—Third round consultation document*. Available from http://www.doverport.co.uk/_assets/client/images/collateral/dover_consultation_web.pdf (accessed July 20, 2011).

European Commission Directorate General Regional Policy. 2008. *Guide to cost-benefit analysis of investment projects*. Available from http://ec.europa.eu/regional_policy/sources/docgener/guides/cost/guide2008_en.pdf (accessed September 7, 2011).

Farrow, S., and Shapiro, S. 2009. The benefit-cost analysis of security-focused regulations. *Journal of Homeland Security and Emergency Management*, 6(1), 1–20.

Guess, G. M., and Farnham, P. G. 2000. *Cases in public policy analysis*, 2nd ed. Washington, DC: Georgetown University Press.

Habbema, J. D. F., Lubbe, J. Th. N., van Oortmarssen, G. J., and van der Maas, P. J. 1987. A simulation approach to cost effectiveness and cost benefit calculations of screening for the early detection of disease. *European Journal of Operational Research*, 29, 159–166.

Hanley, N., and Spash, C. L. 1993. *Cost-Benefit Analysis and the Environment*. Northhampton, MA: Edward Elgar.

Jacobson, S. H., Karnani, T., Kobza, J. E., and Ritchie, L. 2006. A cost-benefit analysis of alternative device configurations for aviation-checked baggage security screening. *Risk Analysis*, 26(2), 297–310.

Kelton, W. D., Sadowski, R. P., and Swets, N. B. 2010. *Simulation with Arena*, 5th ed. New York: McGraw-Hill.

Kim, K. H., Park, Y. M., and Ryu, K. R. 2000. Deriving decision rules to locate export containers in container yards. *European Journal of Operational Research*, 124, 89–101.

Kleindorfer, P. R., and Saad, G. H. 2005. Managing disruption risks in supply chains. *Production and Operations Management*, 14(1), 53–98.

Laughery, R., Plott, B., and Scott-Nash, S. 1998. Simulation of service systems. In *Handbook of Simulation*, edited by J. Banks, pp. 629–644. Hoboken, NJ: John Wiley & Sons.

Pilgrim, H., Tappenden, P., Chilcott, J., Bending, M., Trueman, P., Shorthouse, A., and Tappenden, J. 2009. The cost and benefits of bowel cancer service developments using discrete event simulation. *Journal of the Operational Research Society*, 60(10), 1305–1314.

Robinson, S. 2004. *Simulation: The Practice of Model Development and Use*. Chichester, UK: Wiley.

Siebers, P. O., Sherman, G., Aickelin, U., and Menachof, D. 2011. *Comparing decision support tools for cargo screening processes*. The 10th International Conference on Modeling and Applied Simulation, Rome, Italy.

Turner, K., and Williams, G. 2005. Modeling complexity in the automotive industry supply chain. *Journal of Manufacturing Technology Management*, 16(4), 447–458.

UK Border Agency. 2008. Freight search figures for 2007/2008 (unpublished).

Virta, J. L., Jacobson, S. H., and Kobza, J. E. 2003. Analyzing the cost of screening selectee and non-selectee baggage. *Risk Analysis*, 23(5), 897–908.

Wilson, D. L. 2005. Use of modeling and simulation to support airport security. *IEEE Aerospace and Electronic Systems Magazine*, 208, 3–8.

XJ Technologies. 2005. *AnyLogic User's Guide*. St. Petersburg, Russia: XJ Technologies.

XJ Technologies. 2009. Simulation software and services. Available from http://www.xjtek.com (accessed January 20, 2011).

13

Vehicle Routing in Service Industry Using Decision Support Systems

Buyang Cao and Burcin Bozkaya

CONTENTS

13.1 Introduction

Vehicle routing problems (VRPs) have drawn the great attention of academic researchers and practitioners since the problem was first introduced by Dantzig and Ramser (1959) more than 50 years ago. From a practical point of view, a VRP addresses the important fact that the distribution costs of many firms in the service industry may account for a major portion of the total logistics costs of the firm. Nowadays in the service industry, a company wants to keep the logistics cost as low as possible while offering the best service to its customers, which raises a challenging problem. With the rapid increase in energy costs, to be able to effectively address this problem should eventually bring significant economic benefits.

In general terms, the goal of the VRP is to determine how to deploy a fleet of vehicles or service personnel to distribute products or provide services to customers in the most economic way. Travel time, travel distance, or service quality will impact the cost of such operations. In the service industry, many problems can be modeled as the vehicle routing problem with time windows (VRPTW), an extended version of the VRP, though some of these problems might not involve transportation of physical goods at all. In a VRPTW there are time windows to commit to because customers require that product or service delivery must occur within the imposed time windows. A VRPTW tries to find the most economic way to achieve this while considering other constraints such as vehicle capacities, route lengths, and drivers' (or service persons) working schedules. As usual, in the problems studied here each customer is serviced by one and only one vehicle (person) and each vehicle (person) may have its own schedule defined by a start and end time, break times, and lunch time. A VRPTW can contain one or more depots, and each depot has a time window determined by its working hours. Hence, all vehicles stationed at this depot cannot depart before the start time and must be back by the end time of the depot. In the case of product delivery, one may have to further take into account multiple vehicle capacities including weight, volume, and other units.

In this chapter, we focus on the VRPTW. The VRPTW model is widely applied in the service industry no matter if it is public or private, since planners seek to carry out the most efficient product pickup and delivery or to provide the best service at the lowest possible cost. Some applications in the service industry require simultaneous pickups and deliveries. Therefore, vehicles may have a mixture of pickup and delivery operations at customer sites, such as the case for picking up at or delivering cash to ATMs and bank branches. This problem is structurally a different type of problem requiring somewhat different problem-solving techniques and we will present its application and a corresponding solution methodology in our case study.

With the rapid increase of energy costs, the service industry nowadays is faced with tough challenges such as offering better services while keeping the overall costs as low as possible. Higher gasoline prices, more restrictive traffic rules, and personnel costs put more pressure on providing services efficiently. Real VRPTW instances in the service industry may involve thousands of customers instead of hundreds, requiring hundreds of vehicles instead of tens. Furthermore, companies request that VRPTW solutions be more realistic and feasible rather than optimal. Usually companies want to achieve VRPTW solutions that are accepted by the field people such as drivers, service persons, and dispatching supervisors. A theoretically optimal but practically unrealizable solution is not welcome by end users. When real-time solutions are needed (e.g., when processing a service order over the Web), it is essential that the solution for a VRPTW be obtained in a very short computational time, and be as realistic and implementable as it can be. To this end, modern information technologies such as GIS (Geographic Information System), GPS (Global Positioning System), mobile application techniques, and RFID (radio frequency identification) have been incorporated to solve real-world VRPTW problems. More decision support systems (DSS) for solving various logistics problems including VRPTW that consist of databases, user-friendly graphical interfaces, decision models, and optimization modules are built by many providers. Users are able to take advantage of interactive DSS to solve complicated VRPTW instances. Users may also modify the results created by the system upon their own experiences, and generate more realistic and easy-to-deploy solutions. The modern information technologies and optimization techniques enable these DSS to solve different kinds of large-scale optimization problems more effectively.

This chapter focuses on the practical applications of VRPTW, particularly on utilizing DSS to solve complicated VRPTW in the service industry. Instead of an extensive coverage on the theory behind various VRPTW models and their associated solution methodologies, which has been done in numerous studies in the literature, we have chosen to focus on the adaptation of the VRPTW model and related solution methodologies upon the practical needs, and the DSS implemented to solve problems from the real world based upon our practical experiences. We do not necessarily propose a new methodology for solving the VRPTW, but instead we report successful DSS implementations that employ well-established and proven solution techniques.

After briefly reviewing the status of research on VRPTW and DSS developments, we present the reader with two interesting DSS implementations that aim to assist users to solve real instances of VRPTW: one from tobacco products delivery; the other from the pickup and delivery of cash, coins, and other valuables on a daily basis to and from bank branches, ATMs, and merchant locations. These applications are presented as case studies in this chapter. For each study, we describe the special modeling and system deployment considerations to address the particular business needs. In order to achieve results that are acceptable to practitioners, certain real-world features need incorporating in a VRPTW solution procedure, which are discussed in detail in each case study. Furthermore, we report how to combine modern information technology (IT) techniques such as GIS, GPS, RFID, and optimization techniques to build the corresponding decision support systems to solve large-scale VRPTW instances with ten thousands of nodes (customers) in the service industry. The case studies do not only contain the achievements but also the lessons we learned during the system implementations. Having served as chief technical officer and technical manager in either study, we think the case studies provide the reader with a more realistic and operationally acceptable view of the available solution alternatives. In these case studies, we present the economic benefits of applying various IT techniques combined with optimization methodologies and the system architectures of the underlying DSS.

The rest of this chapter is organized as follows. In the next section, we provide a background on VRPTW and present a brief literature review. Next, we present the basic VRPTW model in mathematical terms and the solution approaches whose enhanced versions were embedded in the DSS built for solving the VRPTW problems in our practice. The following section contains two case studies in which the authors of this chapter were directly and actively involved. The chapter ends with concluding remarks and directions for future studies.

13.2 Background

It is not difficult in the operations research literature to find rich documented applications of VRPTW in the service industry. Some examples include Weigel and Cao (1999) for delivering furniture, appliances, and related services to customer homes; Kim et al. (2006) for solving VRPTW in a waste collection operation; Spada et al. (2005) for scheduling student bus service; and Blakeley et al. (2003) for generating periodical routes for service technicians. The interested reader is referred to Bozkaya et al. (2011) for more details on these and other interesting applications of VRPTW.

Practical VRPTW problems found in the service industry typically involve many business rules that may or may not be completely addressed by the VRP models in the literature. Some of these are listed next (see Bozkaya et al. 2011, for a complete list):

- Multiple objectives—A VRPTW objective function may involve multiple goals, such as minimizing total travel time or distance traveled by the fleet of vehicles, minimizing the fleet size, or minimizing total time window violation (where we define time window violation as the lateness of the product or service delivery).

- Nonhomogeneous vehicles—Vehicles subject to route-building may not be identical in terms of capacities, cost values, working hours, skill(s) required to operate, and so on.

- Vehicle/personnel-customer compatibility—A vehicle or service personnel may be restricted on the type of customers it can feasibly serve. For instance, a service person may have the skill to fix an appliance but not a computer, in which case this person cannot be routed to serve a customer requiring computer repair service.

- Soft time windows—While the basic VRPTW model treats the time window constraints as hard constraints (leading to infeasible solutions at times), in practice, customers may be willing to accept some delays in product or service delivery. This allows the possibility of time window violation, a concept known as the "soft" time window constraint.

- Multiple time windows—For some service companies, such as restaurants, a delivery can only be made in the morning or afternoon hours, excluding the busy lunch hour. In this case the VRPTW model must incorporate multiple time windows.

- Customer preference—A customer may require or request a specific service person for the delivery of a product or service. For example, a customer may ask for a service person who can speak the same language or someone who has serviced him or her before.

It is no doubt that high quality and at times real-time data such as customer orders, street network, and real-time traffic information allow solving VRPTW instances more efficiently and creating more realistic solutions. This is made possible to a great extent by recent advances in information technologies, found in today's enterprise resource planning (ERP), customer relationship management (CRM), and transportation management system (TMS) systems. Such systems typically provide the following input data required for a routing application:

- Vehicle data such as number of available vehicles and their parameters (e.g., capacities, costs, specialties, working hours)

- Customer data including pickup or delivery quantities, skill requirements, time windows, vehicle preferences, service priorities, and so on

- Real-time information such as a vehicle's GPS data, traffic data, weather data for dynamic routing or route adjustment of vehicles

- Geographic data on travel times, distances, and cost, which are typically calculated using GIS tools commercially available

Deployment of decision support systems such as those presented in the case studies section of this chapter relies on utilizing these techniques and data sets mentioned earlier for offering more realistic and acceptable solutions for the VRPTW. Next, we discuss some common solvers and DSS implementation examples for solving practical VRPTW instances.

13.2.1 Solution Methodologies and Implementations

It is well known that research on VRP or VRPTW is motivated by the practical needs. The first paper addressing VRP by Dantzig and Ramser (1959) describes a real application of delivering gasoline to gas stations. The authors also present the first mathematical formulation of the vehicle routing problem. This was followed by Clarke and Wright (1964) who presented a greedy heuristic that is widely known as the Clarke–Wright savings algorithm, and Solomon (1987) who introduced an insertion algorithm. Many other studies followed thereafter, including more recent ones such as Gendreau et al. (1994) and Cordeau et al. (2001) proposing tabu search heuristics, Gendreau et al. (2002) offering a review of metaheuristics for the problem, Cordeau et al. (2002) and Cordeau et al. (2005) proposing heuristic alternatives, Kallenhauge (2008) and Azi et al. (2010) studying the time window version of the problem, Nagata and Bräysy (2008, 2009) proposing memetic and local search heuristics, and most recently Juan et al. (2011) proposing ways to improve the Clarke–Wright savings heuristic.

Literature on exact and heuristic solution techniques for VRP and its variants is vast, so we refer the reader to Laporte (2009) who also, in commemoration of the 50th anniversary of VRP, presented a brief history of the development of VRP, and described a basic model of VRP along with a survey on exact algorithms, classical heuristics, and metaheuristics widely used to solve the problem and some of its variants.

VRP is known to be an NP-hard (nondeterministic polynomial time hard) problem (Lenstra and Kan 1981), meaning that large instances of this problem (with thousands or ten thousands of nodes and hundreds of vehicles) are generally not solvable to optimality in reasonable computational times. For this reason, many heuristic and exact algorithms were developed for VRPTW (see Bodin et al. 1983; Bräysy and Gendreau 2005, for comprehensive surveys).

Even though small instances of VRPTW can be solved by exact algorithms such as branch-and-bound or branch-and-cut approaches (for example, Ropke and Cordeau 2009), simple and effective heuristics are usually preferred for solving real VRPTW instances. There are many different types of heuristics aimed at solving the VRPTW. Two main categories are

- Local search or neighborhood search heuristics that involve route construction and route improvement logic
- Metaheuristic approaches including tabu search, genetic algorithms, simulated annealing, ant colony, particle swarm optimization, hybrid search, and scatter search

These approaches are covered in further detail by Bozkaya et al. (2011) along with many examples from the literature. Thanks to the development of technologies in computer hardware and software, researchers now are able to utilize larger memory capacity, parallel computing capability, and more computational power to solve larger and more complicated optimization problems in acceptable computational times. It is affordable to add more steps (procedures) to these metaheuristic approaches to explore a larger part of the solution space and hence yield better results for the VRPTW.

As observed in the survey by Bräysy and Gendreau (2005), metaheuristics for the VRPTW yield much better solutions than those produced by classical heuristics. While metaheuristics typically require more computational resources compared to classical heuristics, many useful implementation strategies such as parallel processing or multicore hardware utilization help speed up the solution procedures significantly, making it possible to solve real instances of VRPTW effectively.

Recent advances in information technology support the process of solving large and more complicated VRPTW instances found in the service industry. The notion of a "decision support system" has emerged as a computer-aided approach to help users and analysts solve complicated optimization problems. At the same time, a DSS allows users to easily access and manipulate, if necessary, the required data and analytical models while solving problems. Together with optimization techniques and other technologies such as a GIS and vivid graphic user interface, a DSS provides the user a better overview of the model and the problems to be solved (e.g., by displaying the geographic locations of stops to be serviced, the road networks, etc.). Many DSS deployments turned out as systems assisting users to make more effective decisions on vehicle routing and scheduling than before.

Many DSS-based approaches are found in the literature, showing the increasing interest in this technique. Examples include Basnet et al. (1996) for solving the VRP in milk delivery, Ioannou et al. (2002) for intracity VRPTW, Li et al. (2007) for vehicle rescheduling problem, Gayialis and Tatsiopoulos (2004) for routing and scheduling in the oil industry, Mendoza et al. (2009) for routing in the public sector, and Ray (2007) for optimizing routes with oversize and overweight vehicles. In almost all of these studies, a DSS was developed and presented to the users that involved a GIS mapping interface linked with an underlying ERP database for extracting data and an optimization engine for producing routing solutions. Some of these systems also included tools to modify automated solutions generated by the optimization engine and perform what-if analyses.

13.3 Mathematical Model and Solution Methods

VRPTW is a problem widely studied in the literature for which various models have been proposed. Next we present a basic model for VRPTW and its variants, based on which we discuss later the solution methodologies we have implemented in the service industry. As we present our case studies, we also introduce additional features to this model to address the particular business rules imposed by the customers we have been working with. Furthermore, in the solution methodology description of this section, we present only the newly added features of our solution approach in comparison to what is discussed by Bozkaya et al. (2011).

13.3.1 Mathematical Model

Let us introduce first the decision variables and other notation to be used in the mathematical model for VRPTW. Let $i,j \in I$ represent indices of customers from a customer index set I and $k \in K$ be indices of vehicles from set K. The customer index set I, with $|I| = n$, is further extended to $I^+ = I \cup \{0, n+1\}$ to include the single depot location index that marks the start and end point of each route. Furthermore, let

x_{ijk} = 1, if vehicle k drives from customer i to customer j, 0 otherwise
s_{ik} = the time vehicle k starts service at customer i
r_k = total route time (in minutes) of vehicle k
o_k = total overtime (in minutes) of vehicle k

where x_{ijk} is defined only for $i, j \in I^+$; s_{ik} is defined for $i \in I^+$. The rest of the notation is presented as follows:

c_k^d = travel cost per unit distance for vehicle k
c_k^r = regular labor cost per minute for vehicle k
c_k^o = overtime labor cost per minute for vehicle k, assuming $c_k^o > c_k^r$
w_i = amount of capacity used by customer i on any vehicle it is assigned to
p_i = duration of service at customer i location
d_{ij} = travel distance between customer i and customer j
t_{ij} = travel time in minutes between customer i and customer j
a_i, b_i = time window limits (start time and end time respectively) for customer i
R_k = allowable total route time, including any possible overtime, for vehicle k
O_k = allowable overtime for vehicle k
l_k = regular work time for vehicle k beyond which overtime will accumulate
Q_k = maximum capacity of vehicle k

The objective function of the model is the summation of the following cost items: travel cost, route labor cost, and route overtime cost. This objective function, however, may be revised to accommodate other specific business rules and logic, for instance, time window violation penalty, which we do in the following sections.

The basic mathematical model for VRPTW can be listed as follows:

$$\min F = \sum_{k \in K} \sum_{i,j \in I^+} x_{ijk} c_k^d d_{ij} + \sum_{k \in K} [c_k^r r_k + (c_k^o - c_k^r) o_k]$$

subject to

$$\sum_{k\in K}\sum_{j\in I^+} x_{ijk} = 1, \ \forall i \in I \tag{13.1}$$

$$\sum_{i\in I} w_i \sum_{j\in I^+} x_{ijk} \leq Q_k, \ \forall k \in K \tag{13.2}$$

$$\sum_{j\in I^+} x_{0jk} = 1, \ \forall k \in K \tag{13.3}$$

$$\sum_{i\in I^+} x_{ihk} = \sum_{j\in I^+} x_{hjk}, \ \forall h \in I, \ \forall k \in K \tag{13.4}$$

$$\sum_{i\in I^+} x_{i,n+1,k} = 1, \ \forall k \in K \tag{13.5}$$

$$s_{ik} + p_i + t_{ij} \leq s_{jk} + M(1 - x_{ijk}), \ \forall i, j \in I^+, \ \forall k \in K \tag{13.6}$$

$$r_k = s_{n+1,k} + p_{n+1} - s_{0k}, \ \forall k \in K \tag{13.7}$$

$$o_k \geq r_k - l_k, \ \forall k \in K \tag{13.8}$$

$$r_k \leq R_k, \ \forall k \in K \tag{13.9}$$

$$o_k \leq O_k, \ \forall k \in K \tag{13.10}$$

$$a_i \leq s_{ik} \leq b_i, \ \forall i \in I, \ \forall k \in K \tag{13.11}$$

$$x_{ijk} \in \{0,1\}, \ \forall i, j \in I^+, \ \forall k \in K \tag{13.12}$$

$$s_{ik} \geq 0, \ \forall i \in I^+, \ \forall k \in K \tag{13.13}$$

$$r_k, o_k \geq 0, \ \forall k \in K \tag{13.14}$$

In this model, constraints (13.1) indicate that each customer must be assigned to a vehicle, constraints (13.2) ensure that vehicle capacities are not exceeded, constraints (13.3) through (13.5) maintain flow balance at each node of the underlying transportation network, constraints (13.6) calculate the arrival time of a vehicle at each customer location it visits, constraints (13.7) and (13.8) count the total route time (including overtime, if any) and total route overtime, respectively, constraints (13.9) and (13.10) set the required time limits on route time and route overtime, constraints (13.11) make sure the hard time window limitations are not violated, and finally constraints (13.12) through (13.14) enforce the integrality and nonnegativity requirements.

The mathematical model presented is a capacitated VRPTW model with hard time window constraints, limits on the route time, overtime, and capacity constraints. Furthermore, it states that all vehicles start at and return to the same depot location. Nevertheless, we discuss extensions to this model as necessary upon the specific problems in our case studies.

13.3.2 Solution Methodology

Since the VRPTW is a NP-hard problem, heuristic approaches are usually the best choice for obtaining good solutions in reasonable time. Our experience with the real-world instances of the VRPTW confirms this, and we elaborate further in the two case studies presented later in this chapter. To solve the capacitated VRPTW, we take a two-phase approach, which is an implementation of the cluster-first and route-second concept first introduced by Fisher and Jaikumar (1981). These two phases are named "assignment" and "improvement," and are described next.

13.3.2.1 Assignment

During the assignment phase, the primary goal is to produce an initial solution that satisfies all constraints of the VRPTW including vehicle capacities, hard time windows, limits on total route time, and total route overtime, while trying to obtain the initial solution with the minimal overall operational costs (i.e., travel costs and labor costs). To meet different business focuses in the service industry, we introduce a weighted objective function that combines various cost factors. The weighted objective function provides the flexibility of adjusting the weights for different cost dimensions according to the end users' experience and preferences, and thus to yield more acceptable solutions for the practice. Furthermore, this structure also allows us to account for additional cost terms that could be introduced in various extensions of the problem (e.g., time window violations in case of soft time window restrictions, waiting, or idle time).

During the assignment procedure, we build routes one after another, that is, unassigned customers are assigned to the current route whenever it is possible, and a new route will be built if any constraint on the current is violated. The assignment procedure basically consists of three parts: (1) building an initial route for each vehicle, (2) inserting stops into routes, and (3) improving assignment.

This route-building phase with three parts may work effectively for the problems with moderate size; however, it may not work so well for the large-scale problems often encountered in the service industry since the insertion evaluation process will likely be time consuming. Furthermore, routes built via this process may not quite fit the business' desire, for example, some companies may want to service a different area on each weekday. Therefore, before the assignment procedure, it is often necessary to run an additional step to divide the entire service region into several subareas. After these subareas are created, the assignment procedure runs for each subarea. Based upon our experience, building subareas before the assignment step cannot only speed up the assignment procedure but also provide more desirable results.

In our algorithm implementation, we adapted the clustering algorithm proposed by Cao and Glover (2010) to build required subareas. The assignment step is employed to create routes within each subarea. The clustering algorithm proposed by Cao and Glover (2010) is Thiessen polygon based, and it is able to build compact and unoverlapped subareas. In the first case study, the corresponding clustering algorithm is described in detail.

During the assignment procedure, routes are built one after another. It is inevitable that the procedure makes the decision not based upon the global optimal criterion. The improvement steps may be necessary to achieve more satisfactory initial solutions. The improvement step for the assignment consists of two subprocedures with the purpose of creating the assignment result with lower total cost and relatively balanced routes:

- Destroy and rebuild—Similar to what is described by Glover (2000) and Laporte et al. (2010), some stops will be taken away from their routes and be reinserted. Here, we may use the change in the weighted objective function, that is, $\Delta F_{ikt} = \alpha_1 \Delta f_{1,ikt} + \alpha_2 \Delta f_{2,ikt} + \alpha_3 \Delta f_{3,ikt} + \ldots$ as the criterion to evaluate dropping selected stops from routes and reinserting them. This procedure is further elaborated in one of the upcoming case studies.

- Transfer and exchange—Stops may be transferred from their original routes to different (destination) routes or two stops may be exchanged between their respective routes.

As it is known that a weighted objective function is employed in the model, the assignment procedure can adjust the insertion costs by tuning the weights for the cost factors participating in the objective function. In this case, different business rules (e.g., service priority for a customer, soft time windows, etc.) can be addressed effectively.

13.3.2.2 Improvement

Based upon the previous discussion, we know that the assignment procedure terminates when all customers get assigned or no route can accept any more unassigned customers. Next we want to improve the assignment result by applying intraroute and interroute improvement moves. An intraroute move changes the visiting sequence of a single stop within its route to find a better visiting sequence, while an interroute move switches a stop from its current route to a different one or exchanges two stops between their respective routes attempting to find a better solution in terms of total cost (Bozkaya et al. 2011). In our case study, we illustrate how the improvement steps are applied.

The following summarizes the aforementioned two-phase algorithm:

Step 0. Read input data and initialize data structures.

Step 1. (Optional) Build subareas using Thiessen polygon-based clustering algorithm.

Step 2. (Assignment) For each subarea (please note that if step 1 is not performed, then we have only one subarea) repeat until no more customers can be inserted.

   2.1 If there is no empty route, exit step 2. Otherwise, pick an empty route called current route.

   2.2 Evaluate possible feasible insertions for all unassigned customers into the current route.

   2.3 Pick the insertion that has the lowest value of ΔF_{ikt}.

   2.4 Execute the insertion found in step 2.2, and update the route (remaining capacity, etc.) upon the insertion

   2.5 If the current route still has room to accept more customers, go to 2.2, otherwise go to 2.1.

Step 3. (Optional) Perform initial solution improvement procedure (destroy-rebuild, stop exchange, and transfer)

Step 4. (Solution improvement)

   4.1 Initialize tabu lists and tabu tenures.

   4.2 Repeat until total number of iterations is executed.

      4.2.1 Evaluate all feasible intraroute improvement possibilities.

      4.2.2 Evaluate all feasible interroute improvement possibilities.

      4.2.3 Select the best improved tabu or nontabu move. If there is no such move, select the best nonimproved nontabu move.

      4.2.4 Execute the move found in step 4.2.3 and update the corresponding route(s) upon the move.

      4.2.5 Update the tabu lists.

      4.2.6 If new solution is better than the best known solution, record it as the new best solution.

   4.3 Output the best known solution.

This tabu-search-based improvement procedure is reported to be capable of creating better results for VRPTW problems in the service industry (Bozkaya et al., 2011).

13.4 Case Studies

In this section, we present case studies for two real DSS deployments for solving VRPTW in the service sector. These include delivering tobacco products and delivering or picking up cash, coins, and other valuables. Although the nature of the VRPTW is somewhat different in each case study, the general characteristics of the problems are very similar, and each DSS built to solve VRPTW is combined with heuristic optimization approaches, GIS, and RFID techniques.

13.4.1 China National Tobacco Corporation

China National Tobacco Corporation (CNTC) is by sales the largest single manufacturer of tobacco products in the world; the market serviced by CNTC accounts for roughly 30% of the world's total consumption of cigarettes. CNTC is responsible under the close supervision of its administrative organ called the State Tobacco Monopoly Administration or STMA, for marketing, production, distribution, and sales of all tobacco products inside of China. Furthermore, various factories owned by CNTC in different provinces or cities in China fulfill the orders and deliver tobacco products to CNTC's distribution channel or special authorized stores.

Due to the large number of special stores in cities (usually the special stores within a single city will be serviced), the subdivision of CNTC of that city faces a challenge: how to deliver tobacco products more efficiently to their special stores. Every day, the owners of the special stores place their orders over the Web site hosted by CNTC according to estimated demands. When the subdivision of CNTC in a city receives the orders, it will determine the delivery tasks based upon the inventory data and build the corresponding delivery routes for the next day to service these orders. When delivery routes are built in a subdivision of CNTC, in addition to honoring all the imposed business rules, one also tries to:

- Provide delivery truck drivers with consistent routes
- Minimize total delivery costs

Although the delivery problems to be solved here can be modeled as VRPTW, the size of the problems and their practical complexity make them of both theoretical and practical interest. A subdivision of CNTC in a city usually has a fleet of around 50 delivery trucks to service more than 10,000 special stores in total with more than 2000 stores to be delivered on a daily basis. In order to facilitate the decision making on delivery route building, the end user also requests the DSS possess the following features:

- Display all underlying streets and their attributes, speed limit, one-way, etc.
- Display all delivery areas or territories
- Monitor the incoming products
- Track dynamically the delivery status for each truck
- List driving directions for routes
- View all routes and their desired delivery area (seed points) if any
- Manually adjust a route, for instance, select a route and change the stop sequences or reassign stop(s) from one route to another
- Manually assign a delivery (seed point) to a driver based on the business need or customer preference

One of the authors of this chapter was involved in the project and served as a chief technical officer for optimization and system architecture. He experienced the entire procedure of system design and development for CNTC. To assist the daily routing/scheduling and other corresponding daily operation needs, a decision is made to build a DSS for decision making, automatic data processing, and vehicle tracking activities. The DSS combines existing commercial IT products and custom-developed modules that utilize the contemporary IT technologies such as GIS, GPS, and RFID. RFID is employed to track the incoming products, GIS provides necessary data for optimization module (travel time/distance, route displays and adjustment, etc.), and GPS tracks the vehicle movements and delivery statuses. In the following we are going to present the overall system architecture and embedded algorithms for solving VRPTW encountered during building routes for tobacco product delivery operations.

13.4.1.1 System Architecture

Figure 13.1 clearly indicates the purposes of building this DSS to assist the operations in tobacco logistics. The CNTC management decided that a DSS combining various IT techniques and their business data would form a common (information) platform that forms the logistics solution to assist the daily logistics operations and solve related problems. Based on the organizational structure of CNTC, each subdivision of CNTC in a city is responsible for all logistics operations in that city. Therefore, the DSS is built at the province level but is distributed to individual cities of that province.

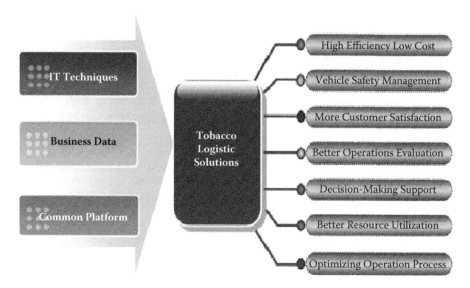

FIGURE 13.1
The purposes of building DSS for CNTC.

The DSS consists of four major components:

- Inventory management—It tracks inbound and outbound products where RFID is utilized, records the inventory, and provides the replacement information. The RFID technique is employed to track the sources, brands, and quantities of incoming products, which is important for CNTC management to decide if the requests from their special stores can be met.

- Optimization module—It builds the daily tobacco product delivery routes, schedules the drivers, which is customized component and will be discussed a bit more later.

- Vehicle tracking—GPS technology is applied to track the vehicles in real-time mode to ensure them to follow planned routes (digital fence), to obey the traffic laws (speed limits, etc.), to provide better estimated arrivals for customers, and evaluate the performance of delivery operations.

- GIS analysis—This component utilizes the operation data collected daily to analyze the operational performances (fuel consumption, mileage per stop, number of stops per vehicle), the market demands, the store performance, and product distributions.

Figure 13.2 depicts the system architecture for the DSS.

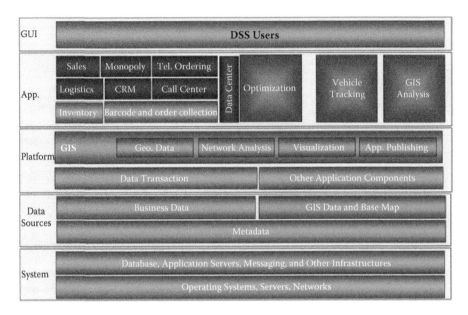

FIGURE 13.2
DSS architecture.

Although certain existing commercial IT products can be used to create certain components in the DSS, the optimization component needs to be developed based on the particular business logic in addition to those in the base model of VRPTW that is also the focus of this section.

13.4.1.2 Optimization Component

The VRPTW to be solved for CNTC usually has thousands of stops and nearly hundreds of vehicles. Furthermore, when the routes are built for tobacco product delivery, we should take the following CNTC business rules into account:

- Stores on the same street segment should be delivered by the same truck whenever possible.
- The side of a street to unload products needs to be considered.
- Whenever a truck encounters a store on its route on the way, it should deliver to that store right away if possible to reduce the weight loaded on the truck. It is referred to as "first-see first-serve" rule. If it is purely based on the objective function value, it is quite possible that a stop is delivered on the back to the depot without changing the cost at all.
- Larger trucks should be utilized first.

In the following discussion, we will focus on the extension to the base VRPTW model and its corresponding solution steps, and will not replicate the contents that are known for a regular VRPTW. Some major additions to the base model and solution methodology are as follows:

- Building daily delivery territories—Since the number of CNTC special stores in a city is huge, and there is not enough capacity in the context of vehicles and manpower to cope with the entire city, the subdivision has to divide the entire city into several daily delivery areas with each being serviced on a single day. Usually the city will be divided into five subareas. However, this delivery area creation is another challenge as it requires:

  - Each subarea should be as compact as possible so that the expected travel time and distance to service the subareas will be minimal.

  - Each subarea must be connected, which means no customer of a subarea can lie within the geographic boundaries of another subarea.

- The subareas should be balanced. Balance criteria can be expressed as bounds on differences in the numbers of customers in different subareas, or in the delivery quantities of different subareas.

  Cao and Glover (2010) propose algorithms for producing such subareas based upon special procedures by exploiting Thiessen polygons. Their methods are able to handle multiple criteria for balancing the subareas, such as the number of customers in each subarea, the service revenue in each subarea, or the delivery/pickup quantity in each subarea. For the details of the algorithms please refer to Cao and Glover's paper. The DSS has incorporated their algorithms, and it achieves very satisfactory results. The delivery territories are planned periodically and may be adjusted upon seasonable demands such as Chinese New Year. Figure 13.3 shows the result of creating five delivery subareas in a city in China, though the user has the flexibility of creating different numbers of subareas upon the business requirements.

- Enhanced assignment procedure—After the subareas are generated, the delivery routes for each day can be built within each individual subarea. We have already presented the base assignment procedure; nevertheless, we need to enhance it to address the concern of CNTC. It states that larger vehicles should be used first before smaller vehicles can be used. In the input data fed to the solver, we sort the vehicles in nonincreasing order of their capacities. Since our assignment procedure builds routes one after another, it copes with this requirement very well. Unfortunately this procedure inherits a weakness, that is, it inserts customers to a single route instead of considering all potential routes; it is unavoidable that the procedure

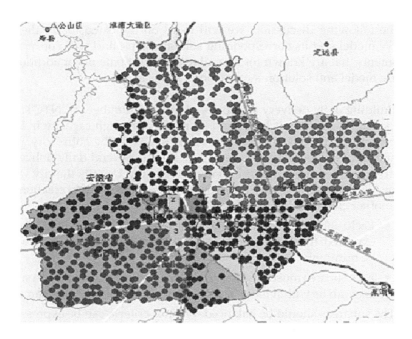

FIGURE 13.3
The result of creating five delivery areas.

is short-sighted. To overcome this problem, we adapt the destroy-and-rebuild strategy similar to removal and insertion heuristics described in Laporte et al. (2010). In the practice, we pick 5%–10% of stops based on the problem size and remove them from the current routes. We then reinsert these removed stops into routes as they were unassigned stops. Specifically, for a stop i, define $\Delta C(r, -i) = C(r) - C(r, -i)$ to be the cost of removing stop i from route r, where $C(r)$ is the current route cost while $C(r, -i)$ is the route cost after stop i is removed. Each time we remove a stop with the largest value of cost of removing and we update the values for the rest of stops until we remove a certain number of stops from their current routes upon the predefined percentage of stops to be removed. All cost factors to be involved in this value calculation could be time consuming, therefore, we use only the travel distance as the sole factor in this value, which can be obtained fairly quickly. This step is optional in the DSS. The user chooses to run it if more computational time is permitted. The procedure, however, demonstrates its effectiveness in fixing some problems in initial solutions.

The assignment needs to address the first-see, first-serve rule. It is relatively easy to implement it because we pick the smallest insertion cost for a stop with the smallest insertion position index in a route. Therefore the result after the assignment meets this rule.

- Enhanced intraroute improvement—We need to take the first-see, first-serve rule into account. This rule is also treated as a *hard* rule. The intraroute improvement procedure needs to add some extra steps to prevent a stop from being moved to an improper delivery sequence though the total cost stays the same. Please note that although we are using TS for our intraroute improvement, this situation may happen. Because the result from the assignment meets the rule, if we do not accept any forward insertion move with zero cost change it will not violate the rule based on the definition of forward insertion. However, we may accept backward insertion with zero cost change. In this case we need to validate the move. Consider a route *r* with a given stop sequence $r = (0, ..., j - 1, j, j + 1, ..., i, ... 0)$, where *j*, *i* are the sequence indices of stops, 0 is the index for the depot, and $i > j$. Define $d(j)$ to be the current travel distance from the previous stop to stop *j* before the move, and d_{ji} to be the travel distance from stop *j* to stop *i*. In case of a backward insertion move (where stop *i* is going to be inserted in front of stop *j*) has the zero cost change, we check if:

$$\Delta D = d(j) - d_{j-1,i} > 0, \text{ if } i = j + 1, \text{ or}$$

$$\Delta D = d(j) - d_{ij} > 0, \text{ if } i > j + 1$$

When the condition is valid, then the rule applied to stop *j* is not violated and the move may be accepted, otherwise we will reject this move.

13.4.1.3 Economic Benefits

The DSS for solving VRPTW has been implemented in many cities where CNTC is responsible for delivering tobacco products to their special stores. The system provides friendly graphic user interfaces and a convenient mechanism to allow users to perform what-if analyses. The DSS is a common information platform on which users can also track vehicles and their working status besides performing routing and scheduling tasks. Rich report generation functionality facilitates the daily management and operations. The DSS developed for CNTC has been deployed in several provinces in China and utilized for more than one year to support their daily tobacco logistics operations (inventory, distribution, tracking, evaluation, etc.). Figure 13.4 shows some screenshots for the system.

The users at subdivision offices of CNTC now are able to build more efficient routes, make better decisions, manage business process more efficiently, and perform more effective analysis tasks. Since the system was deployed a year ago, CNTC has been seeing good ROI (return on investment). Table 13.1 summarizes the economic benefits achieved by deploying the DSS in a city.

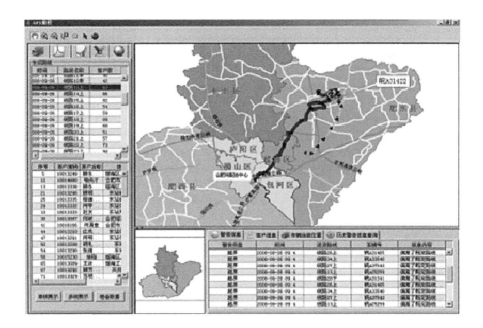

线路名称：线路01单　车牌号：皖AJ5849

送货序号	客户名称	简码	地址	销量	金额	到达时间
1	阚武	50002045	阚集乡阚集站金路百货超市	21	739.5	8:05
2	黄如鹏	10015966	合肥市恒润花园1幢115号	11	252	8:12
3	刘世初	10011894	政务新区汇林湖农贸市场W13	25	1040.5	8:16
4	李翠春	10011887	大铺头下公岗村78号	64	3188	8:29
5	孙启明	60004265	肥西县大柏乡堰墩村先锋组305号	45	2178	9:21
6	王广乐	60003120	高店街道	36	1442	9:53
7	权客元	60003129	肥西县桃花工业园翡翠花园111号	68	3168	9:57
8	梁齐祥	60005126	江夏乡王角街道	41	2442	10:15
9	苏义元	70005429	长丰县岗集镇善岭村	62	2197.5	11:44
10	张在和	70004900	水湖镇张闸乡裴户村	49	1765	13:29
11	阎其学	70005426	长丰县届岗村阎大庄	12	641	13:39
12	阎玉随	70003517	长丰饲料厂	17	454.5	13:47
13	姚兴荣	70003525	长丰饲料厂对面	10	361	13:50
14	胡瑞华	70004478	水湖农场	16	523.5	13:55
15	耿广虎	70003336	朱巷镇	31	1612.5	14:45
16	耿言柱	70003338	朱巷镇	29	766.5	14:48
17	魏孟清	70003341	朱巷镇	11	275.5	14:51
18	梁支付	70003344	朱巷梁山村	14	397	14:54
19	杨梅芳	70003334	朱巷镇	13	361	14:57
20	许庆安	70003329	朱巷镇	40	1519	15:00
21	董朝伟	70001353	石涧小学对面	19	528	16:05
22	孟俭	70001349	双凤镇火车站	13	1009.5	16:17
23	郑台风	10012425	合肥市庐阳区杏花村镇	70	2915.5	16:42
24	江玲	10015965	合肥市界首路宰福人家6幢111号	17	528	16:51
25	李庆元	10014296	合肥市阜南路190号(金清楼隔壁)	11	316	16:57
26	存群百货商店	10010545	来瓶路南侧竹西村朱小郭	12	361	17:14
27	招财商店	10010630	东瓶路南侧竹西村委边	13	355.5	17:17
28	阚武	50002045	阚集乡阚集站金路百货超市	42	1641.5	17:27

FIGURE 13.4

User interfaces (maps) and route manifest.

TABLE 13.1

Economic Benefits Achieved by Using DSS

Before System Deployed	After System Deployed
Number of Vehicles Employed	
40	20
Number of Customers Serviced per Vehicle	
60	93
Number of Products Delivered per Vehicle	
45	75
Average Vehicle Loading Rate	
51%	81%
Time Spent on Planning Territories	
Days	<30 minutes

According to the statistics from CNTC, the logistic operation cost has been reduced by 22% after the DSS was deployed. We also learned that in order to create a DSS that can be relatively easily accepted by the end user, we need to build easy-to-use and easy-to-understand user interfaces. Though the optimality of results is highly desired, reasonable and feasible solutions are more welcome in practice because of problem complexity and lack of computational time for exact solutions. Management and user's involvement during the project is the key to ensure the project success. Analyzing and explaining results jointly with users facilitates the deployment and employ-ment of the DSS. To provide feasible and good quality VRPTW solutions, the quality of map data is extremely important, and more frequent updating of map data is desired as well, particularly for those emerging markets such as China. The management and the user would like to see the newly built roads, bridges, and so forth being incorporated into the map data. They even would like to incorporate real traffic data into the solution procedure. It is, however, a challenge to use real traffic data while solving VRPTW, let alone the consistent availability of high-quality, real-time traffic data. With future advances in sensor technologies and emerging business models for using traffic data, and new optimization techniques utilizing time-of-day travel times, we think these issues will be successfully addressed within the next 3 to 5 years.

13.4.2 Bantas Cash and Valuables Pickup and Delivery Optimization

Our second case study is about a private company named Bantas A.S., based in Istanbul, Turkey, whose main business is to provide cash pickup and delivery services to bank branches, ATMs, and major clients who need cash for their daily operations. The company was formed in 2009 by a joint venture of three major banks operating in Turkey, in recognition for the need to collaborate to achieve economies of scale and increased efficiency. While the immediate intent is to serve the three banks that initiated the joint venture, the company is also interested in expanding its services to different clients across Turkey.

In its short history, Bantas has been a company that has invested in various forms of technology, such as fast money collection and counting devices, GPS-based tracking of its armored trucks, and technological security measures for its cash centers. Although such systems were put in place, Bantas did not have a routing optimization solution. The routing of its trucks was mainly being done manually by a routing specialist based in each one of Bantas' 24 cash centers, serving the territory around it. Each routing specialist was charged with the task of extracting from the ERP system the list of daily service requests from client banks and merchants, and creating a daily visit sequence for each armored truck in his or her jurisdiction. There was no integrated GIS system actively used by specialists in route planning; only a mapping application, named Mobiliz, for tracking armored trucks using GPS was available. Bantas wanted to further automate this system by incorporating a route optimization routine to eliminate potential human errors and inefficiencies due to lack of an overall optimization approach. Increased automation would also mean increased flexibility in handling varying service levels requested by Bantas' clients, and handling extreme situations such as unanticipated cash requests, responding to ATM breakdowns, high service volumes during national holidays, and so on. One of the authors of this chapter acted as the technical manager of the team who developed and implemented a fully integrated GIS-based routing solution. Bantas aimed to launch this route optimization system fast enough to handle the service volume expected to increase soon and realize its benefits.

Bantas' routing business rules were indeed somewhat different from traditional applications of VRPTW. The following information summarizes main operating rules under consideration for the purpose of optimization:

- The country is divided into 24 regions, each containing either a cash center (CC) or a cash in transit (CiT) office responsible for operations in a defined territory using known locations (XY coordinates) of bank branches, ATMs, and Bantas client companies (such as supermarkets, store outlets of major retail chains) who request cash pickup and delivery services of Bantas.

- Each cash center is responsible for timely collection of the required cash in the morning before 10:00 am from the central bank (CB), arrange it into bags and cassettes, and have it delivered to their target destinations throughout the day. Similarly, at the end of day the CC must organize delivery of all excess cash for deposit into the CB before 16:00. CiT offices perform similar tasks, different only in that they do not perform cash storage and processing services; they simply act as transit centers. For cash centers, the portion of the day route before 10:00 am is executed as an independent run and trucks are assumed to have completed this step before they start visiting banks and ATMs.

- Failure to meet the CB's strict morning pickup and afternoon delivery deadlines means either no cash on hand to deliver (and hence failing to meet customers' requests) or facing penalties and lost interest charges from Bantas customers who, by contract, must have their excess funds deposited in the CB every day to earn interest.

- Each CC operates a fleet of armored trucks with known characteristics such as operational costs and daily work schedule. Each truck must start its pickup/delivery route at the CC ensuring that all service requests along its routes are fulfilled. A truck must return to the CC early enough for the CC to arrange on time the transfer of funds to a CB location closest to the CC.

- Three types of pickup/delivery accounts exist: bank branches that require pickup or delivery of cash; ATMs (on-site ATMs at the same locations as a branch and off-site ATMs that are independently served) that require delivery of cash; and corporate clients (e.g., store chains and similar merchants) that require cash collection to be delivered to a specific bank branch for deposit. It is desirable that a truck visits a bank branch location only once if that branch is also a target ATM delivery location or a corporate client's target cash delivery destination. Most of the ATMs visited by Bantas trucks, however, are off-site ATMs.

- Due to service level agreements signed between Bantas and the three banks, a service request must be honored within a specific time frame, which implies a daily time window for picking up or delivering cash under each instance. Hence service requests have associated time windows for visit.

Although there are many other business rules resulting from daily requirements of satisfying various service level agreement terms, our focus is primarily on effectively transferring the aforementioned considerations to a user-friendly decision support system. Next, we describe the details of the GIS-based spatial decision support system we have designed and implemented for Bantas.

13.4.2.1 Proposed System Architecture

Bantas employs an ERP system that is used not only to store and maintain corporate data, but also allows Bantas' customers to post their service requests on a regular basis using the available B2B Web communication tools.

In addition, Bantas maintains a separate vehicle tracking system outsourced to a third-party provider named Mobiliz. This tracking system collects GPS location data from all trucks as well as data from various sensors located on each truck to continuously monitor the security of each truck. Our proposed decision support integrates with this existing system in two ways:

- It provides routing optimization modules that take input data from the ERP system and posts its routing solutions back to the ERP system.
- It offers an analysis and reporting module for comparing and contrasting planned routes versus actual routes.

The interaction of these modules with the current system is depicted as in the system architecture shown in Figure 13.5. This architecture indicates that the main module interacting with the existing ERP system is VisioAnalyst. VisioAnalyst is a Web server system serving a mapping application that provides a Web browser interaction between the user and the decision support system. VisioAnalyst offers two main functionalities: (1) maintaining the input/output interface with the ERP system, and (2) displaying a map that compares and contrasts planned versus actual routes. The former is done by a scheduled service that runs on the VisioAnalyst server at scheduled times to query the ERP Web service for outstanding orders. After the route optimization process

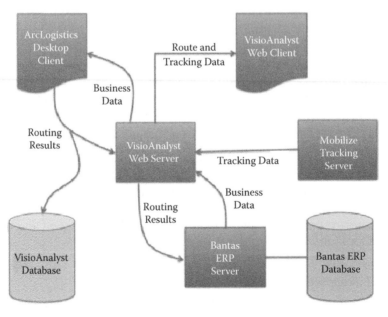

FIGURE 13.5
Bantas routing DSS system architecture.

is completed, VisioAnalyst also provides an outgoing Web service interface for the ERP system to query and request optimized routes.

The second functionality offered by VisioAnalyst is the route tracking analysis and reporting. This is meant for comparing the planned routes with actual routes using GPS track data of each truck, and is accomplished by maintaining a link between the VisioAnalyst system and the third-party tracking system Mobiliz. VisioAnalyst maintains a B2B Web service connection with Mobiliz and frequently (every 2 minutes) queries and stores the location of each truck on its daily route. It then generates, upon user's request, a report comparing the planned route length, duration, and cost against the real data collected from the field.

As a final important component of the decision support system, a spatial database is maintained by VisioAnalyst to store the following types of data:

- Road network for 5 main regions of Turkey covering the 24 offices of Bantas, used mainly by the route optimization engine, and as defined by Bantas staff.
- Base map of Turkey including layers such as administrative boundaries, major and detailed roads, ferry links, and city centers.
- Customer service request data, including attributes such as customer ID, assigned region ID, service request date and session, service duration, service location, pickup/delivery time window(s), and group ID (if service request is linked to other requests that will together be routed).
- Route optimization results, including attributes such as route ID, route date, driver ID and name, number of stops on route, list of stops by service request ID, estimated time of arrival, travel distance and time from previous stop, as well as route statistics such as total cost, total distance (km), and duration (minute), and route optimization parameters such as fixed cost of operating a truck, unit mileage cost, and unit labor cost.

The road network maintained by VisioAnalyst and used by the optimization engine is based on Navteq 2010/Q4 data set and has 1,407,463 street-level road segments covering 71 out of 81 provinces of Turkey (about 81% of country population). The remaining 10 provinces do not have a detailed street database, however, they are still connected to the rest of the Turkish road network via a state highway network. For purposes of efficient network calculations and achieving target computation times, the road network is divided into five subnetworks: Istanbul–Europe, Istanbul–Asia, Aegean–Mediterranean, Ankara–North, and East. There are, however, a total of 24 Bantas cash centers and cash-in-transit offices in Turkey, each one with its own service hinterland, using one of the five regional street subnetworks.

13.4.2.2 Optimization Methodology

The decision support system we implemented for Bantas employs a heuristic route optimization engine, which is used as part of a commercial off-the-shelf software product called ArcLogistics™, purchased and installed by Bantas. ArcLogistics provides a user-friendly graphical interface for managing customer and fleet data, and visualizing optimized routes. More important, ArcLogistics provides a flexible optimization engine capable of handling a variety of vehicle routing problems, including VRP with time windows, and VRP with simultaneous pickup and delivery.

The heuristic optimization algorithm used by ArcLogistics is proprietary, and hence is not revealed to the level of detail that fully describes its workings. However, based on both of the authors' contributions to the development of ArcLogistics' heuristic solver and their experiences in the software's deployment in many industrial projects, and the information available from public sources, we can describe, without violating the nondisclosure of proprietary information, the main logic of the algorithm along the lines described in the previous section of this chapter. Specifically, the ArcLogistics route optimization engine first completes a construction phase, where each service stop is inserted into a suitable route (i.e., assignment), followed by the improvement phase, where the initial feasible solution created in the first phase is improved by a sequence of swaps and exchanges of stops within and between routes. The general execution of this latter improvement logic follows the principles of the tabu search heuristic optimization technique. The algorithm terminates when a predefined total number of iterations or a predefined number of nonimproving iterations are executed and the best routing solution can no longer be improved. During this iterative procedure, the algorithm always maintains hard rules such as capacity restrictions, open and close hours of depot locations, and visit sequence of paired pickups and deliveries, while searching for more economic solutions in terms of total mileage, labor, and fixed vehicle costs. The algorithms handle time windows as a soft constraint, meaning that solutions with time window violations are possible, which, however, are penalized at varying levels that can be chosen in the user interface by the routing specialist. Since time window violations are extremely critical for Bantas' business, the resulting solutions are hence in general free of time window violations, due to the choice of large time window violation penalties.

Bantas uses ArcLogistics in an interactive way to manage daily routes and communicate the resulting routes to the field staff. Initially, after the service requests are downloaded from the VisioAnalyst database and imported into ArcLogistics, a Bantas routing specialist processes each cash center or cash-in-transit office sequentially and generates an automated routing solution. The resulting solution may require manual adjustments by the routing specialist as some dynamic conditions and field knowledge of the personnel (on issues such as traffic congestions and driver preferences for service point assignments) may have to be incorporated. The final solution is then

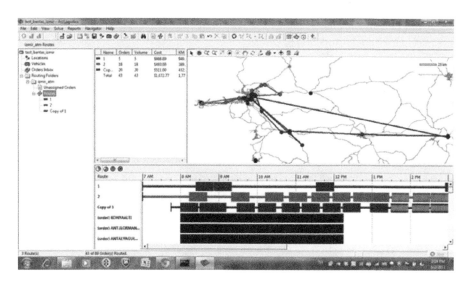

FIGURE 13.6
ArcLogistics screen showing routes in Aegean region.

"exported" to the VisioAnalyst database and also simultaneously to Bantas ERP at the cash center level. The manager of each cash center or CiT office is free to modify the corresponding route plan, however, few modifications are typically made since Bantas staff understand that any manual changes made will reduce the benefits of optimization.

Figure 13.6 shows a sample screenshot of ArcLogistics' main application window, showing the regional map in Aegean Turkey and a routing solution with two routes. Figure 13.7 shows how these solutions appear in relation to GPS coordinates collected from the field. The latter combines views and the reports that can be generated from it are especially important to cash center and CiT office managers and the senior management for monitoring the implementation of planned routes and hence validating the economic benefits of the decision support system.

13.4.2.3 Economic Benefits

Bantas has completed the pilot implementation of the decision support system in two regions of Turkey, namely, Antalya and Istanbul–Europe network. The company is now in the process of deploying the system nationwide in the remaining regions.

The Antalya region is a relatively small region with 12 trucks in total and 200 daily service points on average to visit. Istanbul–Europe, on the other hand, is a relatively larger region with 800 visit points on average serviced by 27 trucks. Because the Antalya region involves more rural coverage compared to the Istanbul–Europe region, its trucks cover greater distances and

FIGURE 13.7
VisioAnalyst mapping application that shows GPS records in comparison to a planned route.

TABLE 13.2

Economic Benefits Achieved by Using DSS for BANTAS

Before System Deployed	After System Deployed
Number of Vehicles Employed	
39	30
Total Daily Kilometers Traveled	
5974	4412
Total Daily Mileage Cost	
7168.80 Turkish Lira	5294.40 Turkish Lira

hence can service fewer points per route. Istanbul–Europe, however, operates in a densely populated area, allowing routes to run in more concentrated areas.

Table 13.2 summarizes the total savings obtained by using the route optimization decision support system in these two regions combined, as part of the pilot study. Further results or time-series based results unfortunately are not yet available, as the company is still in the process of nationwide deployment. The results clearly show a potential cost savings of 26% in these two important regions of the country. As a result, company officials have taken the necessary steps to immediately implement the routes produced in these two pilot areas. Further studies to extend the operation to the remaining regions are under way. Note that these savings are achieved from 39 routes on a single day's worth of route plans. If proportional savings are realized nationwide, Bantas, which currently operates 124 trucks in total, may overall expect to save around 30 trucks and hundreds of thousands of dollars annually. Company officials, however, are reluctant to extrapolate these savings just yet to all regions where they operate, mainly because different regions may have different operating characteristics, which may have a negative impact on the feasibility and applicability of the routes generated by ArcLogistics. The routing engine, however, is fairly flexible in handling a range of operational rules, and the implementation team will nevertheless continue to adapt the system in the remaining regions so as to achieve the maximum possible cost benefits.

Another economic benefit offered by the implementation of this system is due to its scalability. Since Bantas aims to expand its operations beyond the three founding banks as its initial customer base, use of this system will further help Bantas effectively manage its operations. Bantas can not only use the DSS to quickly route the new customer service points that will be added to the system but also use it to simulate different cash center locations and customer–cash center assignments. In other words, Bantas can use the system as a what-if analysis tool in redesigning its nationwide operational structure, in response to business expansion.

Finally, we must note how the implementation of this DSS significantly reduced the labor needs of Bantas. Before the implementation, manual planning of routes took up to 10 hours for all cash center locations combined. With the decision support system in place, the route planning of all regions can now be completed centrally in less than one hour. This means drastically improved efficiency in terms of planning time and fewer errors due to human intervention.

13.5 Conclusions

In this chapter, we present the basic VRPTW model, its corresponding algorithms, and its applications in the service industry. In order to solve real-world VRPTW problems, researchers and practitioners tend to use contemporary IT techniques including GPS, GIS, and RFID together with optimization methodologies to build DSS to facilitate the solution procedures. We report two case studies of using DSS in the service industry. While the VRPTW problems encountered in the service industry possess big challenges such as large number of stops to be visited or serviced, complicated business rules, and restrictive traffic laws, DSS are able to assist the decision makers to make reasonable decisions within relatively short time frames. DSS provide user-friendly interfaces to allow the end users to interactively revise the results created by the system upon their experiences and specific field needs. Our studies show that the DSS described in this chapter have solid enterprise architecture, and they are able to handle the daily decision making in solving VRPTW efficiently. Due to the involvement of the end users during decision making, the results presented for the final deployment address many more real business rules and are welcome by the field people. The economic benefits of using DSS discussed in the chapter are significant.

Of course we are still faced with challenges to help small- to medium-size service companies implement DSS to solve their corresponding VRPTW problems. Because of the investment on hardware and software to build a DSS, these companies are usually hesitant to go forward. Due to their business scales, it may take some time to see the results of a positive ROI. Interestingly enough, cloud-computing platforms are getting more popular. We believe that we can utilize cloud-computing platforms to provide decision-making services to these companies, which only need to pay for the services they use instead of making an initial lump-sum investment. With less investment, small or medium companies are able to achieve better economic results. Nevertheless, deploying a DSS on a cloud-computing platform requires us to address other issues, such as how to configure the DSS to fit the cloud platform, how to develop programs so that they can effectively utilize the flexible computational capabilities, how to address different kinds of VRPTW and business logic by possibly using a library of solvers and switching between solvers to respond to different needs, how to organize the optimization (decision-making) components across multiple physical systems, and what pricing model to use for the customers of such a DSS. These are and will be our ongoing and future research directions aiming to provide better decision-making services to customers in the service industry.

Acknowledgments

We would like to express our gratitude to two anonymous referees for their very useful suggestions to improve our manuscript.

References

Azi, N., Gendreau, M., and Potvin, J. Y. (2010). An Exact Algorithm for a Vehicle Routing Problem with Time Windows and Multiple Use of Vehicles. *European Journal of Operational Research*, 202, 757–763.

Basnet, C., Foulds, L., and Igbaria, M. (1996). FleetManager: A Microcomputer-Based Decision Support System for Vehicle Routing. *Decision Support Systems*, 16, 195–207.

Blakeley, F., Bozkaya, B., Cao, B., Hall, W., and Knolmajer, J. (2003). Optimizing Periodic Maintenance Operations for Schindler Elevator Corporation. *Interfaces*, 33(1), 67–79.

Bodin, L., Golden, B., Assad, A., and Ball, M. (1983). Routing and Scheduling of Vehicles and Crews: The State of the Art. *Computers and Operations Research*, 10, 104–106.

Bozkaya, B., Cao, B., and Aktolug, K. (2011). Routing Solutions for the Service Industry. In *Hybrid Algorithms for Service, Computing and Manufacturing Systems: Routing, Scheduling and Availability Solutions*. J. R. Montoya-Torres et al. (eds.). IGI-Global Publishing, DOI 10.4018/978-1-61350-086-6.

Bräysy, O., and Gendreau, M. (2005). Vehicle Routing Problem with Time Windows, Part I: Routing Construction and Local Search Algorithms. *Transportation Science*, 39(1), 104–118.

Bräysy, O., and Gendreau, M. (2005). Vehicle Routing Problem with Time Windows, Part II: Metaheuristics. *Transportation Science*, 39(1), 119–139.

Cao, B., and Glover, F. (2010). Creating Balanced and Connected Clusters to Improve Service Delivery Routes in Logistics Planning. *Journal of System Sciences and System Engineering*, 19(4), 453–480.

Clarke, G., and Wright, J. (1964). Scheduling of Vehicles from a Central Depot to a Number of Delivery Points, *Operations Research*, 12, 568–581.

Cordeau, J.-F., Gendreau, M., Hertz, A., Laporte, G., and Sormany, J.-S. (2005). New Heuristics for the Vehicle Routing Problem. In *Logistics Systems and Optimization*, A. Langevin and D. Riopel (eds.). Springer, 279–297.

Cordeau, J.-F., Gendreau, M., Laporte, G., Potvin, J.-Y., and Semet, F. (2002). A Guide to Vehicle Routing Heuristics. *Journal of the Operational Research Society*, 53, 512–522.

Cordeau, J.-F., Laporte, G., and Mercier, A. (2001). A Unified Tabu Search Heuristic for Vehicle Routing Problems with Time Windows. *Journal of the Operational Research Society*, 52, 928–936.

Dantzig, G. B., and Ramser, J. H. (1959). The Truck Dispatching Problem. *Management Science*, 6, 80–91.

Fisher, M., and Jaikumar, R. (1981). A Generalized Assignment Heuristic for Vehicle Routing. *Networks*, 11, 109–124.

Gayialis, S. P., and Tatsiopoulos, I. P. (2004). Design of an IT-Driven Decision Support System for Vehicle Routing and Scheduling. *European Journal of Operational Search*, 152, 382–398.

Gendreau, M., Hertz, A., and Laporte, G. (1994). A Tabu Search Heuristic for the Vehicle Routing Problem. *Management Science*, 41, 1276–1290.

Gendreau, M., Laporte, G., and Potvin, J.-Y. (2002). Metaheuristics for the Capacitated VRP. In *The Vehicle Routing Problem*, P. Toth and D. Vigo (eds.). SIAM Monographs on Discrete Mathematics and Applications, 129–154.

Glover, F. (2000). Multi-Start and Strategic Oscillation Methods—Principles to Exploit Adaptive Memory. In *Computing Tools for Modeling, Optimization and Simulation: Interfaces in Computer Science and Operations Research*, M. Laguna and J. L. Gonzales-Valarde (eds.). Kluwer Academic, 1–24.

Ioannou, G., Kritikos, M. N., and Prastacos, G. P. (2002). Map-Route: A GIS-Based Decision Support System for Intra-City Vehicle Routing with Time Windows. *Journal of the Operational Research Society*, 53, 842–854.

Juan, A., Faulin, J., Jorba, J., Riera, D., Masip, D., and Barrios, B. (2011). On the Use of Monte Carlo Simulation, Cache and Splitting Techniques to Improve the Clarke and Wright Savings Heuristics. *Journal of the Operational Research Society*, 62, 1085–1097.

Kallenhauge, B. (2008). Formulation and Exact Algorithms for the Vehicle Routing Problem with Time Windows. *Computers and Operations Research*, 35, 2307–2330.

Kim, B.-I., Kim, S. and S. Sahoo (2006). Waste Collection Vehicle Routing Problem with Time Windows. *Computers and Operations Research*, 33, 3624–3642.

Laporte, G. (2009). Fifty Years of Vehicle Routing. *Transportation Science*, 43(4), 408–416.

Laporte, G., Musmanno, R., and Vocatunro, F. (2010). An Adaptive Large Neighborhood Search Heuristic for the Capacitated Arc-Routing Problem with Stochastic Demands. *Transportation Science*, 44 (1), 125–135.

Lenstra, J., and Kan, A. R. (1981). Complexity of Vehicle Routing and Scheduling Problems. *Networks*, 11, 221–227.

Li, J. Q., Borenstein, D., and Mirchandini, P. B. (2007). A Decision Support System for a Single-Depot Vehicle Rescheduling Problem. *Computers and Operations Research*, 34, 1008–1032.

Mendoza, J., Medaglia, A. L., and Velasco, N. (2009). An Evolutionary-Based Decision Support System for Vehicle Routing: The Case of a Public Utility. *Decision Support Systems*, 46, 730–742.

Nagata, Y., and Bräysy, O. (2008). Efficient Local Search Limitation Strategies for Vehicle Routing Problems. *Lecture Notes in Computer Science*, 4972, 48–60.

Nagata, Y., and Bräysy, O. (2009). Edge Assembly-Based Memetic Algorithm for the Capacitated Vehicle Routing Problem. *Networks*, 54(4), 205–215.

Ray, J. (2007). A Web-Based Spatial Decision Support System Optimized Routes for Oversize/Overweight Vehicles in Delaware. *Decision Support Systems*, 43, 1171–1185.

Ropke, S., and Cordeau, J.-F. (2009). Branch and Cut and Price for the Pickup and Delivery Problem with Time Windows. *Transportation Science*, 43, 267–286.

Solomon, M. M. (1987). Algorithms for the Vehicle Routing and Scheduling Problems with Time Window Constraints. *Operations Research*, 35, 254–265.

Spada, M., Bierlaire, M., and Liebling, T. (2005). Decision-Aiding Methodology for the School Bus Routing and Scheduling Problem. *Transportation Science*, 39, 477–490.

Weigel, D., and Cao, B. (1999). Applying GIS and OR Techniques to Solve Sears Technician-Dispatching and Home Delivery Problems. *Interfaces*, 29(1), 112–130.

Section 4

Decision Making
in Other Service Areas

14

Decision Support for Information Technology (IT) Service Investments under Uncertainty

Magno Jefferson de Souza Queiroz and Jacques Philippe Sauvé

CONTENTS

14.1 Introduction

Although organizations invest substantial financial resources in information technology (IT), they usually struggle to translate these investments into increased business performance. Many studies suggest that the relationship between investment in IT and organizational performance is poorly understood due to the lack of robust and comprehensive methods to evaluate business benefits of IT investments (Lucas 1999; Stratopoulos and Dehning 2000).

The Information Technology Infrastructure Library (ITIL) V3 (Office of Government Commerce 2007) recommends implementing service portfolio management (SPM) to actively manage IT investments. Although SPM provides managers with recommendations to better understand quality requirements for investment in IT services, the SPM part of ITIL provides neither formalism nor illustration to clarify how these recommendations can be undertaken. As such, senior managers continue to struggle with the complex issues involved in deciding how to distribute the corporate IT budget over the portfolio of IT services. This chapter addresses this matter by exploring the prioritization of IT service investment alternatives.

The prioritization of investments in existing and future IT services is difficult and currently done in a highly subjective manner. The problem is difficult due to (1) the often complex infrastructure supporting IT services and the interdependencies that exist between organizational business and IT capabilities; (2) the uncertainties present in the parameters that characterize IT investment scenarios; and (3) the highly subjective nature of the investment decision, which depends particularly on the decision maker's *risk profile*, or, in other words, his or her *attitude toward risk* (Sauvé et al. 2011).

Uncertainty comes from two sources: *epistemic* uncertainty and *aleatory* uncertainty (Vose 2000). The former is the decision maker's lack of knowledge about parameters that characterize the investment scenario, for example, the costs or business benefits of a particular investment alternative. The latter refers to randomness in the variations of these parameters, and comes about because many processes—physical or based on human activity—cannot be completely controlled. Total uncertainty is then the combination of epistemic and aleatory uncertainty.

The behavior of decision makers in situations involving uncertainty is contingent on their attitude toward risk. This behavior varies from risk tolerance to risk avoidance and can heavily impact IT investment analysis. Most decision makers are *risk-averse* and therefore prefer lower investment returns as long as risk is also reduced. Others are *risk-seeking* individuals and aim to maximize investment results even if the degree of risk is high (Keeney and Raiffa 1993).

The problem we address in this study is the prioritization of IT service investment alternatives in order to maximize business results while managing risks and costs. As a partial solution, we present a decision model that takes as input a set of investment alternatives and a set of financial criteria upon which to compare the alternatives, and provides as output a *preference index* for each alternative. Epistemic uncertainty is taken into account by means of interval arithmetic (Gutowski 2003), whereas the risk associated with uncertain scenarios is included in the model by means of probability distributions and by the use of utility theory (UT) (Von Neumann and Morgenstern 2007) to represent the decision maker's risk attitude. Validation of the decision model is performed through a real case study and a sensitivity analysis. Results indicate that the model is useful to solve real problems, easy to use, robust in the presence of small variations in inputs or when assumptions are not fully met, and that the risk attitude of decision makers can crucially affect decisions.

Our approach is novel in that (1) it fully and separately deals with the two types of uncertainty; (2) it factors in the decision maker's risk attitude; and (3) it provides guidance for choosing model parameters through a methodology based on the balanced scorecard (BSC), a tool much used by business executives.

14.2 Previous Research on IT Investment Decision Making

Decision support for investments in IT has been the focus of several studies in the literature. Yet, estimating the business impact of IT investments remains an elusive task in IT research and practice. Most studies concentrate on examining the potential benefits of investments by analyzing financial indicators such as economic value added (EVA) and return on investment (ROI) (Pisello and Strassmann 2003), or techniques such as real options valuation (ROV) and discounted cash flow (DCF) (Devaraj and Kohli 2002; Verhoef 2005). Thus, analysis of intangible business benefits is often neglected.

Decision making for investments in IT is difficult due to the often complex relationships that exist between the business and IT infrastructure and strategies; because investment planning is subject to unexpected changes in the business environment, which often result in new resource, budget or time constraints; benefits from IT investments are usually realized in the long run; and because intangible benefits are inherently difficult to quantify.

Research on IT investment focuses on the potential benefits of IT projects (Im et al. 2001; Pisello and Strassmann 2003; Dutta and Roy 2004), while less emphasis is given to evaluation of investment in operational IT services. For example, Pisello and Strassmann (2003) examine how to maximize IT project investment returns by using EVA to estimate the "value of IT." Other studies have applied techniques such as multiattribute utility theory (MAUT), comparative approaches such as *q*-sort, and analytical hierarchy procedure (Heidenberger and Stummer 1999). The study by Stewart and Mohamed (2002), for instance, applies MAUT to evaluate several tangible and intangible criteria for IT project selection.

Decision support for investment in IT services is particularly challenging because it requires a deep understanding of the service operations as well as the interdependencies that exist both within the IT service portfolio and between the IT service portfolio and the supported business activities.

This study presents a decision model to support IT service investment decisions. Utility theory is used to calculate the expected utility of each investment alternative. Uncertainty is taken into account by a combination of utility theory and interval arithmetic. A methodology based on the BSC framework is used to infer the linkage between IT services and the supported business activities. This linkage may then be used for distribution of investments among the IT services under consideration.

Our approach differs from previous studies on IT investment decision making by considering both epistemic and aleatory uncertainty, by factoring in the decision maker's risk attitude, and by providing guidance for choosing model parameters through a structured methodology. For instance, it differs from the approaches by Chen and Gorla (1998) and Chen and Cheng (2009) that use fuzzy logic to model uncertainty, by considering intervals for the values of a given decision criterion and randomness in the actual value of the criterion by means of a probability distribution. It also differs from Chen and Cheng (2009), Stewart and Mohamed (2002), Gunasekaran et al. (2006), and Lee and Kim (2000) by inferring the linkage between IT services and the supported business activities from the use of the BSC. The case study presented here illustrates our method, which we believe is more easily grasped by decision makers. Moreover, by dealing with explicit service availability-related criteria and service level agreement (SLA) clauses attached to the IT services (this will be seen in a later section), our approach provides better insight into aspects of IT services by linking investment results to service availability. No other method we reviewed considers SLA clauses attached to the IT services, a necessity for IT service management.

14.3 Problem Definition

This section provides both an informal and a formal definition of the technical problem considered here.

14.3.1 Informal Definition of the Technical Problem

The technical problem is specified informally as follows: Given a set of investment alternatives for IT services and a set of criteria on which to compare the alternatives, calculate a preference index to rank alternatives in order of decreasing business impact.

The solution must include the following:

- The preference index should be measured on an *interval scale* rather than on an *ordinal scale*, so that the *distance* between values of the preference index can be interpreted: a larger distance means a larger difference between business results.

- The decision maker's *attitude toward risk* must be taken into account and the *degree* to which the decision maker is risk-averse or risk-seeking must be a model parameter.

- *Epistemic uncertainty* must be taken into account, especially with respect to uncertain information elicited from decision makers.

- *Aleatory uncertainty* must be considered, since the decision processes based on human activity cannot be completely controlled and the data available to the decision maker itself contain some amount of randomness. If a decision process were to be "run again," a different outcome could result, and this variation (randomness) is what characterizes aleatory uncertainty.

14.3.2 Problem Formalization

Let A be a set of IT service investment alternatives. The exact representation of an investment alternative will be described in a further section. Suffice it to say, at this stage, that it includes cost and the effects of investing on service availability.

Let $P(a,T) \in R$ be the preference index of an investment alternative, $a \in A$, with respect to a reference situation. The reference situation reflects the state of IT services before any investment alternatives are introduced. T is the time period over which the alternative is evaluated to establish the comparison with the reference situation. In this study, we will always consider a single evaluation period (T); we therefore refer to $P(a,T)$ as simply $P(a)$. A higher preference index indicates a better alternative and can thus be used to rank alternatives.

Formally, the problem is to find the best alternative, $best \in A$, which is the one with the highest preference index:

$$best = \arg\max_{a \in A} P(a) \tag{14.1}$$

14.4 Proposed Solution

Many different solutions can be found to the aforementioned problem. In this section we identify solution requirements for more realistic analysis and we detail methods and procedures to address them. We then rely on these methods and procedures to formalize a solution model. The model takes as input a set of investment alternatives and a set of financial criteria upon which to compare the alternatives and provides as output a preference index for each alternative. These indexes rank investment alternatives according to the alternatives' estimated business results, which are in turn calculated by analyzing gain and loss rates elicited from the decision maker for each alternative.

14.4.1 Solution Requirements

Taking into account the characteristics that we desire from the solution (see Section 14.3.1), we identify the following solution requirements:

- *A methodology must be used to elicit model parameters from decision makers*—An effective solution model should provide a specification of business activities that use IT services and generate business results (total revenue, for example). Model parameters typically include the possible gains and costs of investment alternatives. However, decision makers often struggle to estimate these parameters due to uncertainties in the investment scenarios. Hence, a methodology should be used to elicit model parameters from decision makers. An effective methodology should recognize the linkage between business activities and the supporting IT services in order to facilitate identification of business activities that will be affected by investments in particular IT services. This linkage may then be used to facilitate the estimation of business results for each investment alternative under analysis.

- *Epistemic uncertainty must be formally represented*—Decision makers are often uncertain about parameters that characterize IT investment scenarios. This leads to large amounts of epistemic uncertainty during the estimation of investments' business results.

- *Aleatory uncertainty must be formally represented*—Aleatory uncertainty refers to randomness in all model data, including independent input variables and parameters. Thus, repeating the evaluation of a given investment alternative may yield different results; aleatory uncertainty captures this fact.

- *The decision maker's risk attitude must be formally represented*—The decision maker's attitude toward risk can heavily impact the analysis of IT investments since the "desirability" of a particular investment alternative depends on both the alternative's potential returns and its degree of risk.

- *A specification of investment alternatives must be provided*—Despite their focus being on improving existing IT services or investing in new ones, alternatives will generally address the level of the service by means of performance metrics governing service quality or service delivery. These metrics are called service level objectives (SLOs) and are usually defined by tangible technical measures, such as "availability," "throughput," and "response time." Thus, an IT service investment alternative should specify (1) the set of IT services considered for investment; (2) the set of business activities affected by the investment, that is, those that will generate business gains or losses; and (3) the desired service level (target values for performance metrics, or SLOs).

- *Multiple criteria must be considered when calculating the preference indexes*—For example, the *cost* and possible *gains* from IT services.

14.4.2 Methods and Procedures

This section details methods and procedures to meet the aforementioned requirements.

14.4.2.1 A Methodology to Elicit Model Parameters from Decision Makers

The methodology to elicit model parameters from decision makers uses the notion of BSC perspectives as a prism to look into the business and to facilitate identification, organization, and estimation of model parameters. These parameters include, for example, gains and losses for each investment alternative. The BSC concepts serve the purpose of facilitating communication between IT and business people by arranging business activities in a structure that is familiar to many business executives. This arrangement is then used to infer the linkage between business activities and the supporting IT services. The four BSC perspectives are

- *Customer perspective* consists of all customer-facing business activities, that is, business activities that assist customers directly in their interactions with the enterprise. These include customer relationship management (CRM), e-commerce, sales, after-sales service, and order tracking business activities.

- *Internal operations perspective* is made up of business activities that support the enterprise's production and delivery of goods or services directly. Typical business activities in this perspective are those supporting service provisioning, manufacturing, supply chain management (SCM), and quality control.

- *Learning and growth perspective* encompasses business activities that assist training of human resources, research and development (R&D) efforts, and planning and acquisition of capital goods for enhanced technological support. These can include collaboration and R&D activities.

- *Financial perspective* includes the business activities used for financial purposes such as those supporting credit checking and approval, invoicing, accounts payable and receivable, and budgeting.

Each business activity is "assigned" to a particular BSC perspective according to how closely it relates to that perspective. For example, a sales business activity is naturally assigned to the customer perspective. This assignment aims to facilitate estimation of gains and losses from business activities affected by IT services considered for investments. To further simplify the estimation process, we present the following guidelines:

- For any business activity, whatever BSC perspective it has been assigned to, the cost of setting up or upgrading a service is included as a loss.
- If a business activity is assigned to the customer perspective, then gains are usually related to revenue and losses are typically related to penalties due to SLA violations.
- If a business activity is assigned to the internal operations perspective, then losses are usually related to SLA penalties or to non-provided services, stocking delays, or nonproduced goods due to business activities with unacceptably degraded performance.
- If a business activity is assigned to the learning and growth perspective, then gains can be estimated as the rate of sale of new (innovative) goods or services. Losses can be SLA penalties or can be due to increased time-to-market or from missed sales opportunities due to the delayed new product, service, or promotion.
- If a business activity is assigned to the financial perspective, gains can be obtained from earned interest and losses arise from not earning interest on paid customer invoices or paying interest on loans to honor accounts payable that were delayed due to cash flow drop caused by late payments.

In order to elicit business gain and loss rates (hourly, say) from the decision maker for each investment alternative we break the task into simpler subproblems, as follows:

1. Identify IT investment alternatives that need to be ranked.
2. Identify business activities; these are activities where business gains or losses can be realized. Only the business activities that are somehow affected by the investment alternatives need to be identified. The decision maker can estimate business gains and losses individually for each business activity.
3. We allow several financial criteria to be used so that the decision maker can focus on each one of them separately. Examples of criteria

are SLA violation penalties, the cost of the investment being considered, and business revenue.

4. Criteria are split into two classes: those that depend on service availability and those that do not. The first class is called *service-availability-dependent criteria* and the second is called *other financial criteria*. There are two reasons for dividing criteria this way: (1) it helps the decision maker to focus while estimating gains and losses, and (2) the calculation of business results is slightly different for each class.

5. Since the decision maker is uncertain about gain and loss rates (and other model parameters), we introduce epistemic uncertainty. All gain and loss rates mentioned earlier are actually estimated as intervals.

14.4.2.2 Handling Uncertainty

Epistemic uncertainty is formally represented by means of interval arithmetic. All model parameters are estimated as *intervals*. An interval is defined between a minimum value (called the *infimum*) and a maximum value (called the *supremum*). The interval notation used here is $R[]$ for an interval named R; the interval infimum is $R[inf]$, and the supremum is $R[sup]$. For an unnamed interval of values, we use the notation [infimum supremum]. Interval arithmetic operators are $\oplus$ (sum of intervals), $\ominus$ (subtraction of intervals), and $\otimes$ (multiplication of intervals). These operators are defined as

$$[a,b] \oplus [c,d] = [a+c, b+d] \,, \; [a,b] \oplus c = [a,b] \oplus [c,c]$$

$$[a,b] \ominus [c,d] = [a-d, b-c] \,, \; [a,b] \ominus c = [a,b] \ominus [c,c]$$

$$[a,b] \otimes [c,d] = [min(ac,ad,bc,bd), max(ac,ad,bc,bd)] \,, \; [a,b] \otimes c = [a,b] \otimes [c,c]$$

As can be seen, when used with interval operators, a scalar c is treated as the interval $[c,c]$. We also define relational operators involving intervals as follows:

$$[a,b] < c \text{ is equivalent to } b < c \text{ and } [a,b] > c \text{ is equivalent to } a > c$$

Now, observe that the impact resulting from an investment alternative over time period T is actually a random variable, since repeated measurements of this metric will yield different values. This leads to aleatory uncertainty, which we represent formally by means of probability distributions and utility theory. Thus, total uncertainty is handled by a combination of interval arithmetic and utility theory.

We explain a bit of utility theory to indicate how it is used to handle uncertainty and to support decisions for alternative investment options.

Most techniques for dealing with decisions under uncertainty allow the decision maker to choose between alternatives with uncertain consequences based on the notion of *utility* (Keeney and Raiffa 1993; Von Neumann and Morgenstern 2007). Invented by von Neumann and Morgenstern (vNM), utility theory is based on the notion of lotteries or distributions of outcomes. It posits that people have utility from lotteries and not from outcomes themselves. For example, consider a lottery with two possible outcomes, a gain of either $100 or $0. Obviously, one prefers $100 to $0. Now, consider two different lotteries: in lottery A, you receive $100 with 90% probability and $0 with 10% probability; in lottery B, you receive $100 with 40% probability and $0 with 60% probability. Obviously, lottery A is better than lottery B and we say that *over* the set of outcomes X = ($100, 0), the distribution p = (90%, 10%) is *preferred to* distribution q = (40%, 60%).

To represent preferences over lotteries, vNM introduce the utility function $u(p)$ where p is a lottery. If lottery p_1 is preferred over lottery p_2, then $u(p_1) > u(p_2)$. In continuous space we can now calculate the "expected utility": $U(p) = \int_x u(x)df$, where x is the random variable representing outcomes, f is the probability support (density function) of outcomes, and $u(x)$ is the utility function on the underlying outcomes. Investment alternatives are then chosen according to the expected utility $U(p)$. If one must represent the risk attitude of the decision maker, then the appropriate choice of a utility function is required.

14.4.2.3 Representing the Decision Maker's Risk Attitude Formally

Utility theory does not explicitly use a risk metric in the decision process. To formally represent the decision maker's risk attitude one needs to elicit his utility function. Please refer to Figure 14.1. The decision maker's utility function is $u(x)$ (the concave curve). Imagine a lottery that can give value 0 or value m with equal probability of 1/2. The expected return of the lottery is $E[x] = \frac{1}{2} \cdot 0 + \frac{1}{2} \cdot m = \frac{m}{2}$. The expected utility of the lottery is $E[u] = \frac{1}{2} \cdot u(0) + \frac{1}{2} \cdot u(m) = \frac{u(m)}{2}$. This is point A in Figure 14.1. Now, how much would the decision maker pay to participate in the lottery? Since there is risk, we must consider the decision maker's risk attitude. If he is risk-seeking, he will pay more than $m/2$ (the expected return) in order to try to win m. If, on the other hand, he is risk-averse, he will pay less than $m/2$; this is the usual case for most people. Let us say that the decision maker is willing to pay at most amount s to participate in the lottery. In other words, the decision maker is indifferent between winning s for certain (point B in Figure 14.1) or participating in the lottery (point A in Figure 14.1). Both points have the

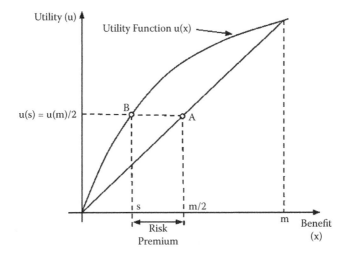

FIGURE 14.1
Utility and the risk premium.

same utility. The difference $\pi = m/2 - s$ is called the *risk premium*, that is, the maximum amount of income that the decision maker is willing to forgo in order to obtain an allocation without risk.

The first step in eliciting the decision maker's utility function is to estimate the risk premium. In what follows we present a three-step procedure to elicit the utility function:

1. Estimate the risk premium.
   a. Estimate the maximum benefit m.
   b. Find the value of s by asking the decision maker the following question (based on the work by Eisenhauer and Ventura 2003): "You are offered the opportunity of acquiring a security permitting you, with the same probabilities, either to gain m or to lose all the capital invested. What is the most you are prepared to pay for this security?" The answer is the value of s. If the decision maker is risk-seeking, he will pay $s > m/2$; if he is risk-averse he will pay $s < m/2$.
   c. Calculate the risk premium: $\pi = m/2 - s$.
2. Choose a utility function shape.
3. Set up the utility function. Find the value of parameters of the function chosen in step 2 to come up with the decision maker's utility function. For illustration purposes, consider that the function $u(x) = m(1 - e^{-\delta x})$ was chosen in step 2. This is a commonly used utility function for risk-averse agents (Keeney and Raiffa 1993); the parameter δ is known

as the *coefficient of absolute risk-aversion*, and can be chosen to match the decision maker's attitude toward risk. In order to set up this utility function:

a. Find the value of δ. This can be done since the values of m and s are known and $u(s) = \dfrac{u(m)}{2}$, or $m(1 - e^{-\delta s}) = \dfrac{m(1 - e^{-\delta m})}{2}$

b. Set up the utility function chosen in step 2 with the values of m (from step 1) and δ.

14.4.2.4 Handling Multiple Criteria

The solution model considers multiple criteria when calculating the preference index. However, it only deals with criteria that can be summed together, criteria that are monetized, for example.

14.4.3 Informal Description of the Solution

In order to compare investment alternatives, the solution model requires that these alternatives be chosen, financial criteria (gains and losses) be decided upon to compare the alternatives, and that the gain and loss rates be identified for each alternative. In order to simplify the estimation of gains and losses, business activities are identified for which gains and losses are individually estimated. Epistemic uncertainty is included in the model by using intervals for all parameters, and interval arithmetic aggregates gains and losses. The risk associated with uncertain scenarios is modeled using probability distributions, a utility function to represent the decision maker's risk attitude, and the use of utility theory to find the expected utility of each alternative. The preference index of an alternative is then simply its expected utility.

14.4.4 Formalization of the Solution

In this section, we formalize a model that calculates the preference index of investment alternatives. We proceed in a bottom-up fashion, starting from the parameters elicited from the decision maker and working "up" the solution model.

Let:

- BA be the set of all business activities.
- C be the set of financial criteria. This set of criteria is divided in two disjoint subsets: C_{SA} (service-availability-dependent criteria) and C_{FIN} (other financial criteria), where $C_{SA} \cup C_{FIN} = C$ and $C_{SA} \cap C_{FIN} = \varnothing$.

- $\gamma_{a,b,c}[]$ be the (monetized) business gain or loss rate of business activity b, under investment alternative a and criterion c, $c \in C_{SA}$. The rates apply when service availability is important. A positive value indicates a gain and a negative indicates a loss.

- $\lambda_{a,b,c}[]$ be the (monetized) business gain or loss rate of business activity b, under investment alternative a and criterion c, $c \in C_{FIN}$. These rates apply when service availability is not a factor. A positive value for $\lambda_{a,b,c}[]$ is a business gain and a negative value represents a loss. An example of a loss is the cost to provision and deploy a service. Since $\lambda_{a,b,c}[]$ is actually a rate, a total cost needs to be converted to a rate (hourly, say), using a time frame over which to spread the investment.

Gain and loss rate parameters $\gamma_{a,b,c}[]$ and $\lambda_{a,b,c}[]$ are elicited from decision makers by using the methodology detailed in the section on methods and procedures. A numerical example will be provided in a further section.

We now proceed to find estimated business result intervals. Let $B_{a,b,c}[]$ be the estimated business result under criterion c for investment alternative a occurring in business activity b. Consider service-availability-dependent criteria and other financial criteria separately. Here, total gains and losses will be calculated from the gain and loss *rates* by summing up over a fixed time period T.

Service-availability-dependent criteria. These criteria have been separated out because gains and losses will here depend on service downtime or, in other words, service availability. A gain is realized only when service is up; as a result, the gain rates will be multiplied by service availability to take downtime into account. On the other hand, a loss is realized only when service is down and the loss rates will be multiplied by (1 – Service availability) to take downtime into account. Let $t_{a,b}$ be the total time duration during which the supporting services violate their SLAs over time period T for business activity b under alternative a. Thus

$$t_{a,b} = (1 - Av_{a,b}) \cdot T \tag{14.2}$$

where $Av_{a,b}$ is the average availability, over period T, of business activity b under alternative a due to the quality of the supporting IT services. Another way of taking epistemic uncertainty into account is through the estimation and inclusion of an *amplification factor*. This is a value specifying how much longer service downtime affects business activities. For example, an IT service may be down for 30 minutes but negatively affect the business for 60 minutes due to activities that must take place in order to come back to normal. Since the amplification factor is itself an interval, we can specify epistemic uncertainty on business activity downtime through the amplification factor. Let the amplification factor for business activity b be $\alpha_b[]$. In other words, the adverse business impact lasts $\alpha_b[]$ longer than just fixing the supporting IT services.

We can now calculate $B_{a,b,c}[]$ for the case of a gain $(\gamma_{a,b,c}[] \geq 0)$ and for the case of a loss $(\gamma_{a,b,c}[] < 0)$. Observe that $\alpha_b[]$ and $\gamma_{a,b,c}[]$ are intervals, allowing us to represent epistemic uncertainty; for this reason, the following calculations use interval arithmetic.

$$B_{a,b,c}[] = \gamma_{a,b,c}[] \otimes (T \ominus (\alpha_b[] \otimes t_{a,b})), \text{ if } \gamma_{a,b,c}[] \geq 0 \text{ and } c \in C_{SA} \tag{14.3}$$

$$B_{a,b,c}[] = \gamma_{a,b,c}[] \otimes (\alpha_b[] \otimes t_{a,b}), \text{ if } \gamma_{a,b,c}[] < 0 \text{ and } c \in C_{SA} \tag{14.4}$$

The following restriction applies in the previous definitions, in order not to have negative uptime for any business activity: $\alpha_b[] \otimes t_{a,b} \leq T$.

Other financial criteria. These criteria are not dependent on service availability. A financial rate $\lambda_{a,b,c}[]$ applies for services used by activity b, under investment alternative a and criterion $c \in C_{FIN}$. A positive value is a gain while a negative rate represents a loss; the cost of developing and deploying a new service is an example that falls in this set of criteria since it does not depend on the availability of existing services. Thus

$$B_{a,b,c}[] = \lambda_{a,b,c}[] \otimes T, \quad c \in C_{FIN} \tag{14.5}$$

To find the net result for an investment alternative, let us treat the reference situation as a special alternative r. Now we find $B_a[]$, the net financial results for alternative a, where $a \in A \cup \{r\}$. The net financial results for investment alternative a can be found by summing individual contributions from business activities over all criteria:

$$B_a[] = \bigoplus_{b \in BA} \bigoplus_{c \in CB} B_{a,b,c}[] \tag{14.6}$$

A positive value for $B_a[]$ is a net gain and a negative value is a net loss.

Remember that the reference situation is a special alternative r and that we have a net result $B_r[]$ for the reference situation. The net expected reference result is simply

$$\bar{B}_r = \frac{B_r[sup] + B_r[inf]}{2} \tag{14.7}$$

We can now find the total estimated *net business impact* of IT services $I_a[]$ for alternative a when compared to the reference situation:

$$I_a[] = B_a[] - \bar{B}_r \tag{14.8}$$

Observe that the term *business impact* is used to refer to the *distance* between a business result and a reference situation. From this interval, we have the lower bound $I_a[inf]$ and the upper bound $I_a[sup]$ of the net business impact for alternative *a*.

To introduce aleatory uncertainty, let us recognize that the impact resulting from alternative *a* over time period *T*—when compared to the reference situation—is actually a random variable $\tilde{I}_a$, since repeated measurements of this metric will yield different values. The probability density function of $\tilde{I}_a$ is $f_{\tilde{I}_a}(x)$, which we assume to be known; this supposition will be investigated in a further section.

To find the expected utility of investment alternatives, let us now bring into the model the decision maker's attitude toward risk by means of his *utility function u(x)*. We provide a quick introduction of utility theory in the section on methods and procedures; a full treatment can be seen in Von Neumann and Morgenstern (2007). Suffice it to say at this point that *u(x)* is a measure of relative satisfaction in receiving benefit *x*. If the decision maker strictly prefers *y* to *z* or is indifferent between them, then $u(y) \geq u(z)$. The utility function is normally concave, due to the *diminishing marginal utility* of benefits exhibited by most decision makers. Such a decision maker is said to be risk-averse.

We can now calculate expected utility, $\bar{E}_a$:

$$\bar{E}_a = \int_{-\infty}^{\infty} f_{\tilde{I}_a}(x)u(x)dx \tag{14.9}$$

If, for example, $\tilde{I}_a$ has a normal distribution $N(\mu, \sigma)$, then:

$$(1)\ \bar{E}_a = \int_{-\infty}^{\infty} N(\mu, \sigma)u(x)dx \cong \int_{\mu-3\sigma}^{\mu+3\sigma} N(\mu, \sigma)u(x)dx \tag{14.10}$$

Since the business impact is a sum of many smaller results (from different activities and different criteria), each with aleatory uncertainty, the central limit theorem leads us to expect a normal distribution for $\tilde{I}_a$. We do not give the form of the utility function *u(x)* here and leave it as a general function, since it will depend on the decision maker and must somehow be elicited from him or her. An example will be discussed in a further section.

The preference index for investment alternative *a* will be set to its expected utility or how "useful" it is to the decision maker:

$$P(a) = \bar{E}_a, \quad a \in A \tag{14.11}$$

The best investment alternative, as defined by Equation (14.1), is simply

$$best = \arg \max_{a \in A} P(a)$$

14.5 Model Validation, Results, and Discussion

In this section, we discuss strategies for model validation and detail a real case study to establish face validity. This is followed by a sensitivity analysis for conceptual and operational validity of the solution model. Finally, we investigate the influence of the decision maker's risk attitude on the model outputs. Additional details on the model validation are available in Sauvé et al. (2011).

14.5.1 Strategies for Model Validation

Validating the decision model includes four tasks to assess whether the model solves the initial problem satisfactorily: (1) validating the conceptual model; (2) verifying the computer implementation of the conceptual model; (3) checking operational validity; and (4) checking data validity. This section describes our strategy in performing these four tasks.

Validating the conceptual model. We must determine whether the assumptions on which the model is based are valid and whether the problem representation (model structure and relationships) is reasonable for the purpose for which the model was developed.

The main assumptions made while developing the model are listed next. We provide either a discussion concerning the reasonableness of the assumption or we indicate how it will be validated in a later section.

- The decision maker can estimate business gains and losses accurately enough. The inclusion of epistemic uncertainty is the way in which the model makes this supposition correct or acceptable.

- Expected gain and loss rates are constant over time period *T*. Recall that the averages are elicited as *intervals*, which can span and absorb expected variations over time.

- The gains and losses considered cover most important financial business result possibilities. *Face validity* of this assumption will be established through a case study.

- Service availability remains constant over any time period *T*. Only a more prolonged case study could validate this assumption.

- The set of business result criteria completely covers one or more aspects of the business. If the decisions to be made can be effectively separated among different orthogonal aspects, then the completeness of the chosen criteria is only necessary for those aspects being modeled. For example, intangible criteria are outside the territory that we cover with the validation example. However, so long as the decision makers clearly understand the separation of concerns between the different classes of possible investments, they will be able to use the tool effectively.

- A known distribution can express aleatory uncertainty. Argument: the net business impacts are a sum of several random variables and the central limit theorem from probability theory leads us to expect a *normal* distribution (Box et al. 2005). However, since these component random variables are probably not independent—as required by the central limit theorem—this question remains a threat to validity and will be investigated through a sensitivity analysis in a further section.

- An adequate utility function can be elicited from the decision maker. Face validity of this assumption will be established through a case study.

The next step in validating the conceptual model is to check the model structure and relationships. This will be done by establishing face validity with a case study and through a *sensitivity analysis* to examine robustness of the model and to verify if it has adequately captured uncertainty and the decision maker's risk attitude.

Verifying the computer implementation of the conceptual model. The conceptual model was implemented in MATLAB. Model verification must ensure that the implementation is correct. This was done through extensive unit and acceptance tests of the MATLAB program using the MATLAB xUnit Test Framework (Eddins 2009). All tests can be run automatically and were run frequently, whenever changes were made to the model code. The model implementation, including unit and acceptance tests are publicly available in Sauvé (2010). The package includes a readme.txt file that explains how to fully reproduce all results.

Checking operational validity. This determines whether the model's outputs have sufficient accuracy for the model's objectives, within the model's domain of validity. We check operational validity through a case study and through a sensitivity analysis.

Checking data validity. This ensures that the data necessary to build, evaluate, and test the model are adequate and correct. We have treated epistemic uncertainty directly and formally in the model parameters to increase (if not ensure) data validity. A remaining threat to validity is capturing the parameters associated with the decision maker's risk attitude. This will be examined in the next section.

14.5.2 A Case Study to Establish Face Validity

In order to establish, at the very least, face validity for the model, we perform a real-life case study to answer the following questions: Can we believe the model's output? What is the model's domain of validity? Is the decision maker satisfied? Observe that the case study did not wait for the winning investment alternative to be implemented to verify whether or not business impacts are similar to the values predicted by the model. This is why we call this case study an exercise in *face validity*.

A furniture manufacturer was contacted to analyze its IT investment alternatives. This company let us use real numbers elicited during the case study, as long as we do not disclose the company's name; we therefore refer to the company as We Make Furniture (WMF). The case study is publicly available and enables the reproduction of all results shown here (Sauvé 2010).

The reference situation for business activities and IT services is as follows (organized by BSC perspective). The business activities included are only those that, according to the decision maker, are affected by the investment alternatives.

- Customer perspective—Sales business activity; this activity is supported by intranet connectivity and by database A (DBA) services, which are subject to 99% availability SLOs.

- Internal operations perspective—Industrial cut of wood for furniture manufacturing (ICFM) activity; this activity is supported by two software applications—"ap1" and "ap2"—and by database B (DBB) services, which are subject to 97% availability SLOs.

- Learning and growth perspective—Collaboration activity that depends on Internet access services with a 97% availability SLO.

- Financial perspective—Accounts payable and receivable (APR) activity that uses a printing spool and the DBA services which are subject to 99% availability SLOs.

There are four IT investment alternatives being considered, as follows:

- Alternative A—Make the sales activity receive 99.9% availability.

- Alternative B—Make sales and collaboration activities receive 99.9% availability.

- Alternative C—Contract a new service to enable a high degree of automation of the ICFM activity. The new service will be responsible for data exchange between three independent software applications (ap1, ap2, and DBB), that are responsible for the control of the ICFM workflow. This communication will avoid manual manipulation of application files and exchange of files among applications. Since much of the downtime of this activity in the reference situation is due to manual errors, the new service will be contracted for 99.9% availability.

- Alternative D—As in alternative C, but also make the APR activity 99.9% available.

Model parameters were elicited from the decision maker at WMF. Recall that most parameters are intervals, since much uncertainty is involved (see

TABLE 14.1

Model Parameters for Case Study

Parameter Name	Value	Parameter Name	Value
γ(Sales, Financial gain)	[1300 2200]	λ(Alternative A, Cost)	[−29 −19]
γ(Sales, Financial loss)	[−13400 −12000]	λ(Alternative B, Cost)	[−29 −19]
γ(ICFM, Financial gain)	[4800 5500]	λ(Alternative C, Cost)	[−10 −6]
γ(ICFM, Financial loss)	[−13100 −11450]	λ(Alternative D, Cost)	[−38 −25]
γ(Collaboration, Financial gain)	[90 120]	Amplification factor (α) (all activities)	[1 3]
γ(Collaboration, Financial loss)	[−500 −260]	T	90 days
γ(APR, Financial gain)	[300 550]	Availability, Av(ICFM)	Reference = 0.970 A = 0.970 B = 0.970 C = 0.999 D = 0.999
λ(APR, Financial loss)	[−1500 −1100]	Availability, Av(Collaboration)	Reference = 0.970 A = 0.970 B = 0.999 C = 0.970 D = 0.970
Availability, Av(Sales)	Reference = 0.990 A = 0.999 B = 0.999 C = 0.990 D = 0.990	Availability, Av(APR)	Reference = 0.990 A = 0.990 B = 0.990 C = 0.990 D = 0.999

Table 14.1). The following evaluating criteria were chosen by the decision maker: Financial gain and financial loss (service-availability-dependent criteria) and Cost (other financial criterion). All financial rates shown in Table 14.1 are expressed in reals/hour (one real, the currency in Brazil, is about US$0.60, in 2011).

As discussed previously, due to the central limit theorem, we expect financial impacts $\tilde{I}_a$ to be approximately normally distributed. We therefore use a normal distribution with mean $\mu_a = (I_a[\sup] + I_a[\inf])/2$. For the standard deviation, we use $\sigma_a = (I_a[\sup] - I_a[\inf])/6$, the reason being that the density of the normal variate drops to almost zero when it reaches 3 standard deviations away from the mean on either side; therefore, these parameter values make the normal distribution span the entire interval.

Although the model allows $\gamma_{a,b,c}[]$ to have a different value for each triple (a,b,c), the parameter does not depend on the alternative a for this particular scenario at WMF; similarly, $\lambda_{a,b,c}[]$ does not depend on the business activity b. The details concerning costs are given in the public package containing the MATLAB model in Sauvé (2010). In summary, costs for investment alternatives A and B include two Internet connections, each with 98% availability (joint availability = 99.96%) plus a "gold support contract" with on-site visits; costs for investment alternative C includes additional software licenses, a

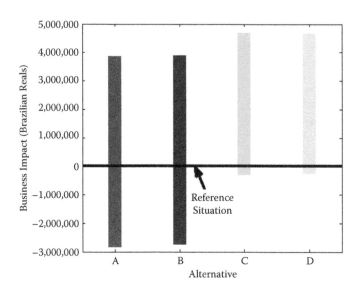

FIGURE 14.2
Business impact intervals for four investment alternatives at WMF.

hardware upgrade depreciated over 2 years, and software maintenance contracts; for investment alternative D, costs are as for alternative C plus a gold support contract.

Now, we find an appropriate utility function representing the decision maker's risk attitude. This is done by following the three-step procedure detailed in the section on methods and procedures. For step 1, we ran the model to find the maximum benefit (m), which turned out to be about 5,000,000 (this will be seen later in Figure 14.3); then we elicited the value of s from the decision maker at WMF; the answer we obtained was $s = 2,000,000$. Our decision maker is clearly risk-averse, with a risk premium of 500,000. For step 2 we used $u(x) = m(1 - e^{-\delta x})$, a commonly used utility function for risk-averse agents; this function represents a decision maker with *constant absolute risk aversion*. From step 3 we have found the value of the coefficient of absolute risk-aversion, $\delta = 0.000000164$. The utility function is thus as follows: $u(x) = 5,000,000(1 - e^{-0.000000164x})$.

The model is now fully identified and we can proceed to the results obtained in the case study.

Case study results. Figure 14.2 depicts the impact $I_a[]$ of the four investment alternatives when compared to the reference situation, while Figure 14.3 depicts the final preference indexes P_a after applying the utility function. All alternatives are an improvement over the reference situation. Alternative D has the best preference index (1,490,000), although alternative C is very close (1,480,000). The business impact of the winning alternative, [–243,000 4,670,000], means that it can be somewhere between 243,000 *worse* than the reference situation and 4,670,000 *better* than the reference situation. The

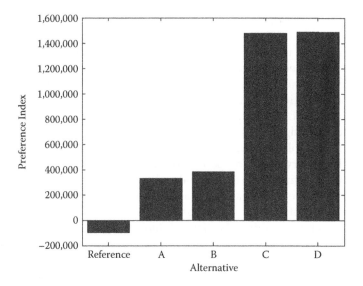

FIGURE 14.3
Preference indexes for four investment alternatives at WMF.

large intervals for business impact are due to the large amount of epistemic uncertainty on the part of the decision maker. If he can somehow reduce this uncertainty, then intervals can be made smaller. The decision maker analyzed the model, the inputs, and outputs, and declared himself quite happy with the results, thus establishing face validity for the model. This, of course, is not a full validation that would involve studies of more scenarios over longer periods of time. We do, however, further establish model validity in the next sections.

14.5.3 Sensitivity Analysis for Conceptual Model Validity and Operational Validity

Sensitivity analysis (SA) is the study of how the variation in the output of a mathematical model can be apportioned, qualitatively or quantitatively, to different sources of variation in the input of the model (Box et al. 2005). SA can be useful to test the robustness of the model as well as to target epistemic uncertainty reduction activities.

The method of sensitivity analysis presented here is through the *design of experiments* (Box et al. 2005). We quickly review the terminology used in experimental design. Some of the independent variables are chosen to be *factors* to be varied in the experiment. Factors can assume one or more values (called *levels*). A particular tuple of levels, one level for each factor, is termed a *treatment*. The model is executed for several treatments and the results are gathered. The choice of factors and treatments is termed an *experimental design*. In a full factorial design, all possible treatments are applied and run, while a partial factorial design only applies a subset of possible treatments.

After gathering results according to the chosen design, the variation in dependent variables (the preference index, in our case) is apportioned to the factors using *contrasts*, a standard statistical technique to combine averages and allocate variation in output to the input variables.

There are a total of 36 independent variables in the model for this case study. A full factorial design is clearly impossible with so many factors. We will not vary certain variables that probably have little impact on output variation (T, costs which are very small compared to gains and losses); we will use a single variable for epistemic uncertainty, make all amplification factors equal, and, finally, fix the value m and vary only δ (either one can be used to vary the risk attitude). We are left with 12 factors, still too many for a full factorial design. We therefore choose to do a *fractional factorial design*, which requires much fewer runs to obtain data. We choose a 2^{12-4} design whereby the 12 factors will be used with 256 experimental runs. The effect of the confounding of factors that always exists when using a fractional factorial design was minimized by the careful identification of *inert factors* (Box et al. 2005). Details about factor confounding and how the treatments for the experimental runs were chosen can be found in the public package available in Sauvé (2010). Each of the factors was varied at two levels as shown in Table 14.2.

After running the 256 experimental runs, the *allocation of variation* for the preference index was calculated using contrasts and is shown in Table 14.3. The factors that explain at least 10% of variation in preference index under any alternative are δ, financial gain (ICFM), amplification factor, spread (overall epistemic uncertainty); and impact distribution.

Between 11% and 24% of the variation in the preference index over all 256 experimental runs are due to the change in δ. This clearly indicates that the decision maker's risk attitude is a very important part of the model. Similarly, epistemic uncertainty (called *spread* in Table 14.3) accounts for 27% to 31% of the variation in preference index and is also a crucial factor. A third conclusion is that this type of sensitivity analysis quickly lets the decision maker see where he must invest time and effort to reduce epistemic uncertainty; in this case, the important factors to investigate are financial gain

TABLE 14.2

High and Low Values for the Fractional Factorial Design

Factor	Low Value	High Value
δ	0.000000164/2	0.000000164 × 2
$\gamma_{a,b,c}[]$	The average value of the case study interval divided by 2	The average value of the case study interval multiplied by 2
Spread on $\gamma_{a,b,c}[]$	0% (no uncertainty)	50%
$\alpha_b[]$	[1 1] (no uncertainty)	[1 5]
Impact distribution	Uniform	Normal

TABLE 14.3

Sensitivity of Preference Index to the Main Model Factors

Factor	A	B	C	D
δ*	23.9%	23.6%	11.1%	10.7%
γ(Sales, Financial gain)	3.50%	3.50%	4.40%	4.40%
γ(Sales, Financial loss)	1.40%	1.70%	0.200%	0.200%
γ(ICFM, Financial gain)*	19.3%	19.4%	16.5%	16.3%
γ(ICFM, Financial loss)	4.20%	4.10%	9.90%	9.60%
γ(Collaboration, Financial gain)	0.600%	0.700%	0.000%	0.100%
γ(Collaboration, Financial loss)	0.500%	0.800%	0.100%	0.100%
γ(APR, Financial gain)	1.80%	1.60%	1.80%	2.00%
γ(APR, Financial loss)	0.300%	0.400%	0.000%	0.100%
Amplification factor α*	3.40%	2.90%	10.2%	11.4%
Spread (epistemic uncertainty)*	27.3%	27.4%	31.0%	30.6%
Impact distribution (aleatory uncertainty)*	13.8%	13.9%	14.5%	14.4%

* Factors that explain at least 10% of variation in preference index under any alternative.

(ICFM) and amplification factor. The other financial gain and loss rates are clearly much less important. As a fourth conclusion, the aleatory uncertainty, through the probability density function of impact, is a major factor. Since there is much sensitivity to the particular distribution chosen, this constitutes a validity threat. Luckily, the central limit theorem leads us to expect a normal distribution; however, since the partial results that are added to form the total impact are not likely to be independent random variables, this must be investigated further. An initial investigation has shown that using different distributions with the same mean and standard deviation results in the same preference index; see the preference_index_with_several_pdfs.m file in the implementation package (Sauvé 2010). As for future work, this result will be further investigated to shed light on the impact (in the model's output) of choosing different probability distributions.

14.5.4 Influence of the Decision Maker's Risk Attitude

We investigate how decisions can be affected by the decision maker's attitude toward risk. There are two alternatives: alternative A has an impact interval of [0 5,000,000]. The value for δ and m are as in the WMF case study (0.000000164 and 5,000,000, respectively). This yields a preference index of 1,650,000. This value is indicated with a solid horizontal line in Figure 14.4. Alternative B is just like alternative A but epistemic uncertainty is varied from 0% to 300% of the uncertainty in alternative A. This yields the dashed, middle curve. Observe that these two lines cross at 100% (value on the x-axis) since A and B are identical. Observe that B is better (higher preference index) when there is less uncertainty than in alternative A and worse when there is more uncertainty.

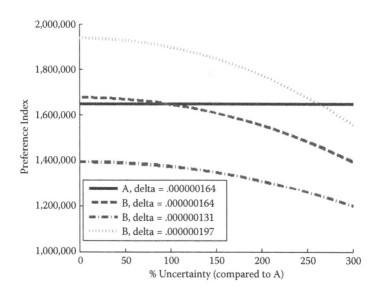

FIGURE 14.4
Decision making for different attitudes toward risk.

Now let us examine alternative B when we change the decision maker's risk attitude. The value of δ is decreased and increased by 20%. When the decision maker is less risk-averse (top curve, Figure 14.4), B remains a better alternative for larger epistemic uncertainty: the decision maker accepts more risk. The opposite happens with a more risk-averse decision maker (bottom curve, Figure 14.4). Alternative B is *never* better than A, indicating that the decision maker accepts much less risk. We conclude that risk attitude can be quite important and thus can heavily affect the prioritization of investment alternatives.

14.6 Conclusions

This study examines the problem of prioritizing IT service investments under uncertainty. We identified solution requirements, detailed methods and procedures to address these requirements, and drew upon these methods and procedures to develop a partial solution model. The model evaluates the business impact of investment alternatives so as to enable a comparison of these alternatives in terms of desirability. Important characteristics such as epistemic and aleatory uncertainties and the decision maker's attitude toward risk are taken into account to calculate a preference index for each alternative. The model was partially validated through a real case study and a sensitivity study. Both allowed us to draw the following conclusions:

- According to the decision maker at WMF, the model is quite easy to use because of its reliance on BSC perspectives, its generality, and its use of intervals to account for uncertainty in parameters.
- The model appears to be quite robust, that is, relatively insensitive to variations in many input parameters; furthermore, robustness indicates that results will not vary significantly if the modeling assumptions are not completely met.
- The case study revealed that the preference indexes are influenced quite significantly by the decision maker's attitude toward risk and by the epistemic uncertainty.
- In the case study, costs had essentially no effect; however, we believe that this is a situation peculiar to the case study and that costs will usually be quite important.
- Face validity of the model was established: the decision maker has found the model "quite useful in practice."
- We believe that the solution model's domain of validity can be characterized as follows: the model should be accurate enough in scenarios where (1) only financial criteria are taken into account when evaluating investment alternatives; (2) service-quality-dependent criteria only deal with penalties associated with service availability; and (3) service availability is reasonably constant in time.

As threats to validity, we cite:

- Few points in the problem space were tested and validated.
- Capturing the decision maker's attitude toward risk uses a potentially crude approach.
- The model is quite sensitive to the model's representation of aleatory uncertainty.
- Service availability over an evaluation period T is a random variable but was represented as a constant average. If there is much variability in availability, it may be better to represent availability as an interval. Since the model already deals with intervals, one can merely replace $Av_{a,b}[]$ for $Av_{a,b}$.

These threats to validity notwithstanding, this study shows that the solution can be useful in practice. Although the case study involved the industrial sector, we expect the solution to be useful in any sector of the economy.

The main contributions of our solution are (1) it fully and separately deals with epistemic and aleatory uncertainty; (2) it factors in the decision maker's risk attitude; and (3) it provides guidance for choosing model parameters through a methodology based on the BSC, a tool much used by business executives.

Future research opportunities include:

- Package the model along with tools to assist with data collection.
- Formally include services in the model, as well as service-to-activity mapping (which services are used by which activities). This treatment is currently done manually "outside the model."
- The model can be extended to nonbinary aspects of service quality, that is, continuous metrics, rather than the current binary metric of service available or not available.
- Consider both tangible and intangible criteria in the model. As it stands, the model only deals with tangible criteria that can be summed together, criteria that are monetized, for example. Other criteria considered by decision makers include intangibles such as company image or technological issues and subjective criteria (that may depend on particular stakeholders). This will, we hope, result in a unified, comprehensive method for supporting investment decisions in IT services. This is currently being investigated using the analytic hierarchy process (AHP) as a base model.
- There appear to be practical situations where cost is more important than future benefits. For such cost-sensitive enterprises, the use of separate preference indexes for cost and business results is appealing. This allows the application of thresholds on costs, which in turn increases the domain of validity for the model since it would be able to capture more scenarios.

Acknowledgments

The authors are grateful to the editors Javier Faulin, Angel Juan, Scott Grasman, Mike Fry, and the two anonymous reviewers for their comments and suggestions.

References

Box, G. E. P., J. S. Hunter, and W. G. Hunter. 2005. *Statistics for Experimenters: Design, Innovation, and Discovery.* New York: Wiley.

Chen, C. T., and H. L. Cheng. 2009. A Comprehensive Model for Selecting Information System Project under Fuzzy Environment. *International Journal of Project Management* 27(4): 389–399.

Chen, K., and N. Gorla. 1998. Information System Project Selection using Fuzzy Logic. *IEEE Transactions on Systems Management and Cybernetics, Part A: Systems and Humans* 28(6): 849–855.

Devaraj, S., and R. Kohli. 2002. *The IT Payoff: Measuring the Business Value of Information Technology Investments.* New York: Prentice Hall.

Dutta, A., and R. Roy. 2004. A Process-Oriented Framework for Justifying Information Technology Projects in e-Business Environments. *International Journal of Electronic Commerce* 9(1): 49–68.

Eddins, S. 2009. MATLAB xUnit Test Framework. http://www.mathworks.com/matlabcentral/fileexchange/22846-matlab-xunit-test-framework (accessed August 13, 2011).

Eisenhauer, J. G., and L. Ventura. 2003. Survey Measures of Risk Aversion and Prudence. *Applied Economics* 35(13): 1477–1484.

Gunasekaran, A., E. W. T. Ngai, and R. E. McGaughey. 2006. Information Technology and Systems Justification: A Review for Research and Applications. *European Journal of Operational Research* 173(3): 957–983.

Gutowski, M. W. 2003. *Power and Beauty of Interval Methods.* Presented at VI Domestic Conference on Evolutionary Algorithms and Global Optimization, Poland, May 26–29.

Heidenberger, K., and C. Stummer. 1999. Research and Development Project Selection and Resource Allocation: A Review of Quantitative Modelling Approaches. *International Journal of Management Reviews* 1(2): 197–224.

Im, K. S., K. E. Dow, and V. Grover. 2001. Research Report: A Reexamination of IT Investment and the Market Value of the Firm—An Event Study Methodology. *Information Systems Research* 12(1): 103–117.

Keeney, R. L., and H. Raiffa. 1993. *Decisions with Multiple Objectives: Preferences and Value Trade-Offs.* Cambridge, UK: Cambridge University Press.

Lee, J. W., and S. H. Kim. 2000. Using Analytic Network Process and Goal Programming for Interdependent Information System Project Selection. *Computers and Operations Research* 27(4): 367–382.

Lucas, H. C. 1999. *Information Technology and the Productivity Paradox: Assessing the Value of the Investment in IT.* Oxford, UK: Oxford University Press.

Office of Government Commerce (OGC). 2007. Information Technology Infrastructure Library, Version 3. Office of Government Commerce (OMG), UK.

Pisello, T., and P. Strassmann. 2003. IT Value Chain Management: Maximizing the ROI from IT Investments. In *Information Economics Press.* New Canaan, CT: Information Economics Press.

Sauvé, J. 2010. The Package to Reproduce the Results Presented in This Work. https://sites.google.com/site/busdrim/documents/matlab-2011-08-10.zip (accessed August 13, 2011).

Sauvé, J., M. Queiroz, A. Moura, C. Bartolini, and M. Hickey. 2011. Prioritizing Information Technology Service Investments under Uncertainty. *IEEE Transactions on Network and Service Management* 8(3): 1–15.

Stewart, R., and S. Mohamed. 2002. IT/IS Projects Selection Using Multi-Criteria Utility Theory. *Logistics Information Management* 15(4): 254–270.

Stratopoulos, T., and B. Dehning. 2000. Does Successful Investment in Information Technology Solve the Productivity Paradox? *Information and Management* 38(2): 103–117.

Verhoef, C. 2005. Quantifying the Value of IT-Investments. *Science of Computer Programming* 56(3): 315–342.

Von Neumann, J., and O. Morgenstern. 2007. *Theory of Games and Economic Behavior,* Commemorative edition. Princeton, NJ: Princeton University Press.

Vose, D. 2000. *Risk Analysis: A Quantitative Guide,* 2nd ed. New York: John Wiley & Sons.

15

An Inventory Criticality Classification Method for Nuclear Spare Parts: A Case Study

Natalie M. Scala, Jayant Rajgopal, and Kim LaScola Needy

CONTENTS

15.1 Introduction

Inventory management is an interesting problem in the service sector, and spare parts management provides an additional challenge. In particular, holding too many spare parts ties up capital, whereas holding too few spare parts or the incorrect ones can lead to costly outages and lost orders if a part failure occurs. This chapter examines spare parts management via a case study of the nuclear power generation industry. This industry is unique and introduces additional nuances to the problem, which are outlined in the next section. However, the general approach outlined in this chapter can be applied to other industries having similar characteristics within the service sector.

15.2 Background: Spare Parts for Nuclear Power Generation

The nuclear industry can be characterized by low probability of parts failing at a plant, but high consequences if a failure does indeed occur. As a result, plants have historically kept excessive amounts of spare parts inventory on hand to address possible part failure issues quickly without having to shut down or reduce the power output of the plant. Some parts, when they fail, can place the plant in a compromised position, based on their locations within the plant where they are installed. These part locations are called "limited condition of operation" (LCO) locations. Parts installed at LCO locations incur preventative maintenance (PM) to reduce the probability of failure. However, if the part does in fact fail, the plant must be shut down or have its power output reduced within a prescribed number of hours. The power reduction, called "derate," and the length of time until it occurs is dependent on the LCO location and how severely the plant is compromised. In the United States, plants typically have 72 hours to remedy the situation before shutdown or derate. A derate or shutdown of a plant results in complete or partial loss in power production. The cost consequences of lost revenues are significant and can be millions of dollars in deregulated operations (Scala 2011). Currently, 23 states and the District of Columbia have some form of deregulation (Quantum Gas and Power Services n.d.). Under the deregulated model, the generation portion of utilities operates as a traditional competitive business, with no guarantee of cost recovery for business expenses.

Under the regulated model that was the norm in the past, electricity rates for generation were set by each state's public utility commission (PUC). These rates included the costs of doing business plus a rate of return (ROR), which could be greater than 10% (Philipson and Willis 2006). Thus, all spare parts related costs—procurement, holding, and part capital—were recovered as costs of doing business, and there was no particular incentive for utilities to be efficient with their spare parts management. Of course, under the deregulated model, this is no longer true. The deregulation process began as early as 1996, with California, Pennsylvania, and Rhode Island as the first states to pass legislation (Lapson 1997; Asaro and Tierney 2005). Utilities whose generation is deregulated are currently transitioning to a competitive market for generation. The spare parts management methods in place at these utilities are, in general, not optimally designed for a deregulated competitive market. Therefore, a need exists to examine and improve the spare parts inventory management in this service industry. Because thousands of parts are stocked, a prerequisite for any inventory management scheme is to determine which parts are critical so that proper attention can be focused. This determination of criticality is the focus of this chapter. In the next section we discuss why this is a challenging problem in the nuclear sector.

15.3 Uncertainty in Spare Parts Management

The problem of controlling spare parts inventories is complicated in the nuclear sector by both a lack of clear supply chain processes at nuclear facilities and a lack of quantitative part data for forecasting. There is considerable uncertainty in part demand data, which cannot be quantified accurately. Moreover, failure rates of plant equipment are not always tracked in the industry. Component maintenance is not based on the equipment condition (Bond et al. 2007), and spare part demands in nuclear generation are highly intermittent. For example, a given part may be used once and then not again for 6 years. The demands are also stochastic because part failure is uncertain. Preventative maintenance is scheduled and routine, so demands for PM parts can be somewhat deterministic. However, the exact extent of the PM work that is needed for all jobs is not obvious; although a specific PM operation can be scheduled, the corresponding parts needed may not be known with certainty. To compensate, utilities typically order all parts that might be needed, contributing to a highly conservative inventory management practice with heavy overordering. Parts that are not used generally cannot be returned to the vendor, because they are uniquely made for the plant. As a result, inventory builds.

The highly intermittent nature of plant demands present two main challenges. First, traditional forecasting methods and intermittent forecasting methods do not produce accurate predictions of demand for nuclear spare parts. Scala, Needy, and Rajgopal (2009) and Scala (2011) present a discussion of why these methods, such as Croston (1972) and Syntetos and Boylan (2001), fail along with a discussion of how the culture at nuclear plants contributes to overordering parts. This heavy overordering causes the plant to return many parts to the warehouse after a maintenance job is complete, thereby heavily confounding the demand data. The authors conclude that practices from the era of regulation have been carried over to current spare parts processes, which should instead be designed for a competitive deregulated business environment. They also present a discussion on how these false demand signals render the demand even more unpredictable and intermittent. A related paper is the bootstrapping approach of Willemain, Smart, and Schwarz (2004); however, the demand data in the nuclear sector does not exhibit any of the three features that this paper addresses, namely, autocorrelation, frequently repeated values, or relatively short series.

Second, the lack of failure rate data preclude a part ordering schedule based on when parts are expected to fail, which is another challenge with spare parts inventory management in the nuclear industry. In a deregulated market, no utility wants to ever be in a situation where a part fails at an LCO location with no part in inventory to quickly fix the issue, leading to a plant shutdown or derate with subsequent revenue loss. The cost of storing critical parts in inventory versus the potential millions of revenue dollars lost is a trade-off that favors storing additional parts. However, this approach can often result in too many additional parts and millions of dollars in inventory; the management policy is not efficient, with the trade-off possibly no longer favoring additional inventory.

As a first step, there is a clear need for a mechanism to determine the importance or criticality of storing a part in inventory to hedge against potential revenue loss. The lack of forecasting models and failure rates coupled with false demand signals lead to high uncertainty regarding which parts are important and how many to store on the shelf. To address this we develop a methodology for identifying critical parts to store in inventory. This methodology uses both influence diagrams and the Analytic Hierarchy Process (AHP), and develops a criticality score for each part for inventory management. The criticality scoring procedure is the focus of this chapter and encompasses all relevant influences to the spare parts process. The current literature is reviewed next, although it should be noted that a shortcoming in this area of study is that most researchers consider only a limited set of influences or criteria for part grouping.

15.4 Literature Review

The technique of using multiple criteria in evaluating an inventory classification is called MCIC (multicriteria inventory classification); similar studies have been presented in the literature, including some studies employing the use of the AHP. However, these studies are limited in scope or consider traditional inventory that has regular turns. This chapter addresses the limitations of the literature and develops criticality evaluations for spare parts inventory.

ABC part classification can be done via a multitude of characteristics, but annual dollar volume is most common (Zimmerman 1975). The main motivation for the MCIC literature using the AHP is that annual dollar volume alone is not always enough to classify parts (Zimmerman 1975; Flores, Olson, and Dorai 1992; Ng 2007). Papers in this research area argue that many criteria are relevant in inventory classification, or using dollar volume alone to classify parts may be limiting and misleading (Flores and Whybark 1987; Flores, Olson, and Dorai 1992; Partovi and Burton 1993; Ramanathan 2006). In particular, Flores, Olson, and Dorai (1992) argue that relative importance and number of criteria used in classification may vary by firm, and the use of the AHP will enable incorporation of multiple criteria. These authors use four criteria—average unit cost, annual dollar usage, criticality, and lead time—in the AHP part classification hierarchy and give sample pairwise comparisons for the criteria. These criteria are consistently used in the MCIC literature when using the AHP. The criteria are then combined with relevant part data to form a criticality factor for each part, after transforming the part data to a 0 to 1 scale.

Gajpal, Ganesh, and Rajendran (1994) also use the AHP to determine part criticality for placing parts into three groups: vital, essential, and desirable (VED). The authors argue that use of the AHP is systematic and not subjective and that they present the first use of AHP in spare parts criticality inventory management. Their AHP hierarchy is based on three criteria—availability of production facility when the part fails, type of spares required, and lead time—and includes alternatives for each criteria. Composite weights are found by combining priorities from the AHP synthesis; parts are assigned to a VED group based on the sum of the composite weights as well as calculated cutoffs for each group. However, the use of only three criteria in the AHP model is limiting and restrictive.

The four criteria from Flores et al.'s (1992) hierarchy are also used in Partovi and Burton (1993). Their approach to the AHP classification of parts uses the ratings mode of the AHP for pairwise comparisons as each stock keeping unit (SKU) to be classified is considered an alternative in the model. AHP priorities of the four criteria are also combined with characteristic values for part data. However, these characteristic values are qualitative; examples include very high, average, short, low, and so forth. Decimal values on a 0 to 1 scale are assigned to each qualitative part characteristic. Parts are placed into groups using a traditional ABC assignment—15% in group A, 35% in group B, and 50% in group C. Braglia, Grassi, and Montanari (2004) extend the AHP classification method and present a multiattribute spare tree analysis specifically for spare parts classification. The tree has four criticality classes combined with strategies for managing parts in each group. The AHP is used to evaluate the decision at each node in the tree. Although specially designed for spare parts, the model is complicated and appears to be difficult to execute in practice.

Other MCIC inventory classification methods are reviewed in Rezaei and Dowlatshahi (2010) and Huiskonen (2001). In particular, Huiskonen (2001) notes that spare parts management research is split into veins of mathematical models and ABC based analyses. Rezaei and Dowlatshahi (2010) present other methods to use besides the AHP for classification and present their own model, which is based on fuzzy logic. For comparison, a classification of a sample set of parts is done with both their fuzzy model and an AHP analysis. The fuzzy model is based on four criteria—unit price, annual demand, lead time, and durability—which is also limiting and not comprehensive of all the relevant factors to the spare parts process.

Flores and Whybark (1987) use a matrix approach with MCIC to classify parts based on two criteria. Although their work was novel and opened a new area of research, the model becomes complicated and cannot be easily extended to more than two criteria. This is considered to be a serious limitation by both Ng (2007) and Flores et al. (1992). Partovi and Burton (1993) note that Flores and Whybark's (1987) matrix approach also assumed that the weights of the two criteria must be equal. Nonetheless, the model combines dollar usage with part criticality, as subjectively defined by managers, and the authors argue that the costs of both holding parts and the implications of not having parts are important in a classification model.

Cluster analysis and weighted linear optimization have also been used to classify parts. The cluster analysis models in Cohen and Ernst (1998) and Ernst and Cohen (1990) compare costs of management policies defined for a cluster of parts to costs of management policies defined for each individual part. The cost trade-offs in clustering the parts are a penalty for placing parts in groups and the resulting loss of discrimination. Their case study is the auto industry. However, Rezaei and Dowlatshahi (2010) argue that this model needs a lot of data to run and can be impractical in practice; furthermore, the model must be rerun when new parts are added, causing a potential change in parts' classifications. The weighted linear optimization models use optimization to solve for each criterion's weight or part's score, and multiple criteria are considered. The part score presented in Ramanathan (2006) is found through a weighted additive function, considering the four criteria commonly used in the literature. This model must be resolved for each part, yielding a score for each part, one at a time. Although this can be time consuming, the author suggests the model is simple for managers to understand. In fact, Huiskonen (2001) argues that managers do not feel comfortable if they do not understand the basics of individual models. Ng (2007) employs a weighted linear optimization model, where the decision maker ranks importance of criteria only; no relative importance such as a pairwise comparison is made, only a ranking. Weights on a scale of 0 to 1 are generated in the linear optimization, parts are scored, and ABC classification is used to place parts in groups. The author concedes that the model can be limited by many criteria, as ranking by the decision makers becomes difficult. Hadi-Vencheh (2010) extends the model presented in Ng (2007) by using a nonlinear programming model, with the four criteria repeatedly used in the literature.

Overall, criticality is important in spare parts logistics (Huiskonen 2001) and is considered in many classification models. Huiskonen (2001) defines *process criticality* as consequences caused by a failure of a part with a spare not readily available and identifies it as a strong factor. However, the assessment of criticality can be subjective, as downtime costs can be hard to quantify.

This chapter discusses criticality of spare parts through the part scoring system and extends and improves the current literature in MCIC inventory classification. Furthermore, the criticality determination is more comprehensive than those presented in the current literature. This model considers all 34 influences from the influence diagram, which is detailed in the next section, and is not limited to the four criteria commonly used in the literature. Many criteria are easily considered in this model without it becoming unmanageable. No issues exist with ranking the influences, as experts from multiple areas are employed to rank the influence set of which they have knowledge. This model is fully representative of the spare parts process without limitation; while it is illustrated with the nuclear spare parts data, the method developed in this research can be applied anywhere and is not specific to one industry or company. In contrast, the Flores and Whybark (1987) model was unique to the case study presented. As we will see in this chapter, the classification groups are not limited to three, as with traditional ABC analysis, but rather a number appropriate for the data and parts being considered. We begin the model development discussion with the use of an influence diagram to understand the nuclear spare parts process.

15.5 Influence Diagram of Nuclear Spare Parts Process

An influence diagram is a visual map of a process or decision problem. The influence diagram does not depict the actual process steps but rather all factors and forces that affect the process and corresponding decision making. By visually summarizing all influences, decision makers can gain a better understanding of what affects the process as well as what can and cannot be controlled. Factors that can be controlled are decision nodes represented by squares on the diagram. Factors that are elements of chance and cannot be controlled are shown by circles. A directed arc between two nodes implies that one node influences and is related to another. Figure 15.1 shows an example of the basic influence diagram structure. Influence diagrams have been widely used. Examples from the literature include maritime risk assessment and decision support (Eleye-Datubo et al. 2006), military alternatives for weapons of mass destruction (Stafira, Parnell, and Moore 1997), and pitcher substitution in Major League Baseball (Scala 2008). A further discussion of the theory of influence diagrams can be found in Howard and Matheson (2005).

Influence diagrams are beneficial to use in a decision-support problem because a visual representation of the problem environment helps to reduce uncertainty. In particular, the nuclear spare parts problem is complicated and affected by more than just the actions of a plant's supply chain department. The plant's planning department plans the maintenance work and generates a list of parts. This list is influenced by maintenance schedules,

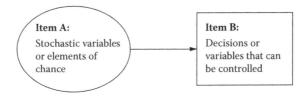

FIGURE 15.1
Influence diagram node example.

vendor availability, scope of work, and so forth. The complexity of the problem quickly becomes clear when building an influence diagram. All plant departments that affect spare parts decision making must be represented on the diagram through identification and mapping of all relevant spare parts influences from each department. Everything that affects spare parts management must be represented on one diagram. Once all the factors, forces, and inputs to the process are depicted, the decision maker can start to identify important relationships and opportunities for continuous improvement.

For the problem previously described, the influence diagram was developed as a collaborative effort between the authors and employees at a test bed nuclear generation facility. This test bed facility is one site in a Fortune 200 utility's generation portfolio. This company manages multiple forms of electricity generation (nuclear, fossil, hydro, etc.) and also holds regulated transmission and distribution companies. The subject matter experts (SMEs) at the facility provided feedback to the model, adding and removing influences as appropriate. This process was iterative and allowed for full discussion of the spare parts problem's decision-making environment. Also, the diagram-development process allowed plant employees, some without technical experience and decision-making backgrounds, to easily participate in the process and contribute vital information. As more information was collected from the employees who work with and make spare parts decisions every day, the diagram became more complete, fully disclosing the decision-making environment along with its complexities and influences. As previously discussed, the lack of detailed and complete part data coupled with highly intermittent demand data leads to high uncertainty regarding the best way to manage the nuclear spare parts process. The influence diagram highlights this uncertainty by visually depicting the decision-making environment and providing a basis for criticality classification of parts based on their influences on the company's overall operations.

All influences in the nuclear spare parts diagram are shown in Table 15.1, along with a description of the nodes to which each influence connects. The relationships between influences also affect the decision-making process, and a clear mapping of these relationships allows for understanding of both how decisions are made and how the process works. Once the basic diagram was completed, the influences were grouped into seven sets based on

TABLE 15.1

Influences in the Spare Parts Ordering Process along with Node Connections

Item	Description	Connects to
0.1	Part score	0.2
0.2	Part criticality	n/a
1	Timeliness of Work Order	0.1
1.1	Work order able to be rescheduled	1
1.2	Reason for work order	1; 1.1; 1.3

TABLE 15.1 *(Continued)*

Influences in the Spare Parts Ordering Process along with Node Connections

Item	Description	Connects to
1.3	Immediacy of schedule - when to complete work order	1; 1.4
1.4	Part demand during lead time	1
2	Part Failure	0.1
2.1	Failure of part leads to LCO or SPV	2
2.2	Lab testing results in predictive maintenance	2
2.3	Surveillance maintenance results	2
2.4	Failure history	2
2.5	System health for related equipment	2
2.6	Normal part life	2
3	Vendor Availability	0.1
3.1	Vendor discontinued or obsolete	3
3.2	History of vendor quality issues	3; 3.3
3.3	Vendor reliability	3
3.4	Vendor lead time	3; 3.3
3.5	Availability on *RAPID*	3
3.6	Quality testing or hold before installation	3
3.7	Ability to expedite the part	3
4	Part Usage in Plant	0.1
4.1	Part is at a single point vulnerability	4
4.2	Installation in a functional location tied to LCO	4
4.3	Usage for equipment train protection and reliability	4
4.4	Number of locations in plant where part installed	4
4.5	Regulatory requirement to keep part on-hand	4
4.6	Usage to prevent equipment forced outage	4
4.7	Open work orders and part demand	4; 1.5
4.8	If the part requested is actually used in the PM work order	4
5	Preventative Maintenance Schedule	0.1
5.1	PM done on the related equipment	5
5.2	Frequency of time the actual PM job is more involved than anticipated	5
5.3	PM done online or only during outage	5
5.4	Associated maintenance rules for the equipment	5
6	Outage Usage	0.1
6.1	Part scheduled for use during outage	6; 6.2
6.2	Actual usage of the part during outage	6
6.3	When the part can be used or equipment accessed	6
7	Cost Consequences	0.1
7.1	Cost of expediting the part	7
7.2	Cost of replacement power	7

Source: Adapted from Scala, N. M., J. Rajgopal, and K. L. Needy, 2010, Influence Diagram Modeling of Nuclear Spare Parts Process, *Proceedings of the 2010 Industrial Engineering Research Conference.*

common characteristics between the influences in each set. In our case, these sets are (1) timeliness of work order, (2) part failure, (3) vendor availability, (4) part usage in plant, (5) preventative maintenance schedule, (6) outage usage, and (7) cost consequences.

Each subset of influences focused on a common theme. For example, all influences in the vendor availability set were related to supply chain activities. All influences on the diagram that were part of the supply chain function were placed in this set. Therefore, an employee in the supply chain department should be familiar with the influences in that set and able to accurately judge the relative importance of each influence in that set. Overall, for all subsets of influences, insight regarding how the influences affect spare parts inventory management and ordering decisions can be gained from the relative importance rankings between the influences in a set.

Once the influences were grouped into sets, subdiagrams were constructed to depict the influences and corresponding node connections in each set. As an example, a subdiagram for the vendor availability set (#3) is shown in the bottom half of Figure 15.2. In the interest of space, subdiagrams for the other influence sets are not included here but were also constructed from the list of influences and node connections in Table 15.1. These subdiagrams can be found in Scala (2011). Each subdiagram expands from the influence diagram node that represents the overall set. An example of how the vendor availability set connects to the main diagram of the overall set of influences is shown via the dotted line in Figure 15.2. The main diagram is shown in the top half of Figure 15.2. Each of the influences, through the overall set, helps to determine the criticality of storing a spare part in inventory. Therefore, all seven sets of influences affect a part's criticality score and corresponding inventory management group, as discussed later in this chapter. Further discussion of the influence diagram for the nuclear generation spare parts process can be found in Scala, Rajgopal, and Needy (2010) and Scala (2011). To our knowledge, this is the first application of influence diagrams to the nuclear generation spare parts process.

Because influence diagrams are widely used and easy to build, the diagram for the nuclear spare parts process can easily be adapted to any decision-making application where there is high uncertainty. The relevant influences can be listed and grouped into sets according to knowledge function and department work area scope. These influences can then be ranked for importance with respect to the decision-making process.

15.6 Ranking the Influences

Any ranking algorithm or decision tool can be used to determine relative importance of the influences in a group. For this problem, the authors used

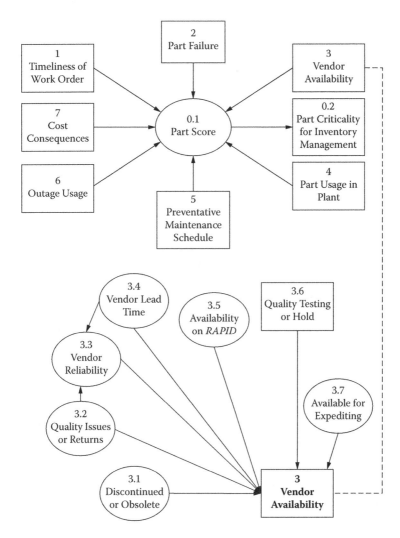

FIGURE 15.2
High level influence diagram, with sublevel influence diagram for Vendor Availability set.
(Adapted from Scala, N. M., J. Rajgopal, and K. L. Needy, 2010, Influence Diagram Modeling of
Nuclear Spare Parts Process, *Proceedings of the 2010 Industrial Engineering Research Conference.*)

the AHP to rank the influences in each subset. The AHP was chosen because
it is an easily accessible decision support tool, and plant employees—even
those with nontechnical backgrounds—can understand and participate in
the ranking process in a meaningful way. The pairwise comparisons of the
AHP force the decision maker to identify which of the two influences being
compared from a subset is more important to him or her in spare parts ordering and management decisions, and then identify how much more important
through the use of Saaty's Fundamental Scale of Absolute Numbers (Saaty

TABLE 15.2

AHP Synthesized Priorities for All Influence Sets

		Set							
		1	2	3	4	5	6	7	0
Influences	1	0.2704	0.4006	0.4566	0.3172	0.4855	0.4074	0.1722	0.0620
	2	0.4151	0.0830	0.0841	0.1693	0.1951	0.4636	0.8278	0.3498
	3	0.1991	0.2375	0.1041	0.1668	0.0941	0.1290		0.0517
	4	0.1155	0.0915	0.1102	0.0448	0.2252			0.0626
	5		0.1281	0.1345	0.1065				0.1013
	6		0.0593	0.0402	0.1242				0.0873
	7			0.0704	0.0334				0.2853
	8				0.0377				

1980, 1990). The pairwise comparison process continues for all pairs of influences within a subset. Synthesizing the pairwise comparisons yield priorities for the influences that are normalized so that they sum to one. Table 15.2 presents the final weights for each influence in the nuclear spare parts inventory management illustration after the AHP prioritization. The table lists each influence set in a column and each influence in rows. For example, set 3, Vendor Availability, has seven influences as shown in Figure 15.2. The priorities for those influences are listed in column 3; the first priority in that column corresponds to influence 3.1 (vendor discontinued or obsolete), the second priority corresponds to influence 3.2, and so on. The first influence is deemed most important in this set, as it has the highest priority (0.4566). The sixth influence is least important to the spare parts process, because it has the lowest priority (0.0402). For a detailed description of the theory and use of the AHP, see Saaty (1980, 1990).

The aging nuclear workforce is another major factor that contributes to the uncertainty in nuclear spare parts management. Based on a study done about 5 years ago (DOE 2006), depending on the specific utility, 11% to 50% of the workforce was eligible to retire in 5 to 10 years. Employing an elicitation process, like the AHP, allows for collection of and recording of employee knowledge, perceptions, and decision-making practices. This information is invaluable to the spare parts process, and, by using this technique, this knowledge will not be lost with employee attrition. Capturing qualitative decision-making practices also helps to reduce uncertainty by providing a quantitative mechanism to record data and determine relative importance of influences. For a full discussion regarding how the AHP was used to rank the relative importance of the influences in each subset along with the interview protocol for the SMEs, the reader is referred to Scala, Needy, and Rajgopal (2010) and Scala (2011).

The next step in determining the importance of storing a part in inventory is to develop a criticality score for each part. Parts can then be grouped based on the corresponding scores. The score combines the influence weights with actual part data, or relevant part information, and assigns a score along an ordinal scale. The remainder of this chapter addresses these steps and includes a definition of and examples of ordinal part data, a definition of the model used to evaluate the part scores, and an example of the model's use.

15.7 Ordinal Part Data

This section describes the development of a common scale for measuring the relative importance of a part with respect to each influence; the parts are diverse and criticality can be measured along different dimensions or in different units (as is the case with the data provided by the test bed company for a random sample of 200 parts that was selected for analysis). This measure of importance is then used as an input to the scoring model.

Actual data from the test bed company for each part with respect to each influence on the influence diagram is used for creating the inputs to the scoring model. The word *data* is used here in a very general sense to encompass information relevant to the part. This data is diverse in nature, with different characteristics that could be measured in different units (e.g., dollars, time, counts, company codes, indicators, etc.) when the part is evaluated with respect to the various influences. For example, cost consequences from lost revenues as a result of shutting a plant down because a replacement part is unavailable after an emergent failure could run in the order of millions of dollars in a deregulated market, while other influences, such as where the part is installed inside the plant, are indicator variables (yes or no, or 1 or 0 to indicate if a part is installed at a given plant location).

To bring this part data to a common set of units for each influence, an ordinal scale of 1 to 5 is adopted for use as a basis for comparison and for distilling the different features of the data across influences. This ensures that all influences are similarly represented and data from one influence does not dominate that from others. The ordinal scale provides a base scale that is unitless, allowing for equal representation of each influence, so that the weights from the AHP prioritization can clearly differentiate the critical parts for inventory management. Thus, the data, or part characteristics, can be easily compared by using a common reference scale for all part data, which are values associated with the characteristics of the parts.

The scale developed herein assigns specific values (or ranges of values) for the data on a 1 to 5 scale. This assignment is done for each influence uniquely. The scale value of 1 is reserved for the least critical values. Part data or data ranges assigned to the value 1 are considered of little importance for storing a part in inventory. This importance or criticality increases with each scale value, with a value of 5 indicating highest importance. The assignments of part data ranges to the ordinal scale are not necessarily linear. Moreover, the assignment of data values to the scale is not necessarily straightforward in all cases due to the diversity of the data in terms of both types and units. The process necessarily involves some degree of subjectivity. When this is the case, assistance for appropriate scaling was solicited from both SMEs and those who work with the part in question on a regular basis. In general, the assignment of part data to the ordinal 1 to 5 scale would be easier when part data is homogeneous, but this is not always the case. In particular, the part data supplied by the test bed company is heterogeneous in units.

Some examples of part data include lists of purchase orders, reorder points, lists of maintenance requests, and safety information. We used the actual historical part data, or values found in the company's enterprise resource planning (ERP) system, in order to evaluate each part and its importance to inventory. An ERP system is an information system used to plan enterprise-wide resources needed to support customer orders (Heizer and Render 2011). For the test bed company, data on all part characteristics are stored in SAP (its ERP system). Numerous characteristics are tracked and updated as time progresses and new data is available. In general, some examples of data or characteristics that are typically stored in the ERP system include vendor information, work orders, lead times, purchase orders, and bills of material. In particular, for nuclear generation, data corresponding to plant refueling outages and safety are also stored and tracked for every spare part. Data for all spare parts are stored in the ERP system, including routine office supplies and parts that are obsolete or retired. The ERP system provides the best and most comprehensive source for part data, and this information is used to test and validate the scoring model. Uncertainty regarding important parts is reduced because the data corresponding to these part characteristics capture the current conditions of the parts and utilizes the limited amount of reliable and accurate data available.

To illustrate how the assignment was done, two examples are given in the following two sections. In general, the item or part characteristic data for every influence are unique. These examples provide an illustration of what that data might be and how it was approached. The assignments for all influences can be found in Table 15.3.

15.7.1 Influence 2.1: Count of Limited Condition of Operation (LCO) and Single Point Vulnerability (SPV) Locations Where Part Is Installed in Plant

Influence 2.1 is the first influence in the Part Failure set, and the characteristic data counts the number of LCO and SPV (single point vulnerability) locations where a part is installed in a plant. Recall that a part failure at an LCO location requires a full shutdown of the plant unless the situation is fixed within a relatively short interval of time following the failure (typically 72 hours). Therefore, the more LCO locations at which a part is installed, the more critical it becomes, because there is an increased probability that a failure of the part may shut down or force a derate of the plant. This probability is based purely on the fact that as the number of LCO locations rises, a failure could happen at more places. This influence does not consider the exact probability of such a failure (because such data does not exist) but does indicate that such a failure could possibly occur.

A nuclear plant typically has two separate production lines that are identical and redundant, called "trains." Generation of electricity can be done with either train, and the plant switches between trains as maintenance is required or if a component failure occurs. Some parts are installed common to both trains and are shared by both lines. Thus, a failure of one of these parts can affect the production of electricity and may cause a derate or loss of production. Locations where these parts are installed are called SPV locations.

To translate this characteristic to the ordinal scale, a list of the LCO and SPV locations at where each part in a sample set of 200 parts was installed was extracted from the test bed company ERP system. Thus, 200 lists were extracted and the number of locations in each list was counted. Reviewing these counts indicated that all 200 sample parts were installed at 24 or fewer LCO or SPV locations. Therefore, 24 became the upper bound of the part data range, with 0 or no LCO/SPV locations as the lower bound. The range of 0 to 24 must then be translated to the dimensionless ordinal scale. Twenty-four was not simply divided by 5, which would assign an equal number of possible counts to each bin (1–5) on the ordinal scale. Rather, the implications associated with various counts of LCO/SPV locations and possible resulting complications were considered. Zero was assigned to ordinal value 1 only because no LCO or SPV installations imply no chance for plant shutdown and no effect on part criticality. Any LCO or SPV installation count greater than 1 implies a nonzero probability of failure and shutdown, which may result in millions of dollars of lost generation revenue. This revenue loss would occur if a part was not in stock to remedy the situation quickly or the plant maintenance team could not fix the issue in 72 hours or less.

Upon reviewing the counts of LCO and SPV locations in the random sample of 200 spare parts at the test bed company, few parts were found to have greater than 7 LCO or SPV installations, so the range corresponding to scale values 4 and 5 was widened. Doing so allows for the parts with the most LCO/SPV installations to be considered most critical to inventory, because these parts would be associated with ordinal scale values 4 and 5. Recall that a scale value of 5 implies highest criticality and a scale value of 1 implies lowest or no criticality. In summary, if the scale was not widened for counts greater than 7 and did not isolate 0 to a scale value of 1, then an equal number of counts would be assigned to each scale value and the complexity and nuances of corresponding inventory criticality associated with counts of LCO or SPV locations would be lost. Table 15.3, row 2.1, shows the final scale assignments for influence number 2.1.

TABLE 15.3

Part Characteristic Data for All Influences

Influence	Influence Description	Characteristic Data	Ordinal Data				
			1	2	3	4	5
1.1	Work order able to be rescheduled	Sum of times work order rescheduled	0–7	8–12	13–28	29–44	45–54
1.2	Reason for work order	Sum of reason codes for work order	0–16	17–37	38–58	59–99	100–109
1.3	Immediacy of schedule, when to complete work order	Maximum priority code over work orders	1	2	3	—	4–5
1.4	Part demand during lead time	Count of future demand during lead time	0–6	7–19	20–29	30–39	40–49
2.1	Failure of part leads to LCO or SPV	Count of installed LCO & SPV locations	0	1–2	3–6	7–13	14–24
2.2	Lab testing results in predictive maintenance	No data available	—	—	—	—	—
2.3	Surveillance maintenance results	No data available	—	—	—	—	—
2.4	Failure history	Count of previous part failures	0	1	2–4	5–7	8–9
2.5	System health for related equipment	Plant system health	0	1	2	3	4
2.6	Normal part life	No data available	—	—	—	—	—
3.1	Vendor discontinued or obsolete	Indicator for vendor discontinued	0	—	—	—	1
3.2	History of vendor quality issues	Count of vendor quality problems	0	1	2–4	5–6	7–9
3.3	Vendor reliability	No data available	—	—	—	—	—
3.4	Vendor lead time	Planned lead time (in days)	0–90	91–180	-	181–364	365–450
3.5	Availability on *RAPID*	No data available	—	—	—	—	—
3.6	Quality testing or hold before installation	Indicator for needed quality inspection	0	—	1	—	—
3.7	Ability to expedite the part	No data available	—	—	—	—	—
4.1	Part is at a single point vulnerability	Count of single point vulnerabilities (SPV)	0	1–2	3–5	6–9	10–14

			0-1	2-3	4-5	6-7	8-9
4.2	Installation in a functional location tied to LCO	Count of LCO locations	0	—	—	—	—
4.3	Usage for equipment train protection and reliability	Indicator for train protection	0	—	—	1	—
4.4	Number of locations in plant where part installed	Count of locations where part installed	0-86	87-144	145-260	261-434	435-551
4.5	Regulatory requirement to keep part on-hand	Regulatory requirement to keep part on-hand	0	—	—	—	—
4.6	Usage to prevent equipment forced outage	Use of part to prevent forced outage	0	—	—	—	—
4.7	Open work orders and part demand	Count of open work order demand	0	1-7	8-14	—	—
4.8	If the part requested is actually used in the PM work order	Total requests for preventative maintenance usage	0-25	26-60	61-100	101-179	—
5.1	PM done on the related equipment	Count of preventative maintenance usage	0	1-12	13-40	41-70	71-87
5.2	Frequency of time the actual PM job is more involved than anticipated	No data available	—	—	—	—	—
5.3	PM done online or only during outage	Count of times used during outage	0	1-4	5-11	12-16	17-19
5.4	Associated maintenance rules for the equipment	Sum of maintenance rules	0-6	7-16	17-30	31-39	40-49
6.1	Part scheduled for use during outage	Count of demand requested for outage	0	1-6	7-15	16-30	31-39
6.2	Actual usage of the part during outage	Count of parts used during outage	0	1-3	4-14	15-22	23-28
6.3	When the part can be used or equipment accessed	Indicator for using part only during outage	0	—	1	—	—
7.1	Cost of expediting the part	Total expediting costs	0	1-200	201-1000	—	—
7.2	Cost of replacement power	Possible revenue loss due to shutdown (in millions)	0	0.01-39.5	39.6-118.7	118.8-197	197.1-251

15.7.2 Influence 1.3: Immediacy of Schedule

Influence 1.3 is the third influence in the Timeliness of Work Order set and addresses the maintenance-related schedule for a work order associated with a part. Every work order for maintenance work is assigned a priority code that identifies how quickly the work needs to be completed. Priority 100 and 200 codes are reserved for work that is likely to have immediate or imminent impact on plant safety or generation output. These codes are reserved for the most critical work and are rarely used. The priority 300 code is reserved for work that is emergent and expedited, a failure in the plant that requires immediate attention. The priority 400 code indicates time-dependent work that must be done within 90 days. Priority 500, 600, 700, and 800 codes are reserved for routine work or work that can be delayed indefinitely.

All historical work orders associated with all 200 parts in the random sample were provided by the test bed company. Then the highest priority code for every part was identified in the data. These priority codes were then matched with the corresponding part characteristic value from the ordinal scale that matched the priority codes. The most critical work orders are clearly priority codes 100 and 200, and priority code 300 and greater are associated with corrective or routine work that cannot immediately shut the plant down. As such, the priority codes were assigned to the ordinal scale values in the following manner: priority 100 and 200 to ordinal scale value 5; priority 300 to ordinal scale value 3; priority 400 to ordinal scale value 2; and priority 500, 600, 700, and 800 to ordinal scale value 1. As a result, priority codes were not assigned the value of 4 from the ordinal scale.

15.8 Scale Assignments

Ordinal scale assignments along the lines of the examples described were done for all 34 influences. Table 15.3 shows, for the nuclear spare parts data, the final mapping of the part data to the ordinal scale along with the corresponding part characteristic data from the ERP system used for each influence. This mapping of values to the 1 to 5 scale was dependent on the range of values stored in the ERP system for a given characteristic. Note the "Influence Description" column repeats the 34 influences to the spare parts process and corresponds to the "Description" column in Table 15.1.

The earlier two examples indicate that the assignment of historical part data characteristic values to the ordinal scale is unique and may not always be linear. Some influences, such as 4.2, follow a linear assignment, but such an assignment is not required. The main goal of the ordinal scale is to capture the importance of the ensuing consequences if a part is not available in inventory when required. Considering consequences in the context of each influence results in a measure of the criticality of storing a part in inventory.

The weights assigned to each influence then scale the criticality in the context of the overall spare parts problem.

In general, the assignment of these data to the ordinal 1 to 5 scale can be somewhat subjective and dependent on the historical values stored in the ERP system or source of part data. However, the strength of the approach lies in the fact that it is data driven; the case study illustrated here is an example of one such assignment. However, this approach could certainly be adapted to other applications through the development of different scales that are appropriate for each application. Ideally, this would be done with the assistance of company SMEs. Our assignments are based on the perceived consequences if a part is not available when requested from inventory. Although they are tied to potential lost revenue associated with a situation that compromises safety at the plant and leads to possible shutdown or derate, these perceptions could be subjective. Therefore, these scales can and should be periodically revisited to ensure that the current part data is effectively being captured by the ordinal scale as well as the most recent data available is being used to evaluate part characteristics. In this case study, data over 69 months was extracted from the ERP system for the sample set of parts provided by the test bed company. Examining a large time window allows for accurate representation of part behavior and demand usage in the spare parts process. Spare parts demands are extremely intermittent, and a large time window is needed to capture all relevant information. The examination of the data for each characteristic led to assignment of historical part data characteristic values to the ordinal scale for every influence.

Finally, it is worth mentioning that the ERP system at the test bed company currently does not track part characteristic data related to every influence. In fact, no relevant data exists for some influences, such as Influence 2.3: Surveillance Maintenance Results. A complete listing of influences for which no characteristic data exists can be found in Table 15.3, denoted by "No data available" in the "Characteristic Data" column. For these influences, a value of 0 was used for the part data for every part. Doing so distinguishes the lack of available data from the ordinal scale values. If a scale value was used, a level of criticality or importance may be assumed for that influence. Such an assumption would be misleading, as no data currently exists for that influence. We recommend that the test bed company begins to track appropriate part characteristic data for the influences in Table 15.3 and incorporate the data into the scoring formulas to increase the accuracy of the part criticality scores. As corresponding part data becomes available, it will replace the current placeholder value of 0 for the influence. The seven influences currently without data have low AHP priority weights, so the current effect of no part data on the overall score is minimal. All influences to the nuclear spare parts process were identified on the diagram, regardless if data was available or not. In order to have a complete representation of all relevant factors to the spare parts process, data must be available for all influences. Therefore, we recommend that the case study company begin to track relevant data for these seven influences and incorporate it into updated parts scores.

The scoring model described in the next section is used to combine the part characteristic data with the AHP priorities to yield an overall part score value in the range of 1 to 5. This procedure is illustrated with an example.

15.9 Criticality Scoring Model

The part criticality scoring model uses the influence weights from the AHP priorities and combines them with part characteristic data, forming a criticality score for each part. This score will identify the importance or criticality of storing a part in inventory. Two equations are used for calculating part scores. For each specific part j, Equation (15.1) combines the weights from a single set k in the influence diagram with the corresponding part characteristic data for part j from the ordinal scale, and then aggregates this across all influences i in the set k to arrive at a subscore for that part with respect to that influence set:

$$\text{Set } k \text{ subscore} = g_{k,j} = \sum_i (p_{k,i} * d_{k,i,j}) \ \forall \ j \tag{15.1}$$

where $p_{k,i}$ is the AHP priority for influence i within set k and $d_{k,i,j}$ is the ordinal scale characteristic data for part j corresponding to influence i within set k. This equation is evaluated for every influence set in the diagram. For the nuclear spare parts influence diagram, we have seven sets of influences, so that for each part, a subscore is found for each set (i.e., seven subscores in all for each part). Once the subscores are found for each part, they are combined with the priority weight for each set of influences and aggregated across the overall set of influences so as to yield an overall part criticality score; this is done by using Equation (15.2):

$$\text{Part score} = s_j = \sum_k (p_k * g_{k,j}) \tag{15.2}$$

where p_k is the AHP priority for set k and $g_{k,j}$ is the subscore for set k and part j. This yields a single criticality score for each part. An example for part *a257* is shown in the next section.

15.10 Scoring Equations Example

Part *a257* is an engineering critical tubing adaptor. "Engineering critical" parts are those that are installed at key locations in the plant and are considered critical to plant safety and operations from an engineering perspective, as defined by the plant. Almost all parts installed at LCO locations have an engineering critical definition. Subscores for each set of influences are found by evaluating Equation (15.1) seven times, once for each set. For influence set 1 (Timeliness of Work Order), the $a257$ set 1 subscore is found by evaluating the set 1 subscore equation, with $k=1$:

$$g_{1,a257} = (p_{1,1} * d_{1,1,a257}) + (p_{1,2} * d_{1,2,a257}) + (p_{1,3} * d_{1,3,a257}) + (p_{1,4} * d_{1,4,a257})$$
$$= (0.2704 * 5) + (0.4151 * 1) + (0.1991 * 3) + (0.1155 * 1) = 2.477 \tag{15.3}$$

Thus, each AHP priority in the Timeliness of Work Order set is combined with the corresponding ordinal part characteristic scale data for each of the four influences in the set. In a similar fashion, the subscores for sets 2 to 7 are found:

$$g_{2,a257} = (0.4006 * 1) + (0.0830 * 0) + (0.2375 * 0) + (0.0915 * 1) + (0.1281 * 5)$$
$$+ (0.0593 * 0) = 1.1326 \tag{15.4}$$

$$g_{3,a257} = (0.4566 * 1) + (0.0841 * 1) + (0.1041 * 0) + (0.1102 * 4) + (0.1345 * 0)$$
$$+ (0.0402 * 1) + (0.0704 * 0) = 1.0217 \tag{15.5}$$

$$g_{4,a257} = (0.3172 * 1) + (0.1693 * 1) + (0.1668 * 1) + (0.0448 * 1) + (0.1065 * 1)$$
$$+ (0.1242 * 1) + (0.0334 * 1) + (0.0377 * 1) = 1.0000 \tag{15.6}$$

$$g_{5,a257} = (0.4855 * 1) + (0.1951 * 0) + (0.0941 * 1) + (0.2252 * 1) = 0.8048 \tag{15.7}$$

$$g_{6,a257} = (0.4074 * 1) + (0.4636 * 1) + (0.1290 * 1) = 1.0000 \tag{15.8}$$

$$g_{7,a257} = (0.1722 * 1) + (0.8278 * 1) = 1.0000 \tag{15.9}$$

The seven subscores are then combined with the overall influence set AHP priority weights to evaluate a final criticality score for part $a257$:

$$s_{a257} = (p_1 * g_{1,a257}) + \ldots + (p_7 * g_{7,a257}) = (2.477 * 0.0620)$$
$$+ (1.1326 * 0.3498) + (1.0217 * 0.0517) + (1 * 0.0626) \tag{15.10}$$
$$+ (0.8048 * 0.1013) + (1 * 0.0873) + (1 * 0.2873) = 1.1213$$

As shown in the next section, part *a257* is subsequently assigned to group III, or the group of parts that is least critical for inventory management. The set 5 subscore for this part is 0.8048, which is low, and that along with the relatively low subscores of 1.00 for influence sets 4, 5, and 6 cause this part to classify as one of the least critical to inventory management. Group III is analogous to group C in an ABC analysis, but recall that with this methodology, we are not limited to just three classifications. This example also illustrates the fact that inventory criticality from a business perspective might be quite different from engineering criticality for plant operation. Part *a257* is actually designated as engineering critical but, based upon this analysis, is not very important to store in inventory.

15.11 Grouping of Parts

Once all 200 parts in the random sample receive a criticality score, they can then be placed into distinct groups for inventory management. This allows for identification of the most critical parts from a management perspective as well as those parts that might not need that much attention. To determine the part groups, the distribution of the 200 criticality scores, one for each part, were graphed in a histogram as shown in Figure 15.3. By visual inspection of Figure 15.3, the histogram might appear to have three natural groups of parts, and a possible grouping is shown in Figure 15.4. However, there is no particularly strong reason to have exactly three groups. Depending on the specific application context, more or less than three groups could well have been appropriate, depending on the data and scores found for the parts.

A more rigorous technique, such as cluster analysis, can be used to determine the number and composition of the groups. Cluster analysis is a technique that breaks up data into meaningful or useful groups; the algorithms used group the data so that the data in each group have common

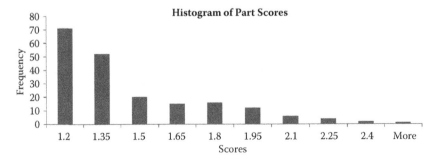

FIGURE 15.3
Histogram of part criticality scores.

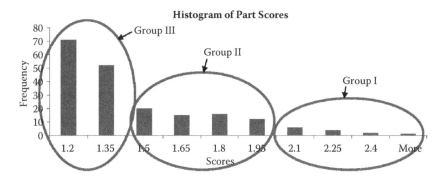

FIGURE 15.4
Histogram of part criticality scores with suggested groups of parts.

characteristics and are distinct from those in other groups (Tan, Steinbach, and Kumar 2006). One common algorithm used in cluster analysis is the k-means algorithm, which groups data into a number of pre-specified groups based on the distance from the centroid of each group. The algorithm checks the value of the distance between a data point and all centroids, and assigns that data point to the group whose centroid is closest to the point. At each iteration, the algorithm assigns all points and calculates a new centroid for each group; the iterations repeat until the centroids converge. For further details on the k-means algorithm and cluster analysis, see Tan, Steinbach, and Kumar (2006) and Everitt, Landau, and Leese (2001).

Three centroids were arbitrarily chosen to seed the k-means algorithm for the part criticality score data, as the histogram appeared to have three natural groups. Because the algorithm is sensitive to the starting centroid values, various combinations of starting centroids were tested with the data. Table 15.4 shows 15 different trials with randomly selected centroids and the resulting number of parts placed in each group. Patterns in the results emerged, and most trials yielded approximately 20 parts in group I, approximately 40 parts in group II, and approximately 140 parts in group III. These similar results provide empirical evidence in support of the natural group boundaries. Therefore, row 13 in Table 15.4 was selected as the final grouping analysis; the algorithm ran 10 iterations for this combination.

Table 15.5 presents descriptive statistics for each group of parts, after the final k-means cluster analysis. Parts with a criticality score of 1.8582 or higher were placed in group I (highest criticality scores, 21 parts); from the remaining parts, those with a criticality score of 1.4612 or higher were placed in group II (moderate criticality scores, 38 parts); the rest of the parts were placed in group III (lowest criticality scores, 141 parts). The cutoff scores for each group ensure the same number of parts in each group as designated by the k-means algorithm are in the actual criticality groups. Based on the criticality scores, all 200 parts in the random sample of parts from the test bed company were thus placed into one of three importance groups for inventory management.

TABLE 15.4

Centroids Tested in k-Means Algorithm

Trial	Starting Centroids			Number of Parts		
	First	Second	Third	Group I	Group II	Group III
1	1.3000	1.2000	1.0000	41	48	111
2	1.8000	1.6000	1.5000	13	29	158
3	2.0000	1.6000	1.0000	25	39	146
4	2.1350	1.6660	1.2670	23	38	139
5	2.1830	1.7140	1.1650	23	39	138
6	2.1860	1.7240	1.2430	23	38	139
7	2.1900	1.7360	1.3400	21	38	141
8	2.2000	1.8000	1.2000	10	30	160
9	2.2600	1.6660	1.2290	23	39	138
10	2.2600	1.8760	1.2290	21	37	142
11	2.2600	1.5500	1.1000	24	40	136
12	2.2600	1.8800	1.2300	21	37	142
13	2.2750	1.7500	1.3310	21	38	141
14	2.3500	1.7000	1.2000	12	28	160
15	2.4120	1.8440	1.4610	16	39	145

TABLE 15.5

Descriptive Statistics for Part Groups

	Group I	Group II	Group III
Number of parts	21	38	141
Mean	2.03	1.6571	1.2132
Standard deviation	0.1552	0.1089	0.098
Minimum score	1.8582	1.4612	1.073
Median score	1.9721	1.6772	1.192
Maximum score	2.4118	1.8436	1.4471
Range	0.5536	0.3824	0.3741

15.12 Conclusions and Future Work

This chapter presents a data-driven method for classifying parts for inventory management based on criticality scores when demand is difficult to characterize because forecasting methods do not work and predictive data do not exist. This methodology is illustrated through a case study from the nuclear power generation industry, using actual spare parts characteristic data from the company's ERP system. The criticality scores are comprehensive, because all relative factors and forces in the spare parts management process are represented in their computation, with influences both being grouped by plant SME knowledge area and ranked for relative importance. The part criticality scores can then be used to identify groups of parts for inventory management, so that the appropriate amount of time and resources can be expended on each part. This is of great value to the decision maker as ultimately it will save significant amounts of money due to the high costs of spare parts inventory. The scores should be periodically updated to ensure accurate representation of actual part usage. This begins with using updated part data from the ERP system and reevaluating both the scoring equations and *k*-means clustering algorithm. Periodically revisiting the criticality scores will ensure that any inventory policy based on the classification is flexible, responsive, accurate, and robust.

The next step after the classification scheme described in this chapter is the development of appropriate inventory management policies for each group of parts. These policies would set a reorder point, order amount, and so forth for every group of parts, with the goal of minimizing inventory costs while maintaining a desired service level. Service levels would be set appropriately high for the more critical parts, while the less critical parts could even be handled using some relatively simple scheme.

Although the classification methodology described herein was developed specifically for nuclear spare parts, the overall approach can be extended to other industries that share similar characteristics. In particular, this approach is especially applicable to environments with (1) a large number of parts with various costs and levels of criticality, (2) unpredictable and intermittent demand in small amounts, and (3) expensive consequences from stockouts. Examples might include military applications, aerospace, and regulated utilities. This approach is data driven, so modification to support the particular data set under analysis is necessary when mapping the part data to an ordinal scale. However, the illustration in this chapter provides a general guideline as to how the ordinal scale assignments might be done. The influence diagram and ranking of influences can be easily applied to other data sets and spare parts decision-making environments.

References

Asaro, C. A., and T. Tierney. 2005. Who's buying who: Mergers and acquisitions of utilities are on the rise. http://www.beechercarlson.com/files/pdf/energyarticle.pdf (accessed January 7, 2011).

Bond, L. J., S. R. Doctor, D. B. Jarrell, and J. W. D. Bond. 2007. Improved economics of nuclear plant life management. http://pmmd.pnnl.gov/program/reports/IAEA-Improved-Economics-100307.pdf (accessed April 14, 2011).

Braglia, M., A. Grassi, and R. Montanari. 2004. Multi-attribute classification method for spare parts inventory management. *Journal of Quality in Maintenance Engineering* 10:55–65.

Cohen, M. A., and R. Ernst. 1988. Multi-item classification and generic inventory stock control policies. *Production and Inventory Management Journal* 29:6–8.

Croston, J. D. 1972. Forecasting and stock control for intermittent demands. *Operational Research Quarterly* 23:289–303.

Eleye-Datubo, A. G., A. Wall, A. Saajedi, and J. Wang. 2006. Enabling a powerful marine and offshore decision-support solution through Bayesian Network Technique. *Risk Analysis* 26:695–721.

Ernst, R., and M. A. Cohen. 1990. Operations related groups (ORGs): A clustering procedure for production/inventory systems. *Journal of Operations Management* 9:574–598.

Everitt, B. S., S. Landau, and M. Leese. 2001. *Cluster Analysis*, 4th ed. London: Arnold.

Flores, B. E., D. L. Olson, and V. K. Dorai. 1992. Management of multicriteria inventory classification. *Mathematical and Computer Modelling* 16:71–82.

Flores, B. E., and D. C. Whybark. 1987. Implementing multiple criteria ABC analysis. *Journal of Operations Management* 7:79–85.

Gajpal, P. P., L. S. Ganesh, and C. Rajendran. 1994. Criticality analysis of spare parts using the Analytic Hierarchy Process. *International Journal of Production Economics* 35:293–297.

Hadi-Vencheh, A. 2010. An improvement to multiple criteria ABC inventory classification. *European Journal of Operational Research* 201:962–965.

Heizer, J., and B. Render. 2011. *Operations Management*, 10th ed. Boston: Prentice Hall.

Howard, R. A., and J. E. Matheson. 2005. Influence diagrams. *Decision Analysis* 2:127–143.

Huiskonen, J. 2001. Maintenance spare parts logistics: Special characteristics and strategic choices. *International Journal of Production Economics* 71:125–133.

Lapson, E. 1997. U.S. electric industry: Understanding the basics. *Proceedings of the AIMR Seminar Deregulation of the Electric Utility Industry*, 5–12.

Ng, W. L. 2007. A simple classifier for multiple criteria ABC analysis. *European Journal of Operational Research* 177:344–353.

Partovi, F. Y., and J. Burton. 1993. Using the Analytic Hierarchy Process for ABC analysis. *International Journal of Operations and Production Management* 13:29–44.

Philipson, L., and H. L. Willis. 2006. *Understanding Electric Utilities and Deregulation*, 2nd ed. Boca Raton, FL: Taylor & Francis.

Quantum Gas and Power Services. n.d. List of energy deregulated states in the United States. http://www.quantumgas.com/list_of_energy_deregulated_states_in_united_states.html (accessed March 5, 2009).

Ramanathan, R. 2006. ABC inventory classification with multiple-criteria using weighted linear optimization. *Computers and Operations Research* 33:695–700.

Rezaei, J., and S. Dowlatshahi. 2010. A rule-based multi-criteria approach to inventory classification. *International Journal of Production Research* 48:7107–7126.

Saaty, T. L. 1980. *The Analytic Hierarchy Process: Planning, Priority Setting, Resource Allocation*. New York: McGraw-Hill.

Saaty, T. L. 1990. How to make a decision: The Analytic Hierarchy Process. *European Journal of Operational Research* 48:9–26.

Scala, N. M. 2008. Decision modeling and applications to Major League Baseball pitcher substitution. *Proceedings of the 29th National Conference of ASEM*.

Scala, N. M. 2011. Spare parts management for nuclear power generation facilities. Ph.D. diss., University of Pittsburgh.

Scala, N. M., K. L. Needy, and J. Rajgopal. 2009. Decision making and tradeoffs in the management of spare parts inventory at utilities. *Proceedings of the 30th Annual ASEM National Conference.*

Scala, N. M., K. L. Needy, and J. Rajgopal. 2010. Using the Analytic Hierarchy Process in group decision making for nuclear spare parts. *Proceedings of the 31st Annual ASEM National Conference.*

Scala, N. M., J. Rajgopal, and K. L. Needy. 2010. Influence diagram modeling of nuclear spare parts process. *Proceedings of the 2010 Industrial Engineering Research Conference.*

Stafira, Jr., S., G. S. Parnell, and J. T. Moore. 1997. A methodology for evaluating military systems in a counterproliferation role. *Management Science* 43:1420–1430.

Syntetos, A. A., and J. E. Boylan. 2001. On the bias of intermittent demand estimates. *International Journal of Production Economics* 71:457–466.

Tan, P., M. Steinbach, and V. Kumar. 2006. *Introduction to Data Mining.* Boston: Pearson Addison Wesley.

United States Department of Energy. 2006. Workforce trends in the electric utility industry: A report to the United States Congress pursuant to Section 1101 of the Energy Policy Act of 2005. http://www.oe.energy.gov/DocumentsandMedia/Workforce_Trends_Report_090706_FINAL.pdf (accessed January 7, 2011).

Willemain, T. R., C. N. Smart, and H. F. Schwarz. 2004. A new approach to forecasting intermittent demand for service parts inventories. *International Journal of Forecasting* 20:375–387.

Zimmerman, G. W. 1975. The ABC's of Vilfredo Pareto. *Production and Inventory Management* 16:1–9.

16

Service Technology, Pricing, and Process Economics in Banking Competition

David Y. Choi, Uday S. Karmarkar, and Hosun Rhim

CONTENTS

16.1 Introduction

Current developments in information technology allow information intensive industry sectors to deliver their services and products electronically (online) to individuals' personal computers or other information appliances at home or work. Electronic delivery channels promise to offer improved convenience and value to customers while reducing service-delivery costs. The continued growth of electronic services is likely to have a profound impact on competition in many of these industry sectors. Moreover, it also provides an alternative channel of service delivery where no service infrastructure has yet been built, for example, in many remote areas of developing countries. In this chapter, we analyze alternative delivery technologies and service processes in information-intensive services, using the example of retail (consumer) banking. Similar methods have been used to analyze industrial services by Karmarkar (2000) and Bashyam and Karmarkar (2004, 2005).

Over the past decade, retail banks have acquired a great deal of experience in their use of new technologies to improve both back and front office processes. However, this has required considerable discussion, and much trial and error. One common discussion has centered on how different service-delivery processes, for example, traditional branch, phone-based and PC-based service, would compete with one another and which would thrive under what conditions (Ernst & Young 1995; Lunt 1995; Choi 1998; Holmsen et al. 1998; Bradley and Stewart 2003). Others have explored the electronic channel as a method of reaching the rural poor in India, Africa, and China (Laforet and Li 2005; Opoku and Foy 2008; Coetzee 2009). Clearly, the banks' service design and pricing policies influence the utilization of each delivery mechanism. These design and pricing decisions may in turn be influenced by competition and market structure. Similar issues and dynamics seem to arise under a wide range of e-business contexts (Starr 2003).

This chapter presents a framework that incorporates process economics and process cost-benefit analysis into an overall analysis of service design, technology impacts, pricing decisions, and competition. As a stylized example, we analyze two alternative service-delivery processes for banking, conventional and electronic, and examine market segmentation and competition between them. The conventional service process refers to the traditional physical branch network, and electronic process refers to online service used through a personal computer (PC). Under the conventional model, customers physically access a branch facility, wait in line if necessary, and make personal contact with a teller to perform certain transactions. In the electronic case, the bank supplies a range of information over telecommunications channels, such as the Internet. Customers choose between the service alternatives so as to minimize their time and costs—financial and other. As mentioned earlier, our analysis in this chapter is simplification of a more complex real world. The analysis is performed under the assumption that the same services can be obtained from either service-delivery processes. Since cash withdrawals and certain other services are currently only feasible at branches and ATMs, we limit our analysis to banking activities that exclude cash transactions. Such banking activities include balance and payment inquiries, fund transfers, bill pay, and loan applications, among others. We also do not include in the current analysis personal financial planning or other in-depth advisory services associated with private banking. Electronic channels have not yet been able to offer adequate replacements for such high-end services.

We start with an analysis of the business processes of banks and the service interaction processes of customers. We define service design for banking services by a measure of service level (or quality of service) and delivery technology. The service level can be measured by customer cost captured as time cost required to complete a service process, that is, high (low) service level implies low (high) time cost. Thus, the process analysis enables cost models for both the service providers and customers as a function of both service level and technology. We utilize this analysis in a process-economics model to describe banking-services markets where customers are indexed by their value of time (using level of income as proxy, as is common in the literature, see Beesley 1965; Thomas et al. 1970; Viauroux 2008), and choose between conventional and electronic service-delivery technologies. We develop market models to obtain insights about market penetration and service levels for banks operating under monopolistic and duopolistic settings.

The research questions we address in this chapter are: How do service-delivery technologies affect pricing and service level decisions of retail banks under competition? How does the delivery process drive customer choices and hence market segmentation?

Our results are as follows: We demonstrate that a market segmentation analysis can reveal the existence or the lack of a market segment for electronic banking customers concerned about time and convenience. In both monopoly and duopoly, we develop an understanding of the factors that affect prices, market segmentation, and service levels of electronic and conventional banking services. For example, a monopolist bank that provides both electronic and conventional processes will offer the lowest conventional service level possible to maximize its profits from electronic processes. In a duopoly, consumers benefit from better service levels at lower prices. We also incorporate in our model how people's reluctance to adopt new technology can significantly reduce market size. In certain situations, bank management may affect customer adoption by improving its service or influencing customer attitudes.

The rest of the chapter is organized as follows. The following section reviews the extant literature. The third section describes a process analysis of the activities that banks and customers perform in creating and using the services, and examine banks' and consumers' costs. The fourth section utilizes a simple market segmentation model to analyze market penetration and service level issues in a monopoly and in several duopoly market settings. We conclude and summarize our results in the final section.

16.2 Literature Review

This chapter presents an analysis of service competition driven by new technologies that impact delivery processes and consumer processes. The analysis ties together the operations issues related to process economics, technology impact, service delivery, cost, and quality with consumer behavior, pricing strategies, and market segmentation. The relevant literature thus extends across operations management, marketing, strategy, technology management, and economics.

Process competition in alternative service-delivery mechanisms is an issue closely related to competition between distribution channels. Competing channels can be retailers (Ingene and Parry 1995; Iyer 1998; Tsay and Agrawal 2000), store versus direct marketer (Balasubramanian 1998; Chiang et al. 2003), pure-direct versus adopted-direct (Yao and Liu 2003), franchisees (Kaufmann and Rangan 1990), and retail versus e-tail (Yao and Liu 2005). Tsay and Agrawal (2004a, 2004b) classify literature on distribution systems, in which they maintain that the class of papers dealing with competition between independent intermediaries in hybrid distribution systems have appeared only recently. Variables used in the analysis of competition have also been limited. While various channel structures are modeled, the majority of papers focus on price competition (Cattani et al. 2004). Viswanathan (2005) is an exception. Viswanathan (2005) examines the impact of differences in channel flexibility, network externalities, and switching costs on competition between online, traditional, and hybrid firms. In our model, we include factors such as service quality and market segmentation as well as pricing. For a comprehensive review of competitive models in marketing and economics, readers are referred to Moorthy (1993).

Research on the impact of the new processes in retail banking has been performed at both conceptual and empirical levels. Choi (1998) introduces the case of Wells Fargo Bank, facing the emergence of the Internet-banking market. Holmsen et al. (1998) recognize the need of a new process to manage multiple channels in retail banking. Geyskens et al. (2002) investigate the impact of adding an Internet channel on a firm's stock market return, using event study methods. Hitt and Frei (2002) compare customers who utilize PC banking to the conventional bank customers. Bradley and Stewart (2003) report Delphi study results in the United Kingdom, in which they predict that Internet banking will contribute as part of a multichannel strategy rather than as a standard-alone strategy in retail banking. Siaw and Yu (2004) use Porter's five forces model to understand the impact of the Internet on competition in the banking industry. Bughin (2004) identifies major differences between online customer acquisition and captive customer conversion, using logistic regression models. Verma et al. (2004) perform conjoint analysis and estimate multinomial logit models to understand customer choices in e-financial services. Customer utilities for online and conventional service features are constructed for various segments. However, the results do not extend to product positioning/design models.

In this chapter, we analytically investigate process competition in the context of retail banking. Byers and Lederer (2001) and Byers and Lederer (2007) are the most closely related articles in this context. These two papers compare various combinations of banking processes, considering entry decisions and customers' attitude about technology. Banks are assumed to be full price-takers, and market segments are given. Since the models presented in these papers are free entry models, focus is on the analysis of total market. Relative market sizes based on total number of banks, channel volumes and outlets, and number of active banks are examined with numerical examples. Equilibrium price is set by equating total demand and supply of bank services. In this chapter, we suppose duopolistic competition. We start from simple process analysis, and provide the intuitive market segmentation scheme endogenously rather than assuming fixed segments. Banks with different processes compete actively with prices and qualities of their services. Thus, we focus on the firm level decisions rather than market level analysis.

16.3 Process Analysis

In establishing and maintaining a service network, whether physical or electronic, the largest cost is typically the initial investment. For a large market size, the conventional service network is likely to be much more expensive. The investment cost of serving customers with a physical branch network is higher due to labor and occupancy costs. Johnson et al. (1995) estimate that to serve a base of a million customers, an electronic bank will have a total headcount of 2000 compared to approximately 3000 to 3500 for a traditional bank. Because of the nature of the jobs involved, the electronic bank will have a total salary bill less than half of its conventional rival. Occupancy costs for an electronic bank can be 80% lower than for a conventional institution, since the electronic bank will have centralized customer servicing and processing centers located in low rent areas, in direct contrast to the geographically dispersed retail bank operating a high cost branch network largely in urban areas.

To a large extent, the initial choice of investment, branch or computer network, determines the level of service offered. That is, the average speed of a service through conventional channel (including driving to a branch) or online channel will not dramatically change over time to the extent that one will overtake the other. We abstract the infrastructure investment costs into a cost function $\hat{I}_C(Q_C)$ and $\hat{I}_E(Q_E)$ with subscripts C and E denoting conventional and electronic options. The values of these functions represent the annualized cost of building and maintaining a service-delivery system over the useful life of the system. $\hat{I}_C(Q_C)$ and $\hat{I}_E(Q_E)$ vary as a function of the respective service levels denoted by Q_C and Q_E. Note that lower Q_C and Q_E indicate higher/faster service levels and that $\hat{I}_C(Q_C)$ and $\hat{I}_E(Q_E)$ are negatively sloped functions because they decrease as Q_C and Q_E increase, and vice versa. We would expect that $\hat{I}_C(Q_C) > \hat{I}_E(Q_E)$ for a wide range of Q_C and Q_E. Note that each of Q_C and Q_E consists of access, pre-service queue, and service stages.

Once the infrastructure is in place, the daily cost of service and transaction processing is also lower for the electronic bank, as described earlier. Paper-based transactions are much more time consuming than an end-to-end electronic transaction mechanism. Thus, the variable operations cost of serving a customer for an electronic bank is usually far less than that of a conventional bank, or $C_E < C_C$.

An example of a pure electronic bank was Security First Network Bank (SFNB), which began operation in 1995 as an Internet bank with one office, no branch, and no ATMs. It was very inexpensive to start but managed to obtain quite a segment of customers (Bauman 1996). Some banks are hybrid service providers, that is, they offer both electronic and conventional processes of service. Most of these banks start as conventional banks and add Internet or other online capabilities (although SFNB actually went in the other direction). Banks make the additional investments if the benefit from the new electronic process exceeds the investment costs.

The process map for customers helps identify the difference in time costs of different service technologies. Cook (1984) found in his survey that customers were most interested in decreasing time involved in carrying out standard but unenjoyable banking activities. We conjecture that each of the major steps of access, pre-service queue, and service take less time with electronic technologies than through conventional means on the average. We note that there may be certain ranges and limits to these quality levels. For example, Q_{Emin} may be bounded by the average speed at which electronic data can be accessed and transported with current networks. Such limits may be determined by a chain of engineering constraints including bandwidth, modem or connection speeds at both ends, database SQL query, information design, and network software, among others. Similarly, conventional service's lower speed limit Q_{Cmin} may be set by certain physical and technological constraints. In addition, Q_{Cmax} and Q_{Emax} may be determined by government or influenced by consumer organizations, since banking services are closely followed by these groups. For example, a complete elimination of branches would be strongly discouraged by them. We hypothesize that $Q_{Emax} < Q_{Cmin}$ based on the average of a large sample of customer experiences. That is, the slowest of the electronic processes is still faster than the fastest of its conventional counterparts in average transaction, mostly because of the travel time required to access the service.

The benefit of electronic banking comes at a cost. The electronic service processes involves a setup cost, F_P, consisting of hardware and software costs and online usage fee, in addition to the learning cost, L_E, (to be discussed later). While the conventional customers incur no initial setup cost in the conventional case other than P_C, the price of a bank account, each service delivery takes longer.

16.4 Market Segmentation Model

This section develops an economic model for determining the market segments choosing electronic and conventional options. The business implications from the model results are summarized in the conclusion. We now present the key variables in our analysis.

P_C, P_E, Q_C, and Q_E are the four decision variables (endogenous) to be set by banks.

P_C, P_E: Yearly (enrollment) fee for electronic and conventional service paid by the customer to the bank

Q_C, Q_E: Average service levels set by banks, measured in time units/ transaction (e.g., minutes), where lower Q_i refer higher service level

q_C, q_E: Average service levels set by banks, measured in cost/salary (to be explained) where lower q_i infers higher service level

The following are the annual operations and investment costs incurred by conventional and electronic banks in providing service.

C_C, C_E: Yearly variable costs of service operations, per customer, incurred by banks

$I_i(q_i)$: Annualized cost of building and maintaining service-delivery system i at level q_i, for $i = \{E, C\}$, decreasing in q_i

S: Describes the annual income level of consumers

$f(S)$: Distribution of consumers by salary levels for $S_{min} \leq S \leq S_{max}$, assumed to be uniform

$T_i(S)$: Total banking cost of customer with salary level S using mechanism $i = \{E, C\}$

$K_i(S, Q_i)$: Variable banking costs of consumer per usage, with salary level S, using mechanism i, at service level Q_i

$L_E(S)$: Annualized learning cost of customer of income S

The variables F_C, F_P, and F_E denote the annualized one-time startup fixed costs incurred by the consumers. Note that F_P represents the fixed startup costs for consumers with a PC.

F_P: Annualized one-time fixed cost to engage in electronic banking, including the initial cost of software, depreciation cost of PC, Internet access cost, and learning needed to start electronic banking

F_C: Annualized one-time fixed cost to engage in conventional banking

F_E: Annualized one-time fixed cost to engage in electronic banking, F_P + P_E

In addition,

m: Total banking market size (in an area) in number of customers

MS_C, MS_E: Market share, measured in terms of values ranging from 0 through 1

N: Usage rate (number of transactions) for a consumer per year

16.5 Framework for Analyzing Process Competition: Segmentation Analysis

We let P_C and P_E represent the enrollment fee for conventional and electronic banking that many banks charge. Many conventional accounts require no routine fees from customers, contingent on maintaining a certain minimum balance level. Although P_C and P_E may be paid monthly or even in shorter time intervals, a rational decision maker, when choosing between conventional and electronic banking, will treat them as an up-front investment (fixed cost) to acquiring services. The startup cost for conventional banking is just its enrollment fee $F_C = P_C$.

We initially assume that the rational user's sole objective is to minimize his financial (startup cost and fee) and time costs involved in his banking activities. Furthermore, the rational customer has neither a preference for personal service nor an inherent dislike for technology (an assumption that will be relaxed later). Given a measure of an individual's value for time, one can estimate the type of service the user will select. We will assume that an individual's salary and value of time are closely related.

Based on the assumptions, the total cost of banking for a conventional customer with an income level S can be modeled as $T_C(S) = F_C + K_C(S,Q_C)N$ and the cost to an electronic customer as $T_E(S) = F_E + L_E(S) + K_E(S,Q_E)N$, where $K_i(S,Q_i)$ represents the customer's cost per service transaction depending on his salary (value of time) and service level. We can simplify these equations further by assuming that $K_i(S,Q_i) = a\,Q_i\,S$ ($a > 0$) and $L_E(S) = l_E\,S$, where l_E is a positive constant, for $S_{min} \leq S \leq S_{max}$, that is, value of time is proportional to the individual's salary rate. Let $q_E = a\,(Q_E N) + l_E$ and $q_C = a\,(Q_C N)$. This gives linear cost models with $T_C(S) = F_C + q_C\,S$ and $T_E(S) = F_E + q_E\,S$, with salary as the segmentation variable, as shown in Figure 16.1. Our new cost coefficients q_C and q_E can conveniently be used as proxy for quality or speed of service delivery. Note that no per-usage charge is incorporated in this model. Linearity may not hold when technology aversion is introduced.

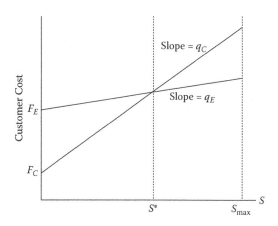

FIGURE 16.1
Customer segments opting for conventional or electronic banking.

In our model, salary is regarded as a proxy of value of time. Using income level as proxy has been common in the transportation economics literature (Beesley 1965; Thomas et al. 1970; Viauroux 2008). The literature shows that when individuals are observed in making decisions that involve a trade-off between money and time, there is evidence that the value of time is an increasing function of income. Moreover, interviews with bank managers have suggested that income level has been a useful variable for technology adoption, with indication that it is correlated with the rate of online banking adoption (see Choi 1998). Hitt and Frei (2002), who collect customer information from four large retail banks, confirm that in all banks the average income levels of PC banking customers are higher than those of conventional banking customers.

Figure 16.1 shows total costs of electronic and conventional processes. To find the point of indifference S^*, equate the costs: $P_C + q_C S^* = F_P + P_E + q_E S^*$, from which one obtains $S^* = (F_P + P_E - P_C)/(q_C - q_E)$. If $S_{min} \leq S^* \leq S_{max}$, the market shares of each mechanism are as follows: $MS_C(P_E, P_C, q_E, q_C) = \int_{S_{min}}^{S^*} f(S)dS = \dfrac{S^* - S_{min}}{S_{max} - S_{min}}$ and $MS_E(P_E, P_C, q_E, q_C) = \int_{S^*}^{S_{max}} f(S)dS = \dfrac{S_{max} - S^*}{S_{max} - S_{min}}$. From Figure 16.1, it is clear that there is at most one intersection (point of indifference) S^* under our linear assumption, so that for S greater than S^* electronic delivery is preferred and S less than S^* conventional service is preferred.

The existence and size of the electronic market depends on the value of F_P, P_E, P_C, q_C, q_E, and the shape of $f(S)$. Uniformity is assumed for consumer distribution $f(S)$ as in Hotelling (1929). Linear market assumption with uniformly distributed customers has been used by many authors (D'Aspremont et al. 1979; Andaluz 2011). Decreasing or increasing q_C and q_E corresponds to lowering or raising the slopes of the lines in Figure 16.1, while decreasing the startup costs implies lowering the y(customer cost)-intercept. Even if P_E is 0, the electronic consumer may pay a substantial startup cost consisting of software, equipment, and learning. The initial learning cost $L_E(S) = l_E S$, which is a function of S, affects the slope q_E, which in turn increases S^*. Thus, lowering the learning cost for the consumer can have a significant effect on the market segment for electronic banking.

We can look at additional market segmentation possibilities by performing a comparative analysis of the cost parameters. The market segmentation in Figure 16.1 is based on the assumption that $F_E > F_C$ and $q_E < q_C$. Even under these conditions, the market for electronic banking will be nonexistent if $S^* > S_{max}$. This situation would imply that either or both of technology (speed) and price of electronic delivery are not marketable.

Consider the situation when $F_E > F_C$ and $q_E > q_C$. Even if electronic delivery is faster, the situation $q_E > q_C$ can occur from the high learning cost l_E involved. Thus, if electronic technology does not save enough time to make up for the learning time lost, then no economically justifiable market exists for electronic banking. An exception would be the segment of customers who for one reason or another have already incurred the learning cost.

Banks may want to price P_E such that $F_E = F_P + P_E < F_C$ to encourage home-banking. If $F_E < F_C$ and $q_E < q_C$, then $S^* < S_{min}$, and all customers in the analysis will engage in electronic banking, strictly from an economic view. As will be discussed, noneconomic variables play an important role in the evaluation of process alternatives. For example, a customer with a high dislike for technology could prefer a conventional bank even if the price of banking is lower for electronic service. If for some reason $F_E < F_C$, but $q_E > q_C$, then the markets will segment, with the customers with salary $S < S^*$ preferring electronic banking, whereas those with $S > S^*$ opt for the conventional process.

In general, if $S^* = (F_P + P_E - P_C)/(q_C - q_E) < S_{max}$, then there is a substantial market for electronic banking. As mentioned earlier, if $S^* > S_{max}$ or is close to S_{max}, then the electronic market is nonexistent or small. The value of S^* and the market size for electronic banking also depends on how and to what extent technology is utilized by the banking industry in servicing their customers. We analyze several different market structures, including two scenarios we have termed "service integration" and "process specialization."

16.6 Monopolist Model: The Service Integration Scenario

With the availability of the electronic process, banks in the industry may integrate it with their existing service line and offer it as an alternative. A simplification of this scenario is a monopolist bank serving customers using both electronic and conventional mechanisms. This scenario is common in the rural areas of many developing countries in which its first local bank has arrived (Laforet and Li 2005; Opoku and Foy 2008; Coetzee 2009). Therefore, consider a scenario in which one bank has both a retail channel and an online banking option.

We assume that customer demand is quite inelastic to service quality or price, and customers choose one of the alternatives. Even though the demand is inelastic to quality and price, as in many market share models (see Anderson at al. 1992), market share itself is elastic to these variables. As mentioned in a previous section, we also assume that P_{Cmax} and P_{Emax}, the maximum conventional and electronic prices, have been set exogenously. Its design and pricing problem can be formulated as a profit maximization problem:

$$\max_{P_E, P_C, q_E, q_C} PR\left(P_E, P_C, q_E, q_C\right) = (P_C - C_C)\, m\, MS_C(P_E, P_C, q_E, q_C) - I_C(q_C)$$
$$+ (P_E - C_E)\, m\, MS_E(P_E, P_C, q_E, q_C) - I_E(q_E)$$

s.t. $\quad q_{Cmin} \le q_C \le q_{Cmax}, \quad q_{Emin} \le q_E \le q_{Emax}, \quad 0 \le P_C \le P_{Cmax}, \quad 0 \le P_E \le P_{Emax}$

In the model, the domains of variables are bounded and closed. Reservation prices and bounded service levels for certain technology have been assumed in many optimization and game theoretic models for compactness (see Anderson et al. 1992). The monopolist bank introduces electronic banking if the increase in revenue exceeds the increase in costs, i.e., if summation of revenue increase, $m(P_E - P_{Cmax})MS_E$, and cost savings, $m(C_C - C_E)MS_E$, is greater than the investment cost, $I_E(q_E)$. Issues related with economies of scope are not considered in this model. Then, the net profit increases by $m(P_E{}^* - P_{Cmax}) MS_E + m(C_C - C_E)MS_E - I_E(q_E{}^*)$ and an interior solution exists. (Proofs of propositions are provided in the Appendix.)

Proposition 16.1
When an interior solution exists (i.e., the monopolist provides both processes), the following hold for the monopoly model.

a. $P_C{}^* = P_{Cmax}$

b. $q_C{}^* = q_{Cmax}$

c. $P_E{}^* = \min \{P_{Emax}, P_{Cmax} + \dfrac{\left[(q_C - q_E)\, S_{max} - (C_C - C_E) - F_P\right]}{2}\}$

d. $q_E{}^* = q_{Cmax} - \sqrt{\dfrac{-m(C_C - C_E + P_E - P_C)(F_P + P_E - P_C)}{I_E'(q_E)(S_{max} - S_{min})}}$ for $q_{Emin} \le q_E{}^* \le q_{Emax}$

It is optimal for the monopolist to move people to the more profitable electronic process: Thus, $P_C{}^* = P_{Cmax}$, the highest price possible for conventional banking, as shown in Proposition 16.1a. Similarly, the bank deliberately sets the physical delivery at the worst level, $q_C{}^* = q_{Cmax}$ in Proposition 16.1b. As expected, P_E^* in Proposition 16.1c is a decreasing function of F_P, that is, if the startup cost is high, P_E^* needs to be low. Similarly, the difference in costs $(C_C - C_E)$ will cause the bank to charge lower P_E^*. The larger the difference of service levels between conventional and electronic service $(q_C - q_E)$, the higher price P_E^* the bank can charge. A closed form expression for q_E^* is found in Proposition 16.1d. The introduction of technology also affects the consumers.

The cost saving to consumers amounts to $\displaystyle\int_{S^*}^{S_{max}} S\,(q_{Cmax} - q_E^*)dS = \frac{\left(q_{Cmax} - q_E^*\right)}{2}$

$[S_{max}^2 - (F_P + P_E^* - P_{Cmax})^2]$ (see Figure 16.1).

Although a monopoly bank would not move as many customers as possible from conventional channels to electronic service, it will not necessarily get rid of its conventional service entirely. Such a move will likely not be optimal for the bank under any realistic assumptions as it would result in a loss of a significant segment of its customers.

16.7 Duopoly Models

16.7.1 Process Specialization

In this section, we consider a market situation in which banks specialize in one of the two service processes. This scenario is in direct contrast with the monopoly (service integration) case, where a single bank controlled both processes. We represent such an industry by a duopoly consisting of a conventional and an electronic bank. This scenario is likely when two banks enter a previously unreached market (e.g., rural market) in a developing country with two different distribution strategies (retail infrastructure versus electronic). Assume that they simultaneously decide on service quality levels, q_E and q_C, and prices, P_E and P_C. Further, assume that electronic delivery is faster than conventional such that $q_E < q_C$. The electronic and conventional bank's design and pricing problems are stated in the following:

Electronic: $\displaystyle\max_{P_E^D, q_E^D} PR_E^D\left(P_E^D, P_C^D, q_E^D, q_C^D\right) = (P_E^D - C_E)\, m\, MS_E(P_C^D, P_E^D, q_C^D, q_E^D) - I_E(q_E^D)$

Conventional: $\displaystyle\max_{P_C^D, q_C^D} PR_C^D\left(P_E^D, P_C^D, q_E^D, q_C^D\right) = (P_C^D - C_C)m\, MS_C(P_C^D, P_E^D, q_E^D, q_E^D) - I_C(q_C^D)$

with $S^* = (F_P + P_E^D - P_C^D)/(q_C^D - q_E^D)$ and $q_E < q_C$, in addition to the same constraints as earlier.

Define C and E as the strategy spaces for the conventional and electronic banks, respectively. We characterize them next:

$$E = \{(q_E, P_E) \mid q_{Emin} \le q_E \le q_{Emax}, -\infty \le P_E \le P_{Emax}\} \subseteq R^{+2}$$

$$C = \{(q_C, P_C) \mid q_{Cmin} \le q_C \le q_{Cmax}, -\infty \le P_C \le P_{Cmax}\} \subseteq R^{+2}$$

Let **e** and **c** denote the strategies of the two players, with $PR_i(\mathbf{e},\mathbf{c})$ denoting the profits for bank i. Banks play a one shot game in which they choose the parameters under their control simultaneously.

Definition 16.1
A Nash equilibrium is a combination $(\mathbf{c}^*,\mathbf{e}^*) \in C \times E$ that satisfies $PR_C(\mathbf{c}^*,\mathbf{e}^*) \geq PR_C(\mathbf{c},\mathbf{e}^*)$ for $c \in C$ and $PR_E(\mathbf{c}^*,\mathbf{e}^*) \geq PR_C(\mathbf{c}^*,\mathbf{e})$ for $e \in E$.

Proposition 16.2
For $S_{min} \leq S^* = (F_P + P_E^D - P_C^D)/(q_C^{D*} - q_E^{D*}) \leq S_{max}$, the markets segment with the following properties in a Nash equilibrium.

a. $P_C^{D*} = \dfrac{\left[-2(q_C^D - q_E^D)\,S_{min} + (q_C^D - q_E^D)\,S_{max} + 2C_C + C_E + F_P\right]}{3}$

b. $P_E^{D*} = \dfrac{\left[2(q_C^D - q_E^D)\,S_{max} - S_{min}(q_C^D - q_E^D) + C_C + 2C_E - F_P\right]}{3}$

c. $q_C^D{}^* = q_E^{D*} + \sqrt{\dfrac{-m(P_C^D - C_C)(F_P + P_E^D - P_C^D)}{I_C'(q_C^D)\,(S_{max} - S_{min})}}$ for $q_{Cmin} \leq q_C^{D*} \leq q_{Cmax}$

d. $q_E^D{}^* = q_C^{D*} - \sqrt{\dfrac{-m(P_E^D - C_E)(F_P + P_E^D - P_C^D)}{I_E'(q_E^D)\,(S_{max} - S_{min})}}$ for $q_{Emin} \leq q_E^{D*} \leq q_{Emax}$

As shown in Proposition 16.2a and 16.2b, we have found analytical expressions for both P_E^{D*} and P_C^{D*} as functions of each other and service levels. Again, the optimal q_C^{D*} in Proposition 16.2c and q_E^{D*} in Proposition 16.2d are either within the minimum and maximum ranges specified or at the boundary points $q_{Cmin}, q_{Cmax}, q_{Emin},$ and q_{Emax}. All the values of $P_E^{D*}, P_C^{D*}, q_C^{D*},$ and q_E^{D*} can be solved for numerically as will be shown in the example and numerical result section.

Corollary 16.1
Comparing the prices between the service integration and process specialization cases:

a. $P_C^{D*} \leq P_C^*$
b. $P_E^D \leq P_E^*$ for $(q_C^{D*} - q_E^{D*}) \leq (q_C^* - q_E^*)$
c. $q_C^{D*} \leq q_C^*$

In Corollary 16.1a, the price of conventional banking under process specialization is less than or equal to the price in a monopoly since $P_C^{D*} \leq P_C^* =$

P_{Cmax}. The price of electronic banking is also lower in a duopoly than monopoly, as specified in Corollary 16.1b. Since the conventional bank's optimal service level may need to be better than q_{Cmax}, $q_C^{D*} \leq q_C^*$, the optimal conventional service level in a monopoly is as shown in Corollary 16.1c. That is, under duopoly, unlike in monopoly, the conventional bank may improve its service level to better compete against the electronic bank.

As will be seen in the following numerical example, cost savings to consumers is larger under duopoly than in the monopoly situation. The total consumer spending in monopoly

$$\frac{P_C + P_C + q_C(S^* + S_{min})}{2}(S^* - S_{min}) + \frac{P_C + q_C S^* + F_P + P_E + q_E S_{max}}{2}(S_{max} - S^*)$$

is larger than that of the duopoly,

$$\frac{P_C^D + P_C^D + q_C^D(S'^* + S_{min})}{2}(S'^* - S_{min}) + \frac{P_C^D + q_C^D S'^* + F_P + P_E^D + q_E^D S_{max}}{2}(S_{max} - S'^*)$$

where S^* and S'^* are the points of indifference for monopoly and duopoly. Thus, as expected, it is more beneficial for the consumer to have two competitive firms each with different processes serving the market than to have a monopoly service provider support both processes.

16.7.2 Duopoly with a Conventional and a Hybrid Bank

The aforementioned results can be extended to the competition between a hybrid and a conventional bank. Consider two banks A and B located in two disparate markets, a and b, with equal size m. Each can be a monopoly conventional service provider in its own market and customers are homogeneous between them. Suppose that bank A now implements an electronic banking service that by nature is accessible to customers in both markets. Assume bank B supports only conventional banking, at least in the short run. Each bank has the same cost function and each needs to maintain a specified range of service level of conventional banking. The hybrid bank A chooses service levels q_C^A and q_E, and prices P_C^A and P_E, while bank B chooses q_C^B and P_C^B. The banks solve the following profit functions:

$$\max_{P_E, P_C^A, q_E, q_C^A} PR_A\left(P_E, P_C^A, P_C^B, q_E, q_C^A, q_C^B\right) = (P_C^A - C_C)\, m\, MS_C(P_E, P_C^A, q_E, q_C^A)$$

$$+ (P_E - C_E)\, m\, MS_E(P_E, P_C^A, q_E, q_C^A)$$

$$+ (P_E - C_E)\, m\, MS_E(P_E, P_C^B, q_E, q_C^B) - I_C(q_C^A) - I_E(q_E)$$

$$\max_{P_C^B, q_C^B} PR_B\left(P_E, P_C^A, P_C^B, q_E, q_C^A, q_C^B\right) = (P_C^B - C_C)\, m\, MS_C(P_E, P_C^B, q_E, q_C^B) - I_C(q_C^B)$$

Bank A's electronic banking segments market a at $S_A^* = (F_P + P_E - P_C^A)/(q_C^A - q_E)$ and market b at $S_B^* = (F_P + P_E - P_C^B)/(q_C^B - q_E)$. Bank A may either price the electronic service differently in each market or offer it at the same price P_E for both. If a differential pricing policy is feasible, it would set its price and service levels in market a and market b according to the results of Proposition 16.1 and 16.2, respectively.

Proposition 16.3

    a. Applying the results of Propositions 16.1c and 16.2c: Whether a differential or single price P_E^* is used, $q_C^A \geq q_C^B$

    b. When the hybrid bank offers its electronic service at the same P_E^H in both markets, then

$$P_E^{H*} = P_{C\max} \frac{(q_C^B - q_E)}{(q_C^A - q_E) + (q_C^B - q_E)} + \frac{C_E}{2} - \frac{F_P}{2}$$

$$+ \frac{P_C^B(q_C^A - q_E) - C_C(q_C^B - q_E) + 2(q_C^A - k_E)(q_C^B - q_E)S_{\max}}{2\left[(q_C^A - q_E) + (q_C^B - q_E)\right]}$$

    for $P_E^{H*} \leq P_{E\max}$

The finding in Proposition 16.3a means that conventional service is worst in market a where bank A offers electronic banking. Market b, where bank B offers only conventional banking, is likely to have a higher level of conventional service. When the hybrid bank offers the electronic service at the same price to both markets, its optimal policy is to price according to the equation in Proposition 16.3b. Note that the price is increasing in C_E and decreasing in F_P and C_C, as one would expect.

16.7.3 Duopoly with Homogeneous Electronic Service Providers

Suppose two banks, A and B, offer electronic banking services only with strategy spaces $E_A = \{(q_E^A, P_E^A) \mid q_{E\min} \leq q_E^A \leq q_{E\max}, -\infty \leq P_E^A \leq P_{E\max}\} \subseteq R^{+2}$ and $E_B = \{(q_E^B, P_E^B) \mid q_{E\min} \leq q_E^B \leq q_{E\max}, -\infty \leq P_E^B \leq P_{E\max}\} \subseteq R^{+2}$. Imagine that they have identical technologies and capabilities, and, consequently, the same cost structure for delivering services. This scenario may be common in many rural areas of developing countries in which more than one bank enters the market with online channels (Yang et al. 2007). In some areas, electronic channels may be the only electronically sensible option for banks (Opoku and Foy 2008). Assume that both banks pick their prices and service levels simultaneously.

Proposition 16.4

If two banks possess identical electronic technologies, then there is no equilibrium in pure strategies in which both providers survive in the market.

The proof requires showing that if both providers survive, they will employ different service levels and prices such that $q_E^A > q_E^B$ and $P_E^A < P_E^B$, and make identical profits. However, such a strategy is not profit maximizing for either player since each benefits by mimicking the other's service levels, and then undercutting his price, thereby gaining a monopolist position.

16.8 Summary and Conclusions

We have examined the cost structure of banks and benefits for consumers through a set of process analyses. Based on these, we analyzed market segmentation under monopolistic and duopolistic conditions. An important aspect of our analysis is that it brings together the effect of technology on processes; the economics of those processes; the consequences for access and service delivery; and issues of pricing, positioning, and segmentation. Our results and business implications are summarized as follows.

The monopolist (or service integration) model indicates that a monopolist bank has a strong incentive to encourage the spread of electronic banking and that it will put very little effort into customer service for branch customers. In fact, the bank may have the incentive to worsen branch customer service. Our results imply that when a bank has little competition (i.e., is in a monopolistic position), its management would consider maximizing its profit by migrating customers onto the electronic channels to the extent possible. In fact, the findings imply that even when the bank has competitors it would encourage its customers to use online banking as long as such migration can take place without the loss of customers.

In a duopoly consisting of an electronic and a conventional bank, prices of banking services are likely to be lower than in a monopoly setting. Also, conventional customer service is better due to competition. As a result, the adoption of electronic banking may be slowed. A similar situation is observed in a duopoly with a hybrid and a conventional bank. These results imply that a bank (conventional or electronic) will need to offer higher service levels in all its methods of delivery when there are competitors, thereby benefiting consumers. Over time, we have seen large banks pushing for electronic banking, whereas other banks, especially smaller regional banks, have been increasing their market share by claiming better conventional customer service (Gumbell 1995).

From a policy perspective, it would be sensible to always encourage the entry of multiple service providers (even though one may seem like an improvement to none in many rural areas). Even if banks pursue a completely different method of delivery (e.g., electronic versus physical), having more than one seems to encourage better service quality.

Appendix

Proof of Proposition 16.1: Monopoly

$$PR(P_E, P_C, q_E, q_C) = (P_E - C_E)\, m \int_{S^*(P)}^{S_{max}} \frac{1}{(S_{max} - S_{min})}\, dS - I_E(q_E)$$

$$+ (P_C - C_C)\, m \int_{S_{min}}^{S^*(P)} \frac{1}{(S_{max} - S_{min})}\, dS - I_C(q_C)$$

a. $\partial PR(P_E, P_C, q_E, q_C)/\, \partial P_C =$

$$= m\, \frac{1}{S_{max} - S_{min}}\, \frac{1}{q_C - q_E}\, (C_C - C_E + P_E - P_C) + m \int_{S_{min}}^{S^*(P)} \frac{1}{(S_{max} - S_{min})}\, dS$$

$$= m\, \frac{1}{S_{max} - S_{min}}\, \frac{1}{q_C - q_E}\, [2(P_E - P_C) + C_C - C_E + F_P - S_{min}(q_C - q_E)] > 0$$

We need to show $[2(P_E - P_C) + C_C - C_E + F_P - S_{min}(q_C - q_E)] > 0$.

As seen in 16.1b in the optimal solution the difference between the prices

is $(P_E - P_C) = \dfrac{\left[(q_C - q_E)S_{max} - (C_C - C_E) - F_P\right]}{2}$. Substituting this back in the

equation we obtain that $S_{max}(q_C - q_E) - S_{min}(q_C - q_E) > 0$. Thus $P_C^* = P_{Cmax}$.

b. $\partial PR(P_E, P_C, q_E, q_C)/\partial q_C = m\, \dfrac{(C_C - C_E + P_E - P_C)}{(S_{max} - S_{min})}\, \dfrac{F_p + P_E - P_C}{(q_C - q_E)^2} - I_C'(q_C)$

As seen in 16.1d in the optimal solution,

$$m\, \frac{(C_C - C_E + P_E - P_C)}{(S_{max} - S_{min})}\, \frac{F_p + P_E - P_C}{(q_C - q_E)^2} = -I_E'(q_E).$$

Thus, $\partial PR(P_E, P_C, q_E, q_C)/\partial q_C = -I_E'(q_E) - I_C'(q_C) > 0$.

c. $\partial PR(P_E, P_C, q_E, q_C)/\, \partial P_E =$

$$= m\, \frac{1}{S_{max} - S_{min}}\, \frac{1}{q_C - q_E}\, (-C_C + C_E - P_E + P_C) + m \int_{S^*(P)}^{S_{max}} \frac{1}{(S_{max} - S_{min})}\, dS = 0$$

Rearranging the terms, we get $P_E^* = P_C + \dfrac{\left[(q_C - q_E)\, S_{max} - (C_C - C_E) - F_P\right]}{2}$

d. $\partial PR(P_E, P_C, q_E, q_C)/\partial q_E = -m \dfrac{(C_C - C_E + P_E - P_C)}{(S_{max} - S_{min})} \dfrac{F_p + P_E - P_C}{(q_C - q_E)^2} - I_E'(q_E) = 0.$

After rearranging the terms, we get $q_E^* = q_C^* -$

$$\sqrt{\dfrac{-m(C_C - C_E + P_E - P_C)(F_p + P_E - P_C)}{I_E'(q_E)(S_{max} - S_{min})}}\ ,\ \text{where } q_C^* = q_{Cmax}.$$

Proof of Proposition 16.2

a. $\partial PR_E(P_E^D, P_C^D, q_E^D, q_C^D)/\ \partial P_E^D =$

$$= m\ \dfrac{1}{(S_{max} - S_{min})}\ \dfrac{1}{q_C^D - q_E^D}\ \left\{ S_{max}\left(q_C^D - q_E^D\right) - \left(F_p + P_E^D - P_C^D\right) - \left(P_E^D - C_E\right)\right\} = 0$$

Rearranging the terms, we get $P_E^{D*} = \dfrac{P_C^D}{2} + \dfrac{\left[(q_C^D - q_E^D)\ S_{max} + C_E - F_p\right]}{2}$

$\partial PR_C(P_E^D, P_C^D, q_E^D, q_C^D)/\partial P_C^D = -(P_C^{D*} - C_C)\ \dfrac{1}{(S_{max} - S_{min})}\ \dfrac{1}{q_C^D - q_E^D} +$

$\dfrac{F_p + P_E^D - P_C^D}{(q_C - q_E)(S_{max} - S_{min})} - \dfrac{S_{min}}{(S_{max} - S_{min})} = 0$

After rearranging the terms, $P_C^{D*} = \dfrac{P_E^D}{2} - \dfrac{\left[(q_C^D - q_E^D)S_{min} - C_C - F_p\right]}{2}$

Substituting P_C^{D*} from the second equation to the first we find

$$P_E^{D*} = \dfrac{\left[2(q_C^D - q_E^D)\ S_{max}\ - S_{min}(q_C - q_E) + C_C + 2C_E - F_p\right]}{3}$$

b. From earlier, $P_C^{D*} = \dfrac{\left[-2(q_C^D - q_E^D)\ S_{min} + (q_C^D - q_E^D)\ S_{max} + 2C_C + C_E + F_p\right]}{3}$

c. $\dfrac{\partial PR_C}{\partial q_C^D} = \dfrac{-m(F_p + P_E^D - P_C^D)}{(S_{max} - S_{min})(q_C^D - q_E^D)^2}(P_C^D - C_C) - I_C(q_C^D) = 0$

After rearranging the terms, we have $q_C^D{}^* = q_E^{D*} + \sqrt{\dfrac{-m(P_C^D - C_C)(F_p + P_E^D - P_C^D)}{I_C'(q_C^D)\ (S_{max} - S_{min})}}$

d. $\dfrac{\partial PR_E}{\partial q_E^D} = \dfrac{-m(F_p + P_E^D - P_C^D)}{(S_{max} - S_{min})(q_C^D - q_E^D)}(P_E^D - C_E) - I_E\ (q_E^D)$

$q_E^D{}^* = q_C^{D*} - \sqrt{\dfrac{-m(P_E^D - C_E)(F_p + P_E^D - P_C^D)}{I_E\ (q_E^D)\ (S_{max} - S_{min})}}$

Proof of Corollary 16.1

a. $P_C^{D*} \leq P_C^*$ since $P_C^* = P_{C\max}$ from Proposition 16.1.

b. We know that at optimality $P_E^{D*} = \dfrac{P_C^D}{2} + \dfrac{\left[(q_C^D - q_E^D)\, S_{\max} + C_E - F_P \right]}{2}$

and $P_E^* = P_C + \dfrac{\left[(q_C - q_E)\, S_{\max} - (C_C - C_E) - F_P \right]}{2}$.

Since $P_E^* = P_C + \dfrac{\left[(q_C - q_E)S_{\max} - (C_C - C_E) - F_P \right]}{2} >$

$\dfrac{P_C}{2} + \dfrac{\left[(q_C - q_E)\, S_{\max} + C_E - F_P \right]}{2}$

$P_E^* - P_E^{D*} \geq 0$ as long as $q_C^* - q_E^* \geq q_C^{D*} - q_E^{D*}$.

c. q_C^* in the monopoly case in Proposition 16.1 needs to be at $q_{C\max}$. It may not be optimal for a conventional bank in duopoly to set $q_C^D = q_{C\max}$. Then $q_C^{D*} \leq q_C^* = q_{C\max}$.

Proof of Proposition 16.3

a. We know from our previous results that the hybrid bank A is a monopolist in its own market. It serves its conventional customers at the lowest possible service level, such as $q_{C\max}$, and electronic customers at some q_E^*. We know from our duopoly results that the conventional bank B, when competing against an electronic bank A, may improve its service level to maximize market share. Thus, $q_C^{A*} \geq q_C^{B*}$.

b. $PR_H(P_E, P_C^A, P_C^B, q_E, q_C) = (P_E - C_E)\, m \displaystyle\int_{S_A^*(P_E)}^{S_{\max}} \dfrac{1}{(S_{\min} - S_{\min})}\, dS$

$+ (P_E - C_E)\, m \displaystyle\int_{S_B^*(P_E)}^{S_{\max}} \dfrac{1}{(S_{\max} - S_{\min})}\, dS - I_E(q_E)$

$+ (P_C^A - C_C^A)\, m \displaystyle\int_{S_{\min}}^{S_A^*(P_E)} \dfrac{1}{(S_{\max} - S_{\min})}\, dS - I_C(q_C)$

$\partial\, PR_H(P_E, P_C^A, P_C^B, q_E, q_C) / \partial P_E =$

$= m\,(P_C^A - C_C^A)\, \dfrac{1}{S_{\max} - S_{\min}}\, \dfrac{1}{q_C^A - q_E} - m\,(P_E - C_E)\, \dfrac{1}{S_{\max} - S_{\min}}\, \dfrac{1}{q_C^A - q_E}$

$$- m \left(P_E - C_E \right) \frac{1}{S_{max} - S_{min}} \frac{1}{q_C^B - q_E} + m \int_{S_A^*(P_E)}^{S_{max}} \frac{1}{\left(S_{max} - S_{min} \right)} dS$$

$$+ m \int_{S_B^*(P_E)}^{S_{max}} \frac{1}{\left(S_{max} - S_{min} \right)} dS$$

After lengthy rearrangement of terms, we get

$$P_E^{H*} = P_C^A \frac{\left(q_C^B - q_E \right)}{\left(q_C^A - q_E \right) + \left(q_C^B - q_E \right)}$$

$$+ \frac{C_E}{2} - \frac{F_P}{2} + \frac{P_C^B \left(q_C^A - q_E \right) - C_C^A \left(q_C^B - q_E \right) + 2 \left(q_C^A - q_E \right) \left(q_C^B - q_E \right) S_{max}}{2 \left[\left(q_C^A - q_E \right) + \left(q_C^B - q_E \right) \right]}$$

where $P_C^A = P_{Cmax}$ from Proposition 16.1.

Lemma

If both electronic banks survive they will employ different service levels and prices s.t. $P_A > P_B$ and $q_E^A < q_E^B$, and make identical profits.

Proof of Lemma

Assume that if both electronic banks use identical service levels and pricing strategies, they split the market equally. Suppose both banks survive in the market but $q_E^A = q_E^B = q_E$, $P_E^A = P_E^B = P$ in equilibrium. Keeping the same service level, bank B can price its service slightly smaller and capture the entire market. Bank A reacts likewise, and the firms undercut each other until one firm changes the smallest price $P_E^0(q_E)$ such that $m \left(P_E^0(q_E) - C_E \right)$ $MS(q_E, P_E^0(q_E)) - I_E(q_E) = 0$, provided q_E and P_E^0 are within the prespecified range. The other firm reaching P_E^0 may not stay in the market, in which case the market reverts to a monopoly situation. A similar argument applies when both attempt to change their service levels while charging identical prices in equilibrium. Thus, if both banks survive, they utilize different pricing and design strategies. Clearly, the higher priced firm will then use a higher service level; otherwise, it will be dominated. In addition, the firms must make identical profits, otherwise the lower profit firm can mimic the decision of the other firm, and undercut its price and do better.

Proof of Proposition 16.4

Suppose the equilibrium described in the lemma exists. Without loss of generality, assume that both firms make zero profits. Then, it must be that $P_E^A >$

$P_E^B > P_0(q_E^B)$, since $P_0(q_E^B)$ is the smallest price at which either firm makes a zero profit as a monopolist with a service level q_E^B. Then, bank A can mimic bank B's service level, undercut his price, and do better. When this happens, firms engage in intense price and size competition till one firm charges $P_0(q_E^{A*})$, while the other firm stays out, reverting the market to monopoly.

References

Andaluz, J. 2011. Validity of the "principle of maximum product differentiation" in a unionized mixed-duopoly. *Journal of Economics* 102: 123–136.

Anderson S. P., A. de Palma, and J.-F. Thisse. 1992. *Discrete Choice Theory of Product Differentiation*. Cambridge, MA: MIT Press.

Balasubramanian, S. 1998. Mail versus mall: A strategic analysis of competition between direct marketers and conventional retailers. *Marketing Science* 17: 181–195.

Bashyam, A., and U. S. Karmarkar. 2004. Usage volume and value segmentation in business information services. In *Managing Business Interfaces: Marketing, Engineering, and Manufacturing Perspectives*, eds. A. Chakravarty and J. Eliashberg. Norwell, MA: Kluwer Academic.

Bashyam A., and U. S. Karmarkar. 2005. Service design, competition and market segmentation in business information services with data updates. In *Current Research in Management in the Information Economy*, eds. U. Apte and U. S. Karmarkar. New York: Springer.

Bauman, L. 1996. Small stock focus: Nasdaq rises amid a broad market decline: Security First Network Bank soars on initial day. *Wall Street Journal*, May 24, C7.

Beesley, M. E. 1965. The value of time spent travelling: Some new evidence. *Economica* 32 (May): 174–185.

Bradley, L., and K. Stewart. 2003. A Delphi study of Internet banking. *Marketing Intelligence and Planning* 21: 272–281.

Bughin, J. 2004. Attack or convert? Early evidence from European on-line banking. *Omega* 32: 1–7.

Byers, R., and P. Lederer. 2001. Retail bank service strategy: A model of traditional, electronic, and mixed distribution choices. *Journal of Management Information Systems* 18: 133–156.

Byers, R., and P. Lederer. 2007. Channel strategy evolution in retail banking. In *Managing in the Information Economy: Current Research*, eds. U. Karmarkar and U. Apte. New York: Springer.

Cattani, K. D., W. G. Gilland, and J. M. Swaminathan. 2004. Coordinating traditional and internet supply chain. In *Handbook of Quantitative Supply Chain Analysis*, eds. D. Simchi-Levi, D. Wu, and M. Shen, 644–677. Norwell, MA: Kluwer Academic.

Chiang, W. Y., D. Chhajed, and J. D. Hess. 2003. Direct marketing, indirect profits: A strategic analysis of dual-channel supply chain design. *Management Science* 49: 1–20.

Choi, D. Y. 1998. Wells Fargo Bank: Information superhighway and retail banking. UCLA Anderson BIT Project Case Study.

Coetzee, J. 2009. Personal or remote interaction? Banking the unbanked in South Africa. *South African Journal of Economic and Management Sciences* 12(4).

Cook, S. 1984. Research finds home banking's key benefit. *ABA Banking Journal* 76: 120–125.

D'Aspremont C., J. J. Gabszewicz, and J.-F. Thisse. 1979. On Hotelling's stability in competition. *Econometrica* 47: 1145–1150.

Ernst & Young. 1995. Managing the virtual bank. In *Special Report on Technology in Banking*.

Geyskens, I., K. Gielens, and M. G. Dekimpe. 2002. The market valuation of Internet channel additions. *Journal of Marketing* 66: 102–119.

Gumbell, J. 1995. Guerrilla tactics to beat the big guys. *Fortune* 132: 150–158.

Hitt, L. M., and F. X. Frei. 2002. Do better customers utilize electronic distribution channels? The case of PC banking. *Management Science* 48: 732–748.

Holmsen, C. A., R. N. Palter, P. R. Simon, and P. K. Weberg. 1998. Retail banking: Managing competition among your own channels. *The McKinsey Quarterly* 1: 82–92.

Hotelling, H. 1929. Stability in competition. *Economic Journal* 39: 41–57.

Ingene, C. A., and M. E. Parry. 1995. Channel coordination when retailers compete. *Marketing Science* 14: 360–377.

Iyer, G. 1998. Coordinating channels under price and nonprice competition. *Marketing Science* 17: 338–355.

Johnson, B. A., and J. H. Ott. 1995. Banking on multimedia. *The McKinsey Quarterly* 2: 94–107.

Karmarkar, U. S. 2000. Financial service networks: Access, cost structure and competition. In *Creating Value in Financial Services*, eds. E. Melnick, P. Nayyar, M. Pinedo, and S. Seshardi. Norwell, MA: Kluwer.

Kaufmann, P., and V. K. Rangan. 1990. A model for managing system conflict during franchise expansion. *Journal of Retailing* 66: 155–173.

Laforet, S., and X. Li. 2005. Consumers' attitudes towards online and mobile banking in China. *The International Journal of Bank Marketing* 23: 362–380.

Lunt, P. 1995. Virtual banking part I: What will dominate the home? *ABA Banking Journal* 6: 36–45.

Moorthy, K. S. 1993. Competitive marketing strategies: Game-theoretic models. In *Handbooks in Operations Research and Management Science* Vol. 5, eds. J. Eliashberg and G. L. Lilien, 143–190. New York: Elsevier Science.

Opoku, L., and D. Foy. 2008. Pathways out of poverty: Innovating banking technologies to reach the rural poor. *Enterprise Development and Microfinance* 19(1): 46–58.

Siaw, I., and A. Yu. 2004. An analysis of the impact of Internet on competition in the banking industry, using Porter's five forces model. *International Journal of Management* 21: 514–523.

Starr, M. K. 2003. Application of POM to e-business: B2C e-shopping. *International Journal of Operations and Production Management* 1: 105–124.

Thomas, T. C., G. I. Thompson, T. E. Lisco, et al. 1970. The value of time for commuting motorist as a function of their income level and amount of time saved. *Highway Research Record* 314: 1–19.

Tsay, A. A., and N. Agrawal. 2000. Channel dynamics under price and service competition. *Manufacturing and Service Operations Management* 2: 372–391.

Tsay, A. A., and N. Agrawal. 2004a. Channel conflict and coordination in the e-commerce age. *Production and Operations Management* 13: 93–110.

Tsay, A. A., and N. Agrawal. 2004b. Modeling conflict and coordination in multi-channel distribution systems: A review. In *Handbook of Quantitative Supply Chain Analysis*, eds. D. Simchi-Levi, D. Wu, and M. Shen, 557–606. Norwell, MA: Kluwer Academic.

Verma R., Z. Iqbal, and G. Plaschka. 2004. Understanding customer choices in e-financial services. *California Management Review* 46: 43–67.

Viauroux, C. 2008. Marginal utility of income and value of time in urban transport. *Economics Bulletin*, 4(3): 1–8.

Viswanathan, S. 2005. Competing across technology-differentiated channels: The impact of network externalities and switching costs. *Management Science* 51: 483–496.

Yang, J., M. Whitefield, and K. Boehme. 2007. New issues and challenges facing e-banking in rural areas: An empirical study. *International Journal of Electronic Finance* 1(3): 336–354.

Yao, D., and J. J. Liu. 2003. Channel redistribution with direct selling. *European Journal of Operational Research* 144: 646–658.

Yao, D., J. J. Liu. 2005. Competitive pricing of mixed retail and e-tail distribution channels. *Omega* 33: 235–247.

Index

Printed and bound by CPI Group (UK) Ltd, Croydon, CR0 4YY

18/10/2024

01776264-0018